Steam Turbines for Modern Fossil-Fuel Power Plants

Steam Turbines for Modern Fossil-Fuel Power Plants

Alexander S. Leyzerovich

THE FAIRMONT PRESS, INC.

CRC Press
Taylor & Francis Group

Library of Congress Cataloging-in-Publication Data

Leizerovich, A. Sh. (Aleksandr Shaulovich)
    Steam turbines for modern fossil-fuel power plants / Alexander S. Leyzerovich.
      p. cm.
    Includes bibliographical references and index.
    ISBN 0-88173-548-5 (print edition : alk. paper) -- ISBN 0-88173-549-3
(electronic edition) -- ISBN 1-4200-6102-X (taylor & francis (distributor) : alk.
    paper)
        1. Steam-turbines. 2. Steam power plants. I. Title.

    TJ735.L395 2007
    621.31'3--dc22

                                                          2007018778

Steam turbines for modern fossil-fuel power plants / Alexander S. Leyzerovich
©2008 by The Fairmont Press. All rights reserved. No part of this publication
may be reproduced or transmitted in any form or by any means, electronic or
mechanical, including photocopy, recording, or any information storage and
retrieval system, without permission in writing from the publisher.

Published by The Fairmont Press, Inc.
700 Indian Trail
Lilburn, GA 30047
tel: 770-925-9388; fax: 770-381-9865
http://www.fairmontpress.com

Distributed by Taylor & Francis Ltd.
6000 Broken Sound Parkway NW, Suite 300
Boca Raton, FL 33487, USA
E-mail: orders@crcpress.com

Distributed by Taylor & Francis Ltd.
23-25 Blades Court
Deodar Road
London SW15 2NU, UK
E-mail: uk.tandf@thomsonpublishingservices.co.uk

Printed in the United States of America
10 9 8 7 6 5 4 3 2 1

0-88173-548-5 (The Fairmont Press, Inc.)
1-4200-6102-X (Taylor & Francis Ltd.)

While every effort is made to provide dependable information, the publisher,
authors, and editors cannot be held responsible for any errors or omissions.

Wherefore I perceive that there is nothing better than that a man should rejoice in his own works; for this is his portion: for who shall bring him back to see what shall be after him?

Ecclesiastes 3:22

Table of Contents

Preface.. xi

Part I   **Main Steam Parameters, Operating Performances,**
            and Design Features of Modern Steam Turbines
            for Fossil-Fuel Power Plants Worldwide 1

Chapter 1   Steam Turbines in the Modern World 3

Chapter 2   Rise of Steam Turbine Output and Efficiency
            with Steam Parameters... 9
            — Steam turbines with subcritical main steam
                pressure of the 1960s-80s... 9
            — Supercritical- and USC-pressure steam
                turbines of the 1960s-80s 14
            — Modern steam turbines with
                elevated steam conditions 23
            — Potential rise of steam conditions
                in the nearest future 34

Chapter 3   Configuration of Modern Power Steam Turbines 43
            — Large power steam turbines 43
            — Profiles for modern power steam turbines of
                a moderate capacity.. 69

Chapter 4   Design of Steam Path, Blading, Gland Seals, and Valves ... 97
            — Features of modern three-dimensional steam
                path design... 97
            — Improvement of turbine gland seals and
                non-bladed areas 119
            — Reduction of losses in control valves 128

Chapter 5   Last Stage Blades and Exhaust Hoods of LP cylinders..... 135
            — Last stage blades 135
            — LP exhaust ports 150

| | — Last stage blade protection against water drop erosion | 155 |

Chapter 6 Thermal Expansion, Bearings, and Lubrication 155

Chapter 6 Thermal Expansion, Bearings, and Lubrication 169
    — Arrangement of thermal expansion 169
    — Improvement of journal bearings 175
    — Prevention of oil fires due to the use of fire-resistant fluids as lubricant 178

Part II   **Steam Turbine Transients and Cycling Operation** ...187

Chapter 7 Operating Conditions and Start-up Systems for Steam-Turbine Power Units 189
    — Typology of operating conditions 189
    — Some features of start-up systems as applied to steam-turbine and CC power units 201

Chapter 8 Experimental and Calculation Researches of Turbine Transients 215
    — Main goals of start-up field tests 215
    — Heat transfer boundary conditions for major turbine design elements 217
    — Metal temperature fields for turbine design elements 234
    — Cooling down characteristics of the turbine 238

Chapter 9 Start-up Technologies as Applied to Different Start-up Systems 247
    — Pre-start warming and rolling up of the turbine 247
    — Start-up loading and raising the steam conditions 258
    — Peculiarities of steam-turbine start-up technologies for CC units 261

Chapter 10 Start-up Instructions for Steam-Turbine Power Units and Their Improvement 271
    — Different approaches to constructing start-up instructions 271
    — Estimation of start-up duration 282
    — Calculated optimization of start-up diagrams 286

Chapter 11   Scheduled and Unscheduled Load Changes within and beyond the Governed Range 295

Chapter 12   Cycling Operation and Its Influence on Turbine Performances..305
— Different approaches to covering the variable part of power consumption 305
— Cost of cycling operation and ways to reduce this cost...315
— Turbine damages caused by or associated with the transients 321

Part III   **Diagnostic Monitoring and Informative Support for Turbine Operators** 349

Chapter 13   Automated Data Acquisition and Control Systems for Modern Power Plants.........................351
— Recent years' evolution of automated control and supervisory systems 351
— Diagnostic monitoring of power equipment 362
— Human-machine interface and informative support for the operator.................................369

Chapter 14   Diagnostic Monitoring of Turbine Heat-Rate and Flow-Capacity Performances 379
— Monitoring of heat-rate performances and deviations from the rated values 379
— Revelation of turbine steam path damages and sources of heat-rate performance deterioration...........394

Chapter 15   Diagnostic Monitoring of Turbine Temperature and Thermal-Stress States......................................403
— General features of this diagnostic function...............403
— The scope of temperature measurements.....................404
— Mathematical modeling of the heating up process for temperature monitoring of turbine rotors 412
— On-line operative support for turbine operators at the transients.................................418

Chapter 16 Post-Operative Analysis of the Turbine's
Operating Conditions 429
— Computerized post-operative ananlysis of
turbine operating conditions...429
— Evaluationn of the control quality at the transients...433
— Post-operative analysis of the transients for
turbines without computerized DAS 434
— Teaching and training the operational
personnel with the use of simulators...........................436
— Specialized program-simulator for teaching
and training operators in running the
turbine transients 445

**Part IV   Lifetime Extension for Aging Steam Turbines
and Their Refurbishment**..453

Chapter 17 Assessment and Extension of Steam Turbine Lifetime 455
— Possible solutions for a problem with mass
steam turbine aging 455
— Steam turbine lifetime limitations................................459
— Assessment of the turbine's residual lifetime 465
— Lifetime extension strategy for sets of the
same type power equipment..470

Chapter 18 Steam Turbine Upgrade 477
— An increase in the turbine efficiency and
other benefits of turbine upgrades 477
— Steam turbine refurbishment by non-OEMs
with the use of alien design solutions 495

Appendix 521
List of Abbreviations and Symbols 521
Acronyms 522
Symbols 524
Subscripts and Superscripts...526
Criteria of Similarity 527
Conversion Table for Main Units Used 528

Index..529

Preface

This book is a continuation of my previous two-volume work *Large Power Steam Turbines: Design & Operation* (PennWell, 1997). Those volumes were conceived as an exposition of steam-turbomachinery fundamentals as they were seen by the end of the 20th century. This new book to a degree rests on its contents, not repeating previous information as far as possible without sacrifice of comprehension. The afore-mentioned work was also supplemented by another book, *Wet-Steam Turbines for Nuclear Power Plants*; PennWell, 2005. The present book considers the newest approaches of the latest decade in design, operation, and refurbishment of steam turbines for fossil-fuel power plants and is designed to be a final part of this trilogy.

At this writing, many people, including some professionals in power engineering, have come to view steam-turbomachinery as a completely matured technology that promises no remarkable achievements in the near future. Indeed, by the early 1990s the efficiency of the best new steam turbines had practically stabilized at the previously attained level and did not grow further. Yet, the mid-1990s brought a new breakthrough in the steam turbine technology, and this progress continues today. As a result, new possibilities can considerably raise power plant efficiency based upon qualitative improvements in the turbine steam path design and gradually applying elevated steam parameters. Dr. Wilfried Ulm of Siemens Power Generation qualified this process as "an almost unnoticed revolution in steam turbine technology" [*VGB PowerTech* 83, no. 1/2 (2003): 1]. These new possibilities can also impact efforts to upgrade old steam turbines in service. New approaches have also been developed and applied to handling the transient operating conditions of steam-turbine-based power units and providing information support for the operational personnel with the use of advanced computerized control and instrumentation (C&I) techniques and friendly human-machine interfaces. I marked some of these trends in all these processes in my above mentioned work, written in the mid-1990s, but what once were novelties have brought their first rich fruits in the first years of the new century. Valuable descriptions of new achievements in steam-turbomachinery and their effects were published by their developers in the late 1990s and early 2000s, but these separate publications have never been gathered together and generalized, to the best of my understanding.

Among modern books on steam-turbomachinery which appeared after the afore-mentioned *Large Power Steam Turbines: Design & Operation*, two noteworthy books were issued just at the turn of the century and addressed primarily to power plant maintenance staff: *Turbine Steam Path Damage: Theory & Practice* by T.H. McCloskey, R.B. Dooley, and W.P. McNaughton (EPRI, 1999) and *Turbine Steam Path: Maintenance & Repair* by W.P. Sanders (PennWell, 2001). These works summarized the current knowledge about damages in the turbine steam path and methods for revealing and repairing them. A general overview of turbomachinery for the power industry at the end of the 20th century is offered in H. Termuehlen, *100 Years of Power Plant Development: Focus on Steam and Gas Turbines as Prime Movers* (ASME, 2001).

The present book, as well as my previous ones, differs from those mentioned above in that it is addressed mainly to power plant operators and operation researchers and was written from their point of view. The book does not try to tell the readers how steam turbines should be designed but rather explains why they were designed as they were and compares different possible design solutions. It absorbs the experience in steam-turbine design and operation accumulated in developed and developing countries throughout the world—in the USA, Germany, Japan, Russia, Denmark, Korea, and India, among others.

In addressing the modern stage of steam turbine design and operation, it might be well to note that in recent years a few advanced and highly efficient large-output steam turbines have appeared, with the utmost steam conditions accessible at the modern technology level. Such machines produce about 1,000 MW in single capacity with main steam pressure of up to 31 MPa (4,500 psi) and main and reheat steam temperatures of up to 600-610°C (1,112-1,130°F). These turbines provide a new benchmark for steam turbine efficiency and herald new possibilities for its further development. Beside these champions, different countries have put into operation some new "ordinary" steam-turbine power units with a rather moderate single capacity (of about 200-300 MW and 500-600 MW) and "common" supercritical steam conditions of about 25 MPa, 540-565°C (3,625 psi, 1,000-1,050°F). Owing to the use of modern technologies and new design approaches, these steam-turbine power units also exhibit good efficiency—much better than power units of the past. Most of these up-to-date power units have been constructed in such countries as Germany, Japan, Korea, China, and others that have a large experience in the use of supercritical-pressure steam-turbine power units. Modern steam-turbine power units are also coming on line after a long recess in the U.S. In parallel, new supercritical-pressure steam-turbine units have been constructed and launched in countries where such units have never been operated before. Noteworthy is that most of these new power units are launched at power plants burning solid fuel, most with coal- and lignite-fired boilers. As a result, thanks to their high efficiency, these units are notable for quite moderate gas emissions to the atmosphere.

Along with new steam turbines for "traditional" fossil-fuel power plants, noteworthy are steam turbines for combined-cycle (CC) units of a new generation. These units have a larger single capacity compared to their forerunners and relatively elevated steam conditions. Many CC units have been designed with single-shaft turbine sets (that is, with one gas turbine and one steam turbine settled at the common shaft with the common electric generator). These circumstances have prompted the use of some unusual design solutions for steam turbines and special technologies for their transients.

Mass implementation of steam-turbine power units in the 1960s, 1970s, and 1980s made it rather difficult to put them out of action even when they achieved and exceeded their rated lifetime. In many countries, these power units formed up a backbone of the power industry and continue to retain their importance. Under these circumstances, it has been considered desirable to extend their lifetimes as far as their reliability allows. In doing so, it is reasonable to retrofit aging turbines by replacing their design elements which accounted for most metal damages. In addition, aging steam turbines, even if relatively young, are as a rule noticeably inferior to those of very recent vintage in their efficiency. From this point of view, it is also advisable to retrofit these turbines by replacing their steam path while retaining secondary design elements that do not affect the turbine's reliability and efficiency. Such partial refurbishment allows not only extending the turbine's lifetime and heightening the operating reliability, but also raising remarkably the operating efficiency and flexibility owing to the use of modern design solutions. It is important to note that involving non-original equipment manufacturers in this job makes it possible to consider a very wide spectrum of possible design solutions and to use some rather unexpected applications to reach optimal results.

This book, as well as my previous ones, is mainly intended for power plant operators, owners, and designers, college students training for work in the power generation industry, as well as the audience of diverse courses and workshops in power engineering, and I believe this book will be useful for its readers. This confidence is to a great degree based on the experience of workshops and seminars I led in the recent years for power plant specialists of different countries. These workshops were accompanied by consultations on diverse operation problems appearing at steam turbines of local power units in service with the single capacity of from 200 to 800 MW, taking into consideration their specific design features and operating conditions. The book also presents some original developments worked up with my participation and implemented at different power plants in service, all of which were approved in their operation practice. Some original materials and overviews given in the book were presented at international power generation conferences in recent years and published in power engineering magazines.

I would like to thank all my co-workers and co-authors for our collaboration. This especially refers to Evgeny Plotkin (presently in Israel), and I would also like to express my deep gratitude to colleagues from various countries that helped me in gathering materials and special artworks for this book. I would like to mention Alan Hesketh, Simon Hogg, Robert Scott, and Don Stephen of ALSTOM Power, Un-Hak Nah of Doosan Heavy Industries and Construction (Korea), the late Tom McCloskey of EPRI, Peter Luby of INGCHEM (Slovakia), Carlos Koeneke and Yoshinori Tanaka of Mitsubishi Heavy Industries, Brajesh Singh of National Thermal Power Corporation (India), Andreas Wichtmann and Wilfried Ulm of Siemens Power Generation, Paul Hurd and Rudy Koubek of Siemens Westinghouse, Hideo Nomoto, Toshihiro Matsuura, and Akira Sakuma of Toshiba. My special thanks are also to editors of The Fairmont Press for their help in preparing and publishing this book.

Dr. Alexander S. Leyzerovich

PART I

MAIN STEAM PARAMETERS, OPERATING PERFORMANCES, AND DESIGN FEATURES OF MODERN STEAM TURBINES FOR FOSSIL-FUEL POWER PLANTS WORLDWIDE

Chapter 1

Steam Turbines in the Modern World

Steam turbines of fossil-fuel power plants make up the bulk of generation capacities for most of industrially developed and developing countries, excepting those whose power industry mainly depends on nuclear and/or hydraulic power generating stations (like, for example, Argentina, Austria, Belgium, Brazil, Canada, France, Lithuania, Norway, Sweden, and some others). So, in the United States by 2002 out of approximately 615 GW of the U.S. generating fleet rated more than 299 MW, the aggregate capacity of non-nuclear steam-turbine-based facilities amounted to 407 GW (approximately 305 GW at coal-fired power plants and 102 GW at oil/gas-fired ones compared to 103.8 GW of nuclear power plants, 62.6 GW of simple-cycle gas turbines, and 40.6 GW of combined-cycle units), not taking into account steam turbines of combined-cycle (CC) units. According to the overview issued by *Power*,[1] 298 US coal-fired steam-turbine power plants, with the individual output of 299 MW and more, referring to their actual production, generate about 58% of the whole amount of the nationally produced electricity, and 121 oil/gas-fired steam-turbine power plants of the same class (with the capacity of 299 MW and more) additionally account for 16.6%, even though many of them are operated in a cycling mode, with a decreased capacity factor.

For China, the world's second-largest country ranked by the electric power capacity and generation, by the end of 2005 the nationally installed power generating capacity reached 508 GW (15% growth in 2005), and the total annual power generation amounted to 2,475 TWh. As this takes place, approximately 80% of the whole electricity production is provided by fossil-fuel power plants (mainly coal-fired steam-turbine power units); about 18% is generated by hydroelectric power plants, and nuclear power plants, with their total capacity of 6.6 GWe, provide 2.1%.[2]

According to *UDI International Directory of Electric Power Producers and Distributors*, Russia, Japan, Germany, and India take the next places after the United States and China by the installed capacity and total annual production of electric power plants.[3]

In Russia, according to the state's Statistic Department,[4] the scope of electric power generation in 2005 made up 952 TWh, including 627 TWh (65.9%) generated by fossil-fuel power and cogeneration plants, whereas hydroelectric and nuclear power stations produced 175 TWh (18.4%) and 150 TWh (15.7%), correspondingly. Of the total installed capacity of about 215 GW, the fossil-fuel power and cogeneration plants make up about 68%. With an insignificant share of gas turbines and CC units, this figure is mainly amounted by steam-turbine-based facilities. Meantime, hydroelectric and nuclear power stations make up approximately 19 and 13%.[5]

The total amount of energy generated by ten electric power utilities of Japan in 1998/9 accounted to 799 TWh, and the maximum aggregate load of their power plants was 168.2 GW. According to predictions, in 2005 the share of fossil fuel in the total annual power production will account to 52.4% (including 23.3% for liquefied natural gas and 19.8% for coal); nuclear power plants will give 34.5%, and hydroelectric stations will contribute up to 9.7%.[6,7] In reality, the share of fossil-fuel power plants, mostly steam-turbine ones, has been even more, accounting to approximately 60%.[8]

In Germany, the net electric power generation in 2004 totaled 570.1 TWh, and major contributions to the electricity supply were based on nuclear energy (27.8%), lignite (25.6%), and hard coal (23.4%), whereas natural gas had a share of 10.4%, and renewables gave 9.4%, including wind power farms—4.4%.[9] In this way, burning fossil fuels provides about 60% of the whole electric power production, mostly owing to steam-turbine power units, whose total installed capacity makes up about 69%.[10]

According to National Thermal Power Corporation (NTPC) of India, by March 2002 the total installed capacity of Indian electric power utilities stood up at 104.9 GW, with the shares of thermal, hydroelectric, and nuclear power plants equal to approximately 71%, 25%, and 2.6%, where the thermal (fossil-fuel) power plants are mainly presented by coal-fired steam-turbine units—about 60% of the total amount of 71%.[11]

So we can see that "old good steam turbines" of fossil-fuel power plants remain a reliable workhorse of the power industry for the very different industrially developed and developing countries, amounting to from 50% to 70% of their total installed capacity and electric power production. With regard to steam turbines of CC units and wet-steam turbines of nuclear power plants, these figures may rise up to 80-90%. A further growth of electric power consumption requires and will require constructing new power plants, and steam-turbine units will continue to make a very great and highly significant contribution to overall power generation everywhere. According to data of the World Energy Council (WEC), the total portion of power contributed around the world by primary fossil fuels and nuclear energy will have remained practically invariable at the current level of approximately 80% until at least 2020.[12]

With rather limited possibilities of mass construction of nuclear power plants (even if the current technical problems of burying nuclear wastes are resolved and existent public prejudices against atomic energy are overcome), the imminent and unavoidable depletion of the world's gas and oil resources, skyrocketing of the gas prices, and striving for desirable independence from oil and gas import, it is obvious that steam-turbine-based power plants burning solid fossil fuels (hard coal, lignite, oil shale, etc.) will take on greater and greater significance. Under these circumstances, of special importance is to raise the efficiency of steam-turbine units to maximally reduce in this way air pollution caused by gas emissions from newly constructed power plants with their taken power output.

Along with constructing new power plants, it is also necessary to upgrade a great fleet of existing aging steam turbines in service with expired (or nearly expired) lifetime. According to data of Utility Data Institute (UDI) of Platt, approximately 200 GW of the installed capacity of the U.S. coal-fired power plants (that is, about 58%) by 2005 were expected to be 30 years or more old; similar data concern oil/gas-fired power plants.[13] As a result, a study carried out by Electric Power Research Institute (EPRI) and published in February 2003 pointed out that 44.6 GW of oil- and gas-fired power units and 15.3 GW of coal-fired units should be retired or refurbished only between 2001 and 2010, and then these processes will go on. Because of growing requirements for the efficiency and environmental safety of power plants, many of them, to remain competitive, should be refurbished prematurely (even before their lifetime terms expire).

A glance to Europe also reveals a clear need for similar immediate actions—the aging power equipment fleet is in need of upgrading, refurbishing, renewing, or retrofitting. According to Siemens Power Generation authorities, in the next ten-to-twenty years European power plants with the total installed capacity of up to 200 GW will have to be replaced or refurbished for reasons of their physical or moral obsolescence, with approximately 40 GW due for the replacement or refurbishment in Germany.[14]

In Russia, by 2002 power equipment of fossil-fuel power plants with the total capacity of 31 GW operated beyond their specific service life, and in 2002-2004 another 23.6 GW of capacities were expected to exhaust their lifetime and would need refurbishment.[15] By the end of 2005, approximately 45% steam turbines of Russian fossil-fuel power plants with the total capacity of 59.3 GW were expected to exhaust their so-called fleet resource (the lifetime term assessed for a certain group of power equipment of the same type operated under similar conditions); by 2010 this figure will increase to 60% (80.5 GW), and by 2015 it can reach 72%, or 94.6 GW.[5] A similar situation arises in many other countries.

Refurbishment, or retrofit, of a steam turbine usually means replacement of its steam path blading, main design components, including rotors, casings, and blading carriers of both the high-temperature and low-pressure cylinders, high-temperature steam-lines and steam admission parts, and so on, retaining, as a rule, the original foundation, bearings, and most of auxiliaries. Technically, the scale of such retrofit is quite comparable with construction of a new turbine.

With a great amount of steam-turbine capacities to be newly constructed or refurbished, it comes as no surprise that main steam turbine producers look quite optimistically into the future.[12] As Wilfried Ulm of Siemens Power Generation wrote, "The steam turbine still has plenty of life left in it. It is set to retain its place in the energy sector of the 21st century. Recent technical advances have contributed considerably to improving its competitiveness. As the use of coal to generate electricity increases, it will regain some of its former dominance. And whether the primary energy sources are now coal, gas, or nuclear power, every new power plant will have at least one steam turbine. There is therefore no need to worry about future market opportunities for this technology." [14]

The world's chief manufacturers of large power steam turbines today are two European-based multinationals: ALSTOM Power, which derived from merging Swiss-German-Swedish ABB Kraftwerke and French-Britain-German GEC Alsthom, and Siemens Power Generation,

Steam Turbines in the Modern World

with its US subsidiary of Siemens Westinghouse Power Corporation. The list of leading manufacturers of large power steam turbines also includes a triad of Japanese concerns—Hitachi, Mitsubishi Heavy Industries (MHI), and Toshiba, then General Electric (GE) of the USA, Leningrad Metallic Works (LMZ) as a branch of the Power Machines of Russia, Turboatom (or Kharkov Turbine Works—KhTGZ) of Ukraine, Ansaldo Energia of Italy, and Skoda Energo of the Czech Republic. Large power steam turbines are also manufactured, mainly under licenses, in China (Shanghai and Dongfang Turbine Works), India (Bharat Heavy Electricals Ltd.—BHEL), and South Korea (Doosan Heavy Industries & Construction). In addition, numerous companies in diverse countries produce power steam turbines of medium and small capacities, as well as driving and marine steam turbines.[16]

For most of these manufacturers, the range of the rated output for large steam turbines designed and produced for fossil-fuel power plants stretches up to 800-1,200 MW, even though MHI marks its readiness to deliver steam turbines with the single capacity of up to 1,400 MW, and ALSTOM Power Turbo-Systems is ready to produce steam turbines with the output of even up to 1,800 MW. Serial steam turbines for CC units envelop the single capacity range up to 280 MW.

References

1. Schwieger B., M. Leonard, S. Taylor, et al. "First Annual Top Plants Survey," *Power* 146, August 2002: 27-70.
2. "Nuclear Power in China," *UIC Nuclear Issues Briefing Paper*, no. 68 (February 2006). *http://www.uic.com.au/nip68.htm*.
3. Bergesen C.A.E., editor. *UDI International Directory of Electric Power Producers and Distributors*, 16th Edition. New York: Platts, 2005.
4. "Russia in 2005 increased electric power production by 2.2%" [in Russian], *http://www.gvc.elektra.ru/show.cgi?news/on10012006.htm*.
5. "Electric power engineering of Russia, 1998-2001" [in Russian], *http://www.raexpert.ru/researches/energy/electric/*.
6. Hoshino K, Y. Otawara, A. Tagishi, and S. Suzuki. "Recent Trends in Thermal Power Generation Technology," *Hitachi Review* 46, no. 3 (1997): 115-120.
7. Leizerovich A. Sh. "Several Current Aspects of the Development of Thermal Engineering in Japan," *Thermal Engineering* 46, no. 10 (1999): 885-892.
8. "Why is nuclear energy necessary in Japan?" *http://www.fepc.or.jp/english/nuclear_power/generation/necessary.html*.
9. "Power Generation, Germany," *http://www.euronuclear.org/info/encyclopedia/p/powgen-ger.htm*.
10. Hille M. and W. Pfaffenberger. "Power Generation in Germany: How to Close the Gap in Generation Capacity in the Context of a Liberalized Energy Market," presented at the Conference on Applied Infrastructure Research, Berlin, 2004, *http://www.wip.tu-berlin.de/workshop/2004/papers/Pfaffenberger_Hille-power_generation_in_*

germany.pdf.

11. "Power Scenario," *http://www.ntpc.co.in/otherlinks/pwrlinks.shtml.*

12. Brunel G. "Why Steam-Turbine Technology Deserves More Attention," *VGB PowerTech* 84, no. 11 (2004): 1.

13. Horton W. and R. Peltier. "Realities restraining North American capacity expansion," *Power* 147, June 2003: 38-44.

14. Ulm W. "What Are the Market Opportunities for the Steam Turbines?" *VGB PowerTech* 83, no. 1/2 (2003): 1.

15. Voronin V.P., A.A. Romanov, and A.S. Zemtsov. "Means for the Technical Upgrading of Electric Power Engineering," *Thermal Engineering* 50, no. 9 (2003): 701-705.

16. "Handbook 2006," *Turbomachinery International* 46, no. 6 (2005): 8-147.

Chapter 2

Rise of Steam Turbine Output and Efficiency with Steam Parameters

STEAM TURBINES WITH SUBCRITICAL
MAIN STEAM PRESSURE OF THE 1960s-80s

Beginning from the early 1950s and up to the early 1990s, the increase of main steam conditions, especially main steam pressure, was the basic and most productive way for raising the efficiency of newly designed power steam turbines. A Mollier diagram with steam expansion lines for some characteristic power steam turbines is presented in Figure 2-1.

It seems reasonable to take a widespread 200-MW reheat steam turbine of LMZ K-200-130 (redesignated later to K-200-12.8) as the starting point for tracing the steam turbine efficiency rise. The first turbine of this type was manufactured in 1958, and this turbine can be considered pretty typical for the 1960s. A great number of these turbines of several modifications were installed at power plants of the former Soviet Union, as well as some European and Asian countries. Under licenses of LMZ, such turbines were also manufactured in China, India, Poland, and Romania. The heat rate of this turbine, with a subcritical main steam pressure of 12.8 MPa (1,856 psi) and both main and reheat steam temperatures of 540°C (1,004°F)—see lines 1 in Figure 2-1,—or, to say more accurately, one of its advanced modifications, with the rated output of 210 MW, designated K-210-12.8-3, was assessed to be equal to 8,045 kJ/kWh (7,626 Btu/kWh), which corresponds to the gross efficiency of 44.7%.[1] With somewhat lower steam temperatures and vacuum in the condenser, depending on features of manufacturing, the actual operating efficiency of these turbines at some power plants was lower. So, for example, at four Indian coal-fired power plants of NTPC (Vinghyachal, Kahalaon, Singrualu, and Badarpur) the rated heat consumption val-

9

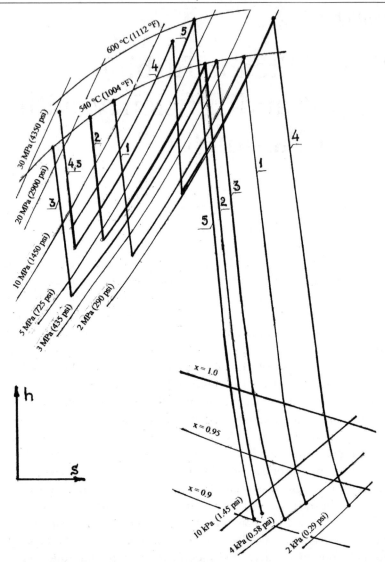

Figure 2-1. Mollier diagram with steam expansion lines for some characteristic power reheat steam turbines.
1 and 2: with subcritical main steam pressure of 12.8 MPa, 540/540°C (1856 psi, 1004/1004°F) and 16.6 MPa, 540/540°C (2407 psi, 1004/1004°F), 3: with "common" supercritical steam conditions of 23.5 MPa, 540/540°C (3414 psi, 1004/1004°F), 4: double-reheat with advanced supercritical steam conditions of 28.5 MPa, 580/580/580°C (4132 psi, 1076/1076/1076°F), and 5: with modern elevated supercritical steam conditions of 27.4 MPa, 580/600°C (3973 psi, 1076/1112°F).

Rise of Steam Turbine Output and Efficiency with Steam Parameters **11**

ues for these turbines, some of which were manufactured by LMZ and some—by BHEL, vary between 8,457 and 8,632 kJ/kWh, corresponding to variations in the turbine efficiency between 42.57 and 41.7%.

Many turbine manufacturers throughout the world produced sub-critical-pressure reheat steam turbines with a higher main steam pressure and approximately the same main and reheat steam temperatures of 530-545°C (986-1,013°F). The rated thermal gross efficiency of such a turbine of Siemens/KWU with the steam conditions of 16.6 MPa, 540/540°C (2,400 psi, 1,004/1,004°F)—see line 2 in Figure 2-1—was assessed equal to 44.6%.[2] With the boiler efficiency of 88%, including pressure drops in the steam-line system, the generator's efficiency equal to 97%, and the efficiency of auxiliaries of 95%, the power unit's net efficiency based on the lower heat value (LHV)* accounts to 36.2%.

The single capacity of turbines with the mentioned steam conditions commonly does not exceed 850-900 MW. The longitudinal section of a characteristic steam turbine of GEC Alsthom (presently ALSTOM Power) with the rotation speed of 3,000 rpm for the single capacity in the range of 500-700 MW, subcritical main steam pressure in the range of 16.5-18 MPa (2,390-2,610 psi), and main and reheat steam temperatures of 540°C (1,004°F) is presented in Figure 2-2. Many such turbines were installed at power plants of different countries, including the UK, France, China, South Africa, and others.

The mentioned output limit refers to tandem-compound (TC), or single-shaft, turbines. With transition to the cross-compound (CC) configuration, steam turbines with such steam conditions reached larger single capacities. Thus, for example, the world's first 1,000-MW steam turbine for the U.S. power plant Ravenswood's Unit 3 was CC with the steam conditions of 16.6 MPa, 538/538°C (2,400 psi, 1,000/1,000°F).

At some power units of this class, mainly applied in Germany, the rated main steam pressure was set at a heightened level of about 19-20 MPa (2,755-2,900 psi). The calculated influence of main steam parameters

---

*It is a general practice in the power industry to calculate the fossil-fuel power facility's performance based on the lower heat value (LHV) of the fuel to be burned, whereas fuel supply requirements and purchase contracts are figured on the basis of the fuel's higher heat value (HHV). The latter is measured as applied to the chemical energy of the fuel which accounts for the total heat given up when the fuel is burned, including, in particular, formation of water vapor, while the LHV measures the finally useable energy. The difference in the efficiency values based on the LHV and HHV depends on the fuel quality, and a typical difference for coal-fired power units is about 5%. So in the considered case, the net efficiency based of the HHV is equal to 36.2%x0.95=34.4%.

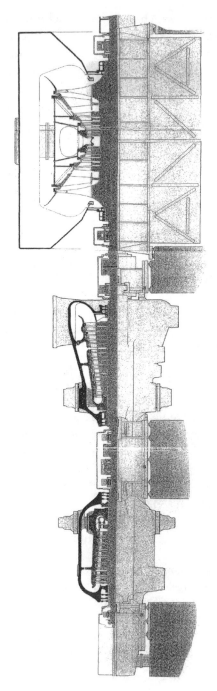

Figure 2-2. Longitudinal section of GEC Alsthom's subcritical-pressure steam turbine with the output of 500-700 MW for the rotation speed of 3,000 rpm (only one of two LP cylinders shown). *Source: By courtesy of ALSTOM*

on the available energy and thermal efficiency of the ideal Rankine cycle is shown in Figure 2-3. It should be noted that for actual steam-turbine cycles, with steam reheat and regeneration, these dependencies occur to be somewhat more complicated.[3-5]

Subcritical main-steam pressure allowed power plant designers to equip such steam-turbine units with traditional drum-type boilers, even though the use of a heightened subcritical steam pressure demanded on special arrangement of the working fluid circulation in the boiler's steam/water circuit. Along with this, some subcritical-pressure power units, especially those with higher main steam pressure values, were equipped with once-through boilers, and these units were used for working out start-up systems and operating technologies that then became obligatory for power units with supercritical steam pressure when drum-type boilers are not acceptable. Once-through boilers, compared to drum-type ones, are more sensitive to feedwater quality, and in this case full-flow condensate polishing is required to protect the turbine from stress corrosion cracking. It is true, but many other prejudices concerning once-through boilers started from early power units of this type with rather primitive start-up systems. In particular, some power plant owners and operators believe

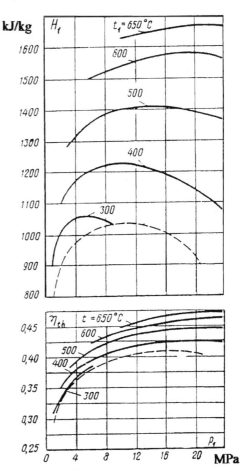

Figure 2-3. Influence of main steam conditions on available energy (a) and thermal efficiency (b) of ideal Rankine cycle.
...saturated steam. Source: A.V. Shcheglyaev[3]

that once-through boilers feature more complicated and time consuming start-up processes, that these boilers should account for more intense solid particle erosion (SPE) of turbine first stage buckets and vanes, that these boilers require higher maintenance costs, and so on. Special analysis of long-term operating experience at power plants carried out by the North-American Electric Reliability Council (NERC), as well as Electric Power Research Institute (EPRI), in the USA and European technical associations for power and heat generation (VGB in Germany and KEMA in the Netherlands) for 1988-1997 did not reveal any significant difference in power plant availability between subcritical-pressure power units (mainly with drum-type boilers) and supercritical-pressure ones.[6,7] Along with this, it should be admitted that the operating performances of power units with once-through boilers depend to a greater degree on the quality of their start-up systems, as well as skills, qualification, and experience of the operational personnel.

SUPERCRITICAL- AND USC-PRESSURE STEAM TURBINES OF THE 1960s-80s

Mass commercial implementation of power units with supercritical steam pressure (more than 21.4 MPa, or 3,108 psi) also commenced in the early 1960s. These units were mainly designed with the same main and reheat steam temperatures of about 535-545°C (995-1,013°F)—see line 3 in Figure 2-1. Owing to the increase of the main steam pressure up to 23.5 MPa (3,414 psi), with the same main and reheat steam temperatures of 540°C (1,004°F), the rated heat-rate of supercritical-pressure steam turbines of LMZ with the output of 300-to-1,200 MW decreased to 7,710-7,626 kJ/kWh, corresponding to the efficiency values varying at the level of 46.7-47.2%,[1,8] compared to the above mentioned data of 8,045 kJ/kWh and 44.7% for the subcritical-pressure 200-MW turbine of the same producer. It is understandable that the increase in the turbine output and a resulted increase in the blading length, the use of longer last stage blades (LSB) providing a decreased energy loss with the exit velocity, as well as some improvements in general design solutions, also played a certain role in this efficiency rise. Of importance is that, since the supercritical-pressure power units made up a significant portion of the installed capacity in the Integrated Power Systems of the former Soviet Union, many of them, especially gas/oil-fired 300-MW units in the European part of that country, had to take participation in covering the

Rise of Steam Turbine Output and Efficiency with Steam Parameters 15

variable part of power consumption graphs with deep unloading and/or regular shut-downs for nights and weekends with subsequent start-ups and catching up the load.[9,10]

Turning to supercritical main-steam pressure made it possible to reach the record-breaking values of the single capacity for both power units as a whole and steam turbines in particular. The greatest single capacity for a fossil-fired steam-turbine power unit became equal to 1,300 MW. The first power units of this type were put into commercial operation in the United States in 1972-73 at the power plants Cumberland and Amos. In the subsequent years, the total number of such 1,300-MW units at U.S. power plants increased to nine: two, commissioned in 1972 and 1973, have operated at the power plant Cumberland of TVA, and seven have been in service at the AEP's power plants: Amos 3 (since 1973), Gavin 1 and 2 (1974 and 1975), Mountaineer 1 (1981), Rockport 1 and 2 (1984 and 1989), and Zimmer (1991).[10] These units, equipped with CC (double-shift) steam turbines of ABB, have remained the largest among the fossil-fuel power units up to now. Somewhat later, there appeared some projects of fossil-fuel power units with the output of up to 1,600 and even 2,000 MW, but they have not come true. The largest TC (single-shaft) high-speed steam turbine, with the rated output of 1,200 MW and maximum accessible output of 1,380 MW, was developed by LMZ and commissioned in 1979 at the Kostroma power plant (Unit 9) in Russia.[1,8,11] It might be well to mention, that later, at the turn of the 21st century, this turbine was excelled in its single capacity by low-speed (1,500 rpm) TC turbines of ALSTOM called the *Arabelle*, with the rated output of 1,500 MW, installed at the French nuclear power plants Chooz and Civaux. Presently, under construction is the nuclear power unit Olkiluoto 3 (Finland) to be launched in 2009 with a low-speed TC turbine of Siemens Power Generation with the rated output of about 1,700 MW.[12] Nevertheless, the 1,200-MW turbine of LMZ remains the largest one among steam turbines for fossil-fuel power plants, even though this title was sometimes given to other machines.[13,14]

The rated steam conditions of the above mentioned 1,300-MW units installed at the U.S. power plants are mainly 24.1 MPa, 538/538°C (3,500 psi, 1,000/1,000°F). These units were preceded and followed by other supercritical-pressure steam-turbine units with approximately the same steam conditions, but lesser single capacities put into operation at U.S. power plants. They are, for example: the Widows Creek Unit 7 (500 MW, 1960), Bull Run Unit 1 (900 MW, 1966), Paradise Unit 3 (1,100 MW, 1969), and others. According to the North-American Electric Reliability Council (NERC), even for the first years of commercial operation the reliability indices (equivalent availability factor and forced outage rate) for super-critical-pressure coal-fired power units with the output of 400-799 MW at U.S. power plants, excepting the early, very first "supercriticals," were quite comparable and even better than those for similar subcritical-pressure units.[10,15] For years, supercritical-pressure units have retained their leading positions among the U.S. most efficient steam-turbine power plants. In the list of top U.S. coal-fired power plants of 2001,[16] ten of 20 power plants with the highest efficiency are completely or partially furnished with supercritical-pressure units, even though all of them were in operation for many years. For example, the operating plant heat consumption of the Bull Run power unit with the net capacity of 870 MW and standard supercritical-pressure steam conditions of 24.1 MPa, 538/538°C (3,500 psi, 1,000/1,000°F) was equal to 8,861 kJ/kWh corresponding to the efficiency of 40.63%, even though this unit was commissioned in 1966 and has been in operation for about 35 years by 2001.

The total number of supercritical-pressure steam-turbine units commissioned at U.S. power plants until 1991 had accounted to 155.[17] Thirteen of them, launched between 1967 and 1972, were double-reheat with the steam temperatures equal to 538/538/538°C (1,000/1,000/1,000°F) or 538/552/566°C (1,000/1,025/1,050°F). Excepting nine cross-compound 1,300-MW turbines and some others delivered by Brown Boveri (later ABB, then ALSTOM Power), the rest of supercritical-pressure steam turbines for the U.S. power plants were produced by GE and Westinghouse.*

Beginning from the early 1960s, supercritical-pressure steam-turbine units were put into operation in some other countries as well, aside from the United States. The greatest number of such units were installed in the former Soviet Union: 180 ones with 300-MW turbines and 15 ones with 500-MW turbines of LMZ and KhTGZ, 14 units with 800-MW turbines, and one with 1,200-MW turbine of LMZ; this list should be also supplemented with 22 cogeneration units with the rated capacity of 250/300 MW with steam turbines produced by Ural Turbine-Engine Works (TMZ). Longitudinal sections of HP and IP cylinders for the 800-MW TC turbine of LMZ (K-800-23.5-5) are shown in Figure 2-4, and Figure 2-5 demonstrates LP cylinders of the 800-MW and 1,200-MW

---

*Another source calls the number of the U.S. supercritical-pressure power units equal to 164, including pilot ultra-supercriticals, as well as some other experimental facilities.[7]

Rise of Steam Turbine Output and Efficiency with Steam Parameters 17

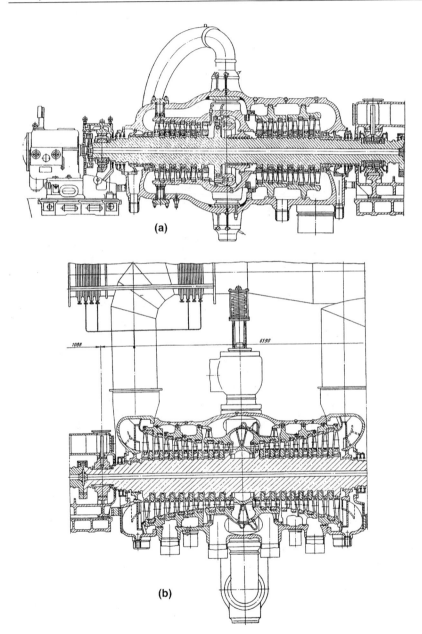

Figure 2-4. Longitudinal sections of HP (a) and IP (b) cylinders for an 800-MW supercritical-pressure turbine of LMZ. *Source: By courtesy of LMZ*

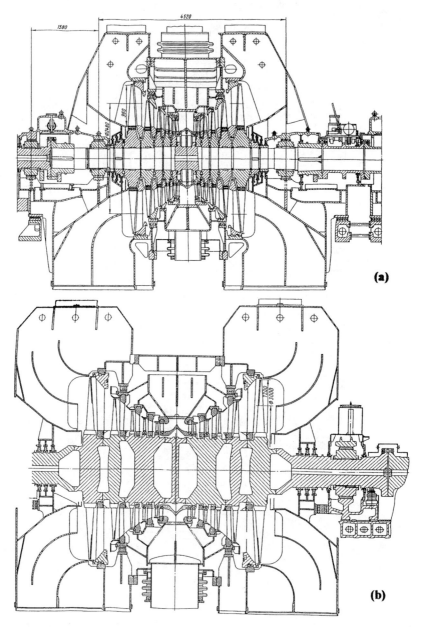

Figure 2-5. Longitudinal section of LP cylinders for 800-MW (a) and 1,200-MW (b) supercritical-pressure turbines of LMZ. *Source: By courtesy of LMZ*

turbines (each one has three LP cylinders). Longitudinal section of the 300-MW supercritical-pressure steam turbine of LMZ (K-300-240) can be seen in Figure 12-15. Supercritical-pressure steam turbines of LMZ were also installed in some other countries, such as Argentina, China, Iran, and Yugoslavia.

Despite the greater number of supercritical-pressure units commissioned in the Soviet Union, it ranked below the United States in their total capacity. Considerable number of supercritical-pressure steam-turbine units were launched in Japan; some units of this class were also constructed and implemented in Germany, Italy, Denmark, the UK, China, the Netherlands, Finland, and some others.

Comparative number and total capacity of supercritical-pressure steam-turbine units put into operation in diverse countries before 1991 is given in Table 2-1. The border of 1991 is chosen because it was the year, when the former Soviet Union collapsed and disintegrated. In addition, just in 1991 the latest until now supercritical-pressure steam-turbine unit was commissioned in the United States (the Zimmer power plant). Even though the U.S. supercritical-pressure steam-turbine units quite assured their high operating efficiency and reliability, impressive developments of the 1990s in gas turbine technologies plus relatively low prices of those years for natural gas provoked a boom in implementing power generating facilities with the use of simple-cycle gas turbines and combined-cycle (CC) units that featured pretty high efficiency, low capital expenditures, and lower gas emissions. They almost completely took over the investments for newly designed and constructed power capacities. It is characteristic that in the list of 488 U.S. power facilities over 50 MW recently completed, being under construction, and planned for constructing in 2001-2005 only 19 ones were steam-turbine-based, including 13 coal-fired power units with the output over 300 MW, 18 wind power farms, 4 hydroelectric facilities, and all the rest were simple-cycle gas turbines and CC units.

In the early 1960s, the United States also pioneered in developing and mastering pilot commercial steam-turbine power units of ultra-supercritical (USC) pressure, that is, with the main steam pressure of 30 MPa and above (over 4,350 psi).* Most of these units were also designed with elevated main steam temperatures of 593°C (1,100°F) and more. In

---

*Sometimes, the term of USC is erroneously applied to supercritical-pressure units with the main steam pressure lower than 30 MPa, but with elevated main and/or reheat steam temperatures—593°C and more.

Table 2-1. The number and total capacity of commercial supercritical-pressure steam-turbine power units (without ultra-supercritical-pressure ones) put into commercial operation in diverse countries until 1991

Country	Number of Units	Total Capacity, GW	Main Turbine Manufacturers
USA	155	106.2	GE, Westinghouse, ABB
Soviet Union	232	79.4	LMZ, KhTGZ, TMZ
Japan	76	56.0	MHI, Hitachi, Toshiba, Siemens, GE
Italy	16	10.56	Westinghouse, Ansaldo
Germany	16	9.19	Siemens, ABB
Denmark	7	2.43	ABB, Siemens
China	5	1.74	LMZ
UK	4	1.5	GEC Alsthom
Others	13	4.94	Siemens, LMZ, Westinghouse, ABB

Mainly based on data of P. Luby[17]

addition, the increased main steam pressure required to increase the reheat steam temperature or use double-reheat to avoid inadmissibly high wetness at the LP exhaust. The most well-known of these pilot USC units were 125-MW Philo Unit 6 with the turbine of GE and steam parameters of 31 MPa, 621/565/538°C (4,500 psi, 1,150/1,050/1,000°F) and 325-MW Eddystone Unit 2 with the turbine of Westinghouse and steam conditions of 34.5 MPa, 650/565/565°C (5,000 psi, 1,200/1,050/1,050°F). Some experimental power units were operated in Germany: for example, the Hattingen Units 2 and 4 of a 107-MW output each, with a low supercritical main steam pressure of 22.1 MPa (3,205 psi), but elevated main steam temperature of 600°C (1,112°F); they were preceded by a few experimental steam turbines of a small capacity with advanced main steam temperatures of up to 650°C (1,200°F). In 1967, a semi-commercial unit with a 100-MW back-pressure turbine of KhTGZ for steam conditions of 29.4 MPa, 650/565°C (4,350 psi, 1,200/1,049°F) was implemented at the Kashira power plant near Moscow, in the USSR; the outlet of this turbine was connected to three existing older 50-MW condensing turbines of medium steam conditions.[10] The use of "ultra-high" main steam temperatures required the use of austenitic steels for the most-stressed high-temperature components of boilers and turbines, as well as hot steam-lines. One of main disadvantages of such steels is the fact that they are ill-welded with common Cr-steels. In addition, austenitic steels are more vulnerable to unsteady thermal stresses arising at the transient operating conditions. Shortcomings of austenitic steels were mainly responsible for temporarily abandonment of the use of advanced steam temperatures in steam-turbine units, and in this case the USC steam pressure unavoidably required the use of double-reheat to get an acceptable wetness degree for the LP last stage blades.

In 1989-91 two USC-pressure double-reheat 700-MW 3,600-rpm steam-turbine units with the steam conditions of 31 MPa, 566/566/566°C (4,495 psi, 1,050/1,050/1,050°F) were put into operation at the Japanese power plant Kawagoe.[10,18-20] The turbine of such a unit (Figure 2-6), developed by Toshiba, consists of the integrated super-HP-and-HP cylinder, double-flow IP one, and two double-flow LP cylinders. Most of turbine features, with exception of the integrated SHP-HP cylinder, are very similar to those of preceding supercritical-pressure 700-MW units for Japanese power plants. Field tests of the Kawagoe Unit 1's turbine completely confirmed its expected high performances. The turbine's gross (thermal) efficiency was found equal to 48%, as compared to 46.1% for conventional supercritical 700-MW turbines of the same producer; the unit's net efficiency at the rated operating conditions amounted to 41.9%. Despite the USC steam conditions, the Kawagoe units were designed to be capable of operating in a cycling mode.

For a time, transition to the USC steam pressure and double-reheat scheme with a simultaneous increase of both the steam pressure and temperatures was considered the general way for the further increase of the power generation efficiency that should also decrease the environmental impact of newly designed fossil-fuel power plants. The diagram of assumed progress in the power plant efficiency of supercritical- and USC-pressure power units, according to Hitachi, is shown in Figure 2-7.[21] According to Hitachi and Toshiba, the next design evolution step was supposed to be presented by USC double-reheat 1,000-MW turbines with the steam pressure of 31 MPa (4,495 psi) and steam temperatures of 593°C (1,100°F). These turbines were expected to have a configuration similar to the Kawagoe turbines. However, such units have not appeared, and it can be said that this notion was defeated by another approach to raising the steam turbine efficiency as applied to newly de-

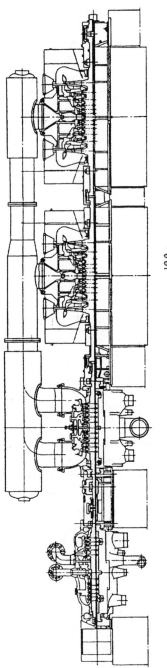

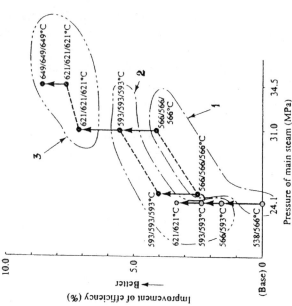

ABOVE: Figure 2-6. Longitudinal section of a 700-MW USC steam-pressure double-reheat turbine of Toshiba for the Japanese power plant Kawagoe's Units 1 and 2. Source: H. Mimuro et al.[19]

RIGHT: Figure 2-7. Expected improvement in thermal efficiency for USC double-reheat fossil-fuel power units compared to common supercritical-pressure ones, according to Hitachi. 1: first stage with conventional steam temperatures, 2: zone of steam conditions accessible with the use of present technologies, 3: zone of technologies to be developed. Source: S. Moria, M. Haraguchi, and Y. Yamazaki[21]

MODERN STEAM TURBINES WITH
ELEVATED STEAM CONDITIONS

signed fossil-fuel power units, a new paradigm in steam-turbomachinery bringing about a new remarkable increase in the turbine efficiency.

Speaking only of turbines, not touching on boilers and high-temperature steam-lines, a new breakthrough primarily ensued from two main factors. The first one was the development of new heat-resistant high-chromium-percentage steels that enable steam turbines to reach pretty elevated steam temperatures without the use of austenitic steels. And the second one was the development and mass implementation of new advanced approaches to the turbine steam path design. As well, noteworthy were remarkable achievements in developing longer LP last stage blades (LSBs) that further decreased exit energy losses and increased the steam turbine efficiency. As a result, there appears a possibility to increase considerably the turbine efficiency without resorting to the use of USC double-reheat steam conditions, which, requiring a special super-HP (SHP) turbine section, highly-stressed hot USC steam-lines, and additional steam-lines of the second reheat, would make the power unit's scheme and turbine design excessively complicated. These circumstances gave birth to a new trend in setting the steam conditions for the most advanced steam-turbine units.

Of interest is an experience of developing and mastering fossil-fuel power units with advanced steam conditions launched in Denmark just at the turn of the century. After a few conventional supercritical-pressure steam turbine units, with the steam conditions of 25 MPa, 540/540°C or 560/560°C (3625 psi, 1,004/1,004°F or 1,040/1,040°F) commissioned in 1973-92, in 1997-8 the Danish electric power utility ESLAM launched two double-reheat 410-MW units Skærbæek 3 and Nordjyllandsværket 3 with steam turbines of ALSTOM for the steam conditions of 29.0 MPa, 582/580/580°C (4,205 psi, 1,080/1,076/1,076°F)—see lines 4 in Figure 2-1.[22,23] The turbine is a five-cylinder machine with a single-flow very-high pressure (VHP) cylinder, integrated HP-IP cylinder with single-flow sections after the first and second steam reheat, additional double-flow IP cylinder with asymmetrical flows IP1 and IP2, and two double-flow LP cylinders separately fed with steam from the IP1 and IP2 sections. The use of a double-reheat scheme was, in particular, desirable because of a deep vacuum in the condensers owing to the use of cold sea water in the circulating contours. Schematic flow diagram of the turbine and its general view are given in Figure 2-8. The unit's net efficiency based on the LHV makes up to 47%.

The Skærbæek 3 and Nordjyllandsværket 3 units were followed by the USC single-reheat cogeneration unit Avedøre 2 with the steam conditions of 30.0 MPa, 580/600°C (4,350 psi, 1,076/1,112°F)—see lines 5 in Figure 2-1. The unit entered commercial operation at the end of 2001.[24] The Avedøre Unit 2 consists of three main modules: the USC power unit burning mainly gas or coal with a small amount of heavy oil as supplementary fuel, two gas-turbines providing peak load generation and used

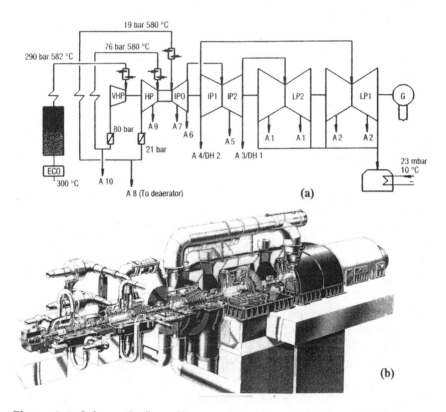

Figure 2-8. Schematic flow diagram (a) and outside view (b) of the supercritical-pressure double-reheat 410-MW steam turbine of ALSTOM for Danish power unit Nordjyllandsværket 3

VHP—"very-high pressure" cylinder. *Source: V. Jensen*[23]

to preheat feedwater to the USC boiler, and biomass-burning boiler, providing additional steam to the USC steam turbine. The turbine, delivered by Ansaldo Energia, can provide the maximum electric output of 535 MW and maximum heat output of 620 MJ/s. Because of significant controllable steam bleedings, the turbine has a rather complicated five-cylinder configuration, with separate single-flow HP and IP1 cylinders, additional double-flow IP2 cylinder (with asymmetrical left-hand and right-hand flows, each feeding its "own" LP cylinder), and two double-flow LP cylinders with butterfly-type valves installed in the crossover pipes before the cylinders. The turbine steam path was designed with the use of an advanced three-dimensional (3D) methodology. All the mentioned circumstances resulted in a pretty high efficiency of the unit. In a pure condensing mode of operation, without steam extractions for district heating, without the biomass boiler and gas turbines, according to the data of actual operating experience, the unit's net efficiency based on low heat value (LHV) reaches up to 49.5% when firing gas and 48.0% for coal, making this unit the most efficient in the world.[25]

A comparative diagram of the efficiency for the most advanced Scandinavian supercritical- and USC-pressure fossil-fuel power units against some relatively new subcritical-pressure units of the UK and Hong Kong is given in Figure 2-9. Of importance is a difference in the efficiency between Avedøre 2 and Nordjyllandsværket 3 in favor of the former, even despite the double-reheat scheme of the latter and quite insignificant difference in their main steam pressure. The further raising of steam conditions up to 38.3 MPa, 702/720/720°C (5,554 psi, 1,296/1,328/1,328°F), according to the project called the *Termie 700*, would make it possible to reach the net efficiency of up to 55%. However, this project was postponed.

The latest experience of designing and mastering the newest USC- and supercritical-pressure steam-turbine units put into operation in 1992-2005 in diverse countries throughout the world are given in Table 2-2, as if a continuation of Table 2-1.

It stands out that since 1992 the United States has constructed and put into operation no supercritical and USC power units. This fact, as well as some others (in particular, no new nuclear power units ordered and constructed since 1978), gave ground to Jason Makansi, former editor-in-chief of *Power*, to write, "The US quickly losing its renowned leadership in the commercial application of advanced technologies for power generation [that] are being applied commercially at large scale, or

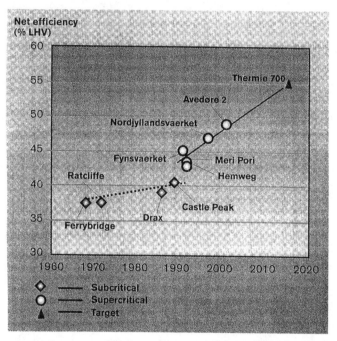

Figure 2-9. Progress in the net efficiency for modern Scandinavian supercritical and USC fossil-fuel steam-turbine units supercritical and USC units.
...subcritical units of the UK and Hong Kong. *Source: G. Welford*[26]

will be soon, but at offshore locations."[27] A similar note can be addressed to Russia as the Soviet Union's scientific and technological successor; it has kept on the implementation of supercritical-pressure steam-turbine units, mainly developed and renewed in the past century, with their quite moderate steam conditions, without any progress. True, from time to time, there appears information about projects of new power units with advanced steam conditions in both the USA[28,29] and Russia[30-33], but all of them seem to be rather far from materialization.

Korea and China are carrying out their impressive programs of constructing and implementing new large fossil-fuel capacities. Even though most parts of the Korean and Chinese facilities presented in Table 2-2 are the units with a rather conventional supercritical steam pressure of about 25 MPa (3,625 psi) and main and reheat steam temperatures of about 540°C (1,004°F),[34-38] some newest power units are supposed to operate with more advanced steam conditions. The turbines for these

Rise of Steam Turbine Output and Efficiency with Steam Parameters

Table 2-2. The number and total capacity of commercial supercritical and USC steam-turbine power units put into operation in diverse countries in 1992-2005

Country	Number of Units	Total Capacity, GW	Main Turbine Manufacturers
Japan	25	20.06	Hitachi, MHI, Toshiba, Siemens, GE
Korea	24	13.54	GE, Doosan
China	18	11.6	Siemens, LMZ, MHI, ABB
Germany	13	9.4	Siemens, ABB
Russia	4	2.7	LMZ
Denmark	4	1.8	Siemens, ABB, Ansaldo
Australia	4	1.72	Toshiba, Ansaldo
USA	—	—	—
Others	11	5.77	Siemens, MHI, LMZ, ALSTOM

Mainly based on data of P. Luby[17]

projects are to be manufactured by foreign companies in cooperation with domestic ones: GE and Doosan Heavy Industries & Construction for Korean 500-MW units with steam conditions of 24.2 MPa, 566/593°C (3,510 psi, 1,050/1,100°F) to be put into operation at the Tangjin Units 5 and 6 in 2005-6[39] and Siemens Power Generation with participation of Shanghai Turbine Works for Chinese 1,000-MW units with steam parameters of 26.2 MPa, 600/600°C (3,800 psi, 1,112/1,112°) to be commissioned at the Yuhan beginning from 2006 and then at the Wajgaoqiao-2 power plants.[14,40]

Taking in consideration all these facts, it can be said that nowadays the main future development trends in the fossil-fuel power industry and steam turbomachinery are mostly shaped and worked out in Germany and Japan, playing leading roles in these fields. In particular, this can be seen from consideration of the design, first operation experience, and field test results for two noteworthy power units commissioned at the turn of the century in these two countries. They are the 907-MW unit Q of the power plant Boxberg in Germany and 1,050-MW Tachibana-wan Unit 2 in Japan.[41] The efficiency figures for these two units provides a benchmark for newly developed power plants. Meaningful is that, unlike the Cawagoe power plant, which burns natural gas, both of the considered power units burn solid fuel. As to the Tachibana-wan power plant, with commissioning its second unit, it reached the installed capacity of 2,100 MW and became the largest, along with the Hekinan power plant (3x700 MW), and most efficient coal-fired power plant of Japan.

The lignite-fired power unit Boxberg Q with a turbine of Siemens (Figure 2-10) went on line in June 2000 and passed acceptance tests in October of the same year. The turbine's gross efficiency was found equal to 48.5% with the unit's net efficiency of 42.7%; the acceptance tests also demonstrated the internal efficiencies for the HP and IP sections equal to 94.2% and 96.1%, respectively.[42,43] Its steam conditions of 26.6 MPa, 545/581°C (3,860 psi, 1,013/1,078°F) differ only in a somewhat higher reheat steam temperature from other German power units of the recent years, such as other lignite-burning power units Schkopau 2x400 MW, Schwarze Pumpe 2x800 MW, and Lippendorf 2x930 MW, as well as hard-coal-fired supercritical units, such as the 510-MW Staudinger Unit 5, 550-MW unit for power plant Rostock, 750-MW Bexbach Unit 2, 910-MW Heyden Unit 4, and some others. Higher steam conditions were taken only for the Hessler's 720-MW unit: 27.5 MPa, 578/600°C (3,990 psi, 1,072/1,112°F).[44-47]

One of the most characteristic German power plants, preceding the Boxberg Q unit, is the Schwarze Pumpe (2x800 MW), with supercritical steam conditions of 26.8 MPa, 547/565°C (3,886 psi, 1,017/1,049°F) and four-cylinder steam turbines of Siemens (single-flow HP and double-flow IP cylinders and two double-flow LP ones). It was the first ever experience of using supercritical-pressure steam conditions at lignite-fired boilers. The unit's flow chart is presented in Figure 2-11.[48] The plant's feature is a discharge of cleaned flue gas into the atmosphere through the cooling tower. (This design solution was earlier worked up at the Staudinger Unit 5.) Before being desulphurized and entering the cooling tower to be discharged, flue gas is cooled by the turbine's main condensate. The turbines are made with huge controllable steam extractions for the adjacent briquette factory and district heating. The turbine's two condensers are connected in sequence with the cooling water. The plant was completed in 1998. The acceptance tests demonstrated the net efficiency of 41.1% relating to the plant as a whole; the net efficiencies of over 41% are also supposed for other new power plants to be built in Germany.[49]

The Tachibana-wan Unit 2 was put into commercial operation in December 2000. With the gross efficiency of 49.0%, this turbine of MHI (Figure 2-12) has been acclaimed the most efficient worldwide.[50] The unit's steam conditions of 25 MPa, 600/610°C (3,625 psi, 1,112/1,130°F)

Rise of Steam Turbine Output and Efficiency with Steam Parameters   29

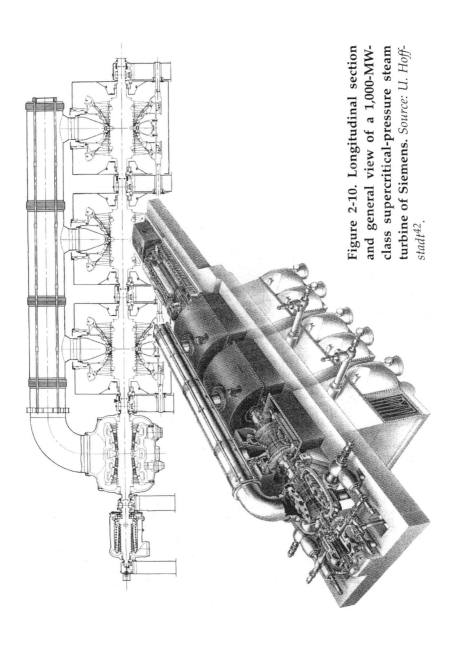

Figure 2-10. Longitudinal section and general view of a 1,000-MW-class supercritical-pressure steam turbine of Siemens. *Source: U. Hoffstadt[42].*

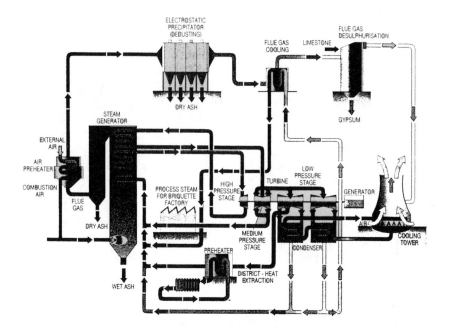

Figure 2-11. Flow chart for power units of Schwarze Pumpe. *Source:* "Schwarze Pumpe: a new era..."[48]

represent the next step in the staircase of raising the main and reheat steam temperature level shown in the diagram of Figure 2-13. In this process, the main steam pressure has remained approximately invariable at the level of about 24-25 MPa (3,480-3,625 psi), whereas the main and reheat steam temperatures considerably grew. The preceding power units had the main and reheat steam temperatures at the level of 593 and 600°C (1,100 and 1,112°F)—beginning from the 700-MW Hekinan Unit 3 and up to the 1,000-MW Matsuura 2 and Misumi 1 Units.[51–53] To a great degree, this steam temperature growth has been also substantiated by practice gained at an experimental 50-MW steam turbine of the Wakamatsu experimental plant with the main and reheat steam temperatures of 593°C at the first investigation stage (1986-90) and main steam temperature of 650°C at the second stage, after 1990.[54]

A similar process of raising the steam temperatures went at power plants with steam turbines of other Japanese turbine producers. Thus, in 1998 Hitachi implemented the 1,000-MW Haramachi Unit 2, with the steam conditions of 24.5 MPa, 600/600°C (3,550 psi, 1,112/1,112°F),

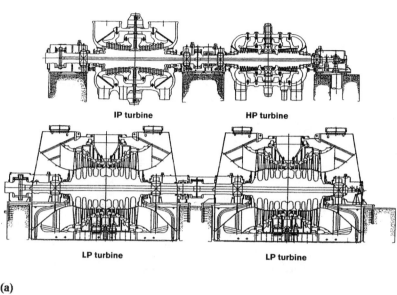

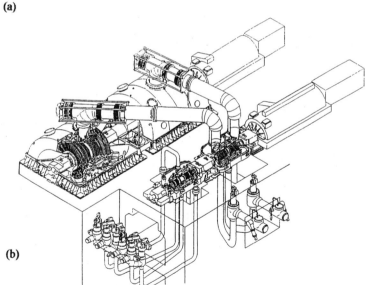

Figure 2-12. Longitudinal section (a) and outside view (b) of a 1,000-MW-class CC steam turbine of MHI for a supercritical steam pressure and elevated main and reheat temperatures. *Source: By courtesy of Mitsubishi Heavy Industries*

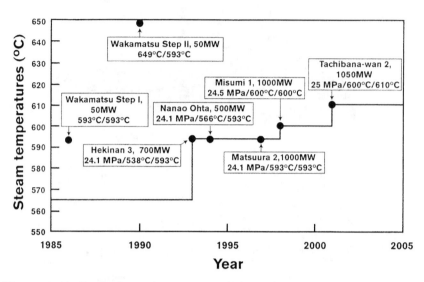

Figure 2-13. Evolution of steam conditions for MHI steam turbines.
Source: By courtesy of Mitsubishi Heavy Industries

and supposed to launch in 2005 the first power unit with the main and reheat steam temperatures equal to 630°C (1166°F).[55-57] According to Hitachi, the raise of steam conditions from 24.1 MPa, 538/566°C (3,495 psi, 1,000/1051°F) to 24.5 MPa, 600/600°C (3,552 psi, 1,112/1,112°F) as applied to the 1,000-MW power units Shinchi 1 and Hitachinaka 1 (commissioned in 2003) should increase the thermal efficiency from 41.89% to 43.78%.[58] Similarly, after two USC-pressure double-reheat 700-MW turbines of Kawagoe, Toshiba produced and implemented a supercritical-pressure 700 MW turbine with the main and reheat steam temperatures of 593°C (1,100°F) at the Nanao power plant and two supercritical-pressure 1,000 MW TC turbines of the Hekinan Units 4 and 5 with the steam temperatures of 566/593°C (1,050/1,100°F); in the nearest future, it is supposed to raise the steam temperatures for new power units to 610°C (1,130°F).[59]

The gross efficiency value for the most efficient steam turbines of German and Japanese power plants put into operation at the end of the 20th century, including those of the Boxberg Unit Q and Tachibana-wan Unit 2, are brought together in Table 2-3.

Of significance is that these turbines achieved their close efficiency values at materially different steam conditions. According to MHI, a

Rise of Steam Turbine Output and Efficiency with Steam Parameters 33

Table 2-3. The gross efficiency data for the most efficient German and Japanese steam turbines of the recent years

Power Unit, Country	Rated Output, MW	Steam Turbine Producer	Steam Conditions, MPa-°C (psi-°F)	Turbine Efficiency, %	Put in Operation in (year)
Hekinan Unit 3, Japan	700	MHI	24.0-538/593 (3,480-1,000/1,100)	47.4	1993
Heßler, Germany	720	ABB	27.5- 578/600 (3,990-1,072/1,112)	47.6	1997
Kawagoe Units 1 &2, Japan	700	Toshiba	31.0- 566/566/566 (4,495-1,050/ 1,050/1,050)	48.4	1989
Boxberg Unit Q, Germany	907	Siemens	26.6, 545/581 (3,860-1,013/1,078)	49.0	2000
Tachibana-wan Unit 2, Japan	1,050	MHI	25.0- 600/610 (3,625-1,112/1,130)	49.2	2000

steam temperature increase from 538/593°C (1,000/1,100°F) to 600/600°C (1,112/1,112°F), with the same steam pressure, makes a turbine more efficient by 2.2%;[50] that is, its efficiency rises by approximately 1.1%. According to calculations of Siemens, raising the steam conditions from 25 MPa, 540/560°C (3,625 psi, 1,004/1,040°F) to 27 MPa, 585/600°C (3,915 psi, 1,085/1,112°F) should increase the turbine efficiency by 1.1%.[60,61] So, closeness of the actual efficiency values for the same capacity class turbines with remarkably different steam temperatures, and with regard to differences in the condenser vacuum, feedwater heating, and so on, says that at least some of these turbines have noticeable reserves for increasing their efficiency by reducing losses in the steam path and using more advanced design solutions. What is more, all these turbines have potentials for improving their internal efficiency further. (A comparative consideration of main design features for modern large power steam turbines, including the above mentioned ones, will be given in the next sections.)

POTENTIAL RISE OF STEAM CONDITIONS
IN THE NEAREST FUTURE

Even though the efficiency values reached at the considered power units are quite impressive, they likely represent only interim, temporary benchmarks to be surpassed in the nearest future. A potential champion could be, for example, the largest German lignite-fired Niederaußem Unit K, commissioned in 2002.[62-64] Compared to the Boxberg Unit Q, the Niederaußem Unit K, equipped with a Siemens turbine similar to the Boxberg Q one, has more advanced steam conditions of 27.4 MPa, 580/600°C (3,990 psi, 1,076/1,112°F), lower condenser pressure, and a 25% larger exhaust area thanks to the use of longer LSBs. This units, with its optimized engineering (called BoA), at the first stage of its operation achieved an efficiency level of 43%,[65] but it is expected to have 45.2%. For comparison, the previous 600-MW units of the same plant (Units G and H), with the steam conditions of 17.4 MPa, 530/530°C (2,525 psi, 986/986°F), commissioned in the mid-70s, have the net efficiency of 35.5%.

The next lignite-fired 670-MW power unit is to be built at the same Boxberg power plant site (Unit R); it should have steam conditions of 28.5 MPa, 600/610°C (4,132 psi, 1,112/1,130°F) and reach the net efficiency of almost 44%. Two 820-MW hard-coal-fired power units with the same steam conditions are planned to be constructed at the site of the former Moorburg power plant.[66] An even higher net efficiency value is targeted for the power plant Westfalen's newly projected Unit D, with the single capacity of 350 MW and highest in Germany steam conditions of 29 MPa, 600/620°C (4,210 psi, 1,112/1,148°F).[47] The same steam conditions are set as applied to a so-called "reference power plant" for the North Rhine-Westphalia (RPP NRW) project.[67,68] The gross plant capacity is set equal to 600 MW (the net output of 556 MW). The RPP NRW is expected to reach the net efficiency of 45.9%.

High performances are also expected from new Japanese power units, due to further heightening their steam conditions—up to 30 MPa, 630/630°C (4,350 psi, 1,166/1,166°F) and combined employment of developed advancements in the turbine steam path.[57] True, the next power units commissioned in Japan after the Tachibana-wan Unit 2 had the main and reheat steam temperatures not exceeding 600°C (1,112°F)—for example, the coal-fired 700-MW Tomatoh-Atsuma Unit 4 (commissioned in 2002), 1000-MW Hitachinaka Unit 1 (2003), 600-MW Hirono Unit 5 and 900 MW Maizuru Unit 1 (both commissioned in 2004), and others.[58,69-71]

Heat rate improvements in the past, present and nearest future for fossil-fuel power plants, due to raising their steam conditions as applied to technologically developed and economically and ecologically driven markets, are shown in Figure 2-14.[72] According to ALSTOM Power, steels suitable for a supercritical process with a steam pressure up to 27 MPa (3,915 psi) and main and reheat steam temperatures of 580 and 605°C, respectively (1,076/1,121°F), are now successfully in the use at many power plants and have completely met with expectations. A further rise of steam temperatures needs steels with higher long-term creep strength. Such an innovation of design materials has advanced, in particular, thanks to the COST program (Co-Operation in the field of Scientific and Technical researches) supported by the European Union. Different types of steel with high long-term creep strength at high process temperatures are being developed and tested.

The next chart of Figure 2-15 depicts the key influence of developments in high-temperature resistant materials and component optimization on the net power plant efficiency as applied to a hypothetical 700-MW power unit.[61,73,74] According to Siemens, the existent chromium steel P91 has permitted the use of steam conditions of up to 27 MPa, 580/600°C (3,915 psi, 1,046/1,112°F), whereas the further transition to

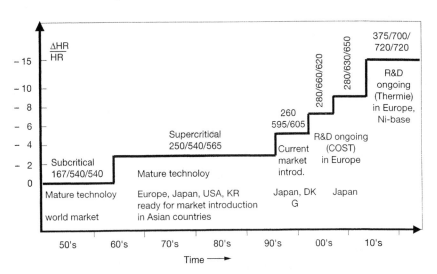

Figure 2-14. Gradual advancement of materials for steam-turbine power plants and its influence on power generation efficiency. *Source: A. Tremmel and D. Hartmann*[72]

the wolfram-alloyed steel NF12 should enable steam parameters of 30 MPa, 625/640°C (4,350 psi, 1,157/1,184°F). Research and development efforts have been started to qualify nickel-based alloys for the hottest design components, making it possible to raise the steam temperatures up to 700/720°C (1,262/1,328°F). The right-hand side of the chart shows potentials of component developments in heat cycle optimization. With all these measures, the net power plant efficiency of 50% seems to be quite reachable and even visible.

The steam condition increase really enables a considerable rise of the turbine's and power unit's efficiency. Along with this, as said above, remarkable reserves in the turbine efficiency increase lie in improvements of the internal turbine design, and these possibilities can be materialized with much lesser expenditures than those needed for the steam condition increase up to their limit values. The considered data of Table 2-3 lead to the conclusion that a very favorable level of the thermal efficiency, quite comparable to that attainable with the use of extreme steam conditions and rather complicated scheme and design solutions, could be reached and kept even presently and in the nearest future with the use of rather moderate supercritical steam pressure and moderately elevated steam temperatures—up to 593-600°C (1,100-1,112°F), thanks to

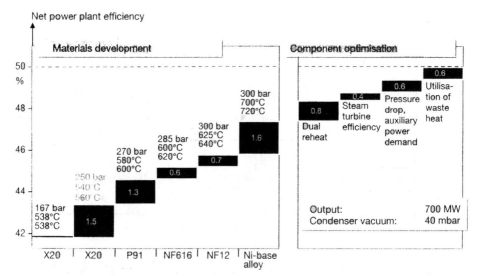

Figure 2-15. Potential efficiency increase for hard-coal-fired power plants. *Source: B. Rukes and R. Taud[74]*

an advanced design of the turbine's steam path. It is especially important for electric power utilities with a lack of experience in operating supercritical-pressure units and once-through boilers.

From this point of view, it seems absolutely reasonable that, for example, the first Canadian supercritical-pressure unit at the power plant Genesee Phase 3, commissioned at the end of 2004, has pretty conservative steam conditions of 25.1 MPa, 570/568°C (3,640 psi, 1,058/1,054°F).[75-78] With the rated output of 495-MW and efficiency of 43.6%, or a gross heat-rate of 8,524 kJ/kWh, it is considered to be 18% more efficient than an average existent coal-fired unit of Alberta, Canada. The unit project is based on a proven 500-MW Hitachi reference plant, capable of operating at sliding steam pressure. The unit's turbine, as well as its other main power equipment, is produced and delivered by Hitachi. It is a very compact two-cylinder TC machine with the integrated HP-IP cylinder and double-flow LP one.

Similarly, the first after 14 years U.S. supercritical-pressure unit for the Council Bluffs Energy Center (CBEC) is constructed for the steam conditions of 24.7 MPa, 565/593°C (3,675 psi, 1,050/1,100°F) and with Hitachi's power equipment.[79,80] The net output of the turbine is to be 790 MW; according to other sources, the rated net output is 870 MW. This will be the largest Hitachi steam turbine produced for service outside of Japan. The project was supposed to be completed in 2007. In the same year, it is assumed to begin construction of a 950-MW USC-pressure coal-fired unit called the Red Rock Generating Facility for Public Service Company of Oklahoma to be on-line in 2011.[81]

By the same token, for example, the first supercritical-pressure power units of Thailand (the Ratchaburi Units 1 and 2, with the gross output at the generator terminals of 735 MW, in operation since 1999-2000) were designed by Mitsubishi Heavy Industries with very moderate steam parameters of 24.2 MPa, 538/566°C (3,510 psi, 1,000/1,050°F).[82] The same steam conditions were set for the first two supercritical-pressure steam turbines produced by MHI for Taiwan and put in operation in 1999. In Australia, first supercritical-pressure power units commissioned in 2001-3 (Tarong North—450 MW, Callide C—2x420 MW, and Millmerran—425 MW), with steam turbines produced by Hitachi (first two projects) and Ansaldo, have main and reheat steam temperatures at the level of 566°C (1,050°F), and the new supercritical-pressure power unit of 750 MW for the power plant Kogan Creek, to be the largest in the country, is also projected for the very conservative steam parameters of 25 MPa, 540/560°C (3,628 psi, 1,004/1,040°F). The scheduled commercial operation date is August 2007; the unit's turbine is delivered by Siemens.[83,84]

It was mentioned before that all the first 500-MW supercritical-pressure units of Korea have conservative steam conditions of 24.1 MPa, 538/538°C (3,495 psi, 1,000/1,000°F), the 800-MW units have main and reheat steam temperatures of 566°C (1,050°F), and only after about ten-year experience of these supercritical-pressure units the steam temperatures for the newest 500-MW units were raised up to 566/593°C (1,050/1,100°F).[34] Similarly, all the early Chinese supercritical-pressure power units, with steam turbines of the single capacity of 320, 500, 600, 800, and 900 MW produced by ABB, LMZ, MHI, and Siemens, have the main and reheat steam temperatures not exceeding 540/566°C (1,000/1,050°F). The flowchart diagram for the 800-MW LMZ turbine as applied to the coal-fired power plant Suijun (the first two units of four went into commercial operation in 2000) is shown in Figure 2-16.[37] As applied to the power plant Waigaoqiao Phase II (2x900 MW) with steam turbines of Siemens, put into commercial operation in 2004, with approximately the same steam conditions of 24 MPa, 538/566°C (1,004/1,050°F), it is supposed to reach the net efficiency based on the LHV a little over 42%.[35] And only after accumulating some experience in operating these early (for that country) supercritical units, the first supercritical-pressure power plants with really advanced steam temperatures of 600°C (1,112°F) were ordered: 1,000-MW units for the power plants Yuhan, also with Siemens turbines.[38,40] The power plant is projected to use relatively warm sea water for cooling the condensers. With the resulted relatively low vacuum in the condensers, the expected power plant efficiency is assessed at the level of somewhat over 44%, and the turbine is designed as a four-cylinder machine—with two double-flow LP cylinders (as distinct from the above mentioned 1,000-MW-class turbines of Siemens with three LP cylinders).[14]

References

1. Trukhnii A.D. *Stationary Steam Turbines* [in Russian], 2nd edition, Moscow: Energoatomizdat, 1990.
2. Termuehlen, H. *100 Years of Power Plant Development. Focus on Steam and Gas Turbines as Prime Movers*, New York: ASME Press, 2001.
3. Shcheglyaev A.V. *Steam Turbines* [in Russian], 2 vols., 6th ed., revised and expanded by B.M. Troyanovskii, Moscow: Energoatomizdat, 1993.
4. Elliot T.C., K. Chen, and R.C. Swanekamp, editors. *Standard Handbook of Powerplant Engineering*, 2nd ed., New York: The McGraw-Hill Companies, 1998.
5. El-Wakil M.M. *Powerplant Technology*, New York: McGraw-Hill Book Co., 1984.

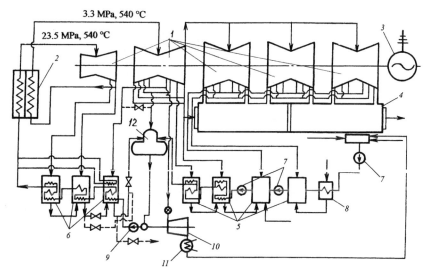

Figure 2-16. Steam/water flowchart for supercritical-pressure 800-MW steam turbine of LMZ for Chinese power plant Suijun
1: steam turbine K-800-240-5, 2; boiler, 3: generator, 4: condensator, 5: LP feedwater-heaters, 6: HP feedwater heaters, 7: condensate pumps, 8: gland steam condenser, 9: boiler feed pump (BFP), 10 and 11: BFP's driving steam turbine and its condenser, 12: deaerator. *Source: V.A. Kuznetsov et al.*[37]

6. Schimmoller B.K. "Supercritical Technology Looks to Trump Subcritical Designs for New Coal Capacity," *Power Engineering* 105, no 2 (2001): 34-37.
7. Viswanathan R., A.F. Armor, and G. Booras. "A critical look at supercritical power plants," *Power* 148, April 2004: 42-49.
8. Ogurtsov A.P. and V.K. Ryzhkov, editors. *Steam Turbines of Supercritical Pressure of LMZ* [in Russian], Moscow: Energoatomizdat, 1991.
9. Plotkin E.R. and A.Sh. Leyzerovich. *Start-ups of Power Unit Steam Turbines* [in Russian], Moscow: Energiya, 1980.
10. Leyzerovich A. *Large Power Steam Turbines: Design & Operation*, 2 vols., Tulsa (OK): PennWell Books, 1997.
11. Ryzhkov V.K. "LMZ K-1200-240 High-Speed Tandem Turbine," *Thermal Engineering* 32, no. 6 (1985): 91-93.
12. Leyzerovich A. *Wet-Steam Turbines for Nuclear Power Plants*, Tulsa (OK): PennWell Corporation, 2005.
13. Busse L. and K.-H. Soyk. "World's highest capacity steam turbosets for the lignite-fired Lippendorf power station," *ABB Review*, no. 6 (1997): 13-22.
14. "Yuhuan: a Chinese milestone," *Modern Power Systems* 25, no. 6 (2005): 27-31.
15. Curley G.M. "Supercritical Units—How Are They Performing?" *Proc. of the American Power Conference* 46, Chicago, 1984: 295-299.
16. Schwieger B., M. Leonard, S. Taylor, et al. "First Annual Top Plants Survey," *Power* 146, August 2002: 27-70.
17. Luby P. "Supercritical Systems," *Modern Power Systems* 23, no. 8 (2003): 27-32.
18. Iwanaga K., A. Ohji, and H. Haneda. "The construction of 700 MW units with

advanced steam conditions," *Proc. of the Institute of Mechanical Engineers* 205, Part A, no. 4 (1991): 249-252.

19. Mimuro H., H. Nomoto, and M. Fujii. "The Development and the Operational Experiences of the Steam Turbine with Advanced Steam Conditions," *Proc. of the American Power Conference* 53, Chicago, 1991: 709-716.

20. Fushimi T. "Operational Experience of Power Plant Kawagoe of Chubu Electric Power Utility" [in German], *VGB Kraftwerkstechnik* 74, no. 3 (1994): 213-219.

21. Moria S., M. Haraguchi, and Y. Yamazaki. "High Efficiency Technology for Steam Turbines," *Hitachi Review* 42, no. 1 (1993): 31-36.

22. Kjær S. and F. Thomsen. "Status of Advanced Super-Critical Fossil-Fired Power Plants in the ELSAM Area," *VGB PowerTech* 79, no. 6 (1999): 24-27.

23. Jensen V. "Experience Gained from Commissioning a New 411 MW Coal-Fired Plant with Advanced Steam Data," *VGB PowerTech* 80, no. 5 (2000): 54-59.

24. "Avedøre 2 sets new benchmarks for efficiency, flexibility and environmental impact," *Modern Power Systems* 20, no.1 (2000): 25-36.

25. Noppenau H. "Concept and First Operating Experience with Avedøre 2," *VGB PowerTech* 83, no. 5 (2003): 88-91.

26. Weldorf G. "Vertical tubes improve supercritical systems," *Modern Power Systems* 20, no. 5 (2000): 31-41.

27. Makansi J. "Major new technology applications located offshore leave void in US," *Power* 139, no. 11 (1995): 15-28.

28. Armor A.F., R. Viswanathan, S.M. Dalton, and H. Annendynk. "Ultrasupercritical Steam Turbines: Design and Materials Issues for the Next Generation," *VGB PowerTech* 83, no. 10 (2003): 48-53.

29. "Program Will Develop Ultra-Supercritical Coal-Gen Materials," *Power Engineering* 106, no 10 (2002): 8, 19.

30. Trukhnii A.D., A.G. Kostyuk, and B.M. Troyanovskii. "Ways for Improving Steam-Turbine Installations in Russia and the Advisability of Creating a Pilot Power-Generating Unit Having Extra-High Steam Parameters," *Thermal Engineering* 44, no. 1 (1997): 1-7.

31. Trukhnii A.D. "A New Power-Generating Unit for Ultra-Supercritical Steam Conditions," *Thermal Engineering* 45, no. 5 (1998): 418.

32. Avrutskii G.D., V.V. Lysko, A.V. Shvarz, and B.I. Shmukler. "About creating coal-fired power units with USC steam conditions" [in Russian], *Elektricheskie Stantsii*, no. 5 (1999): 22-31.

33. Avrutskii G.D., I.A. Savenkova, M.V. Lazarev, et al. "Development of Technical Solutions for Creating a Supercritical Steam-Turbine Power Unit" [in Russian], *Elektricheskie Stantsii*, no. 10 (2005): 36-40.

34. "Koreans set a standard for supercritical systems," *Modern Power Systems* 22, no. 5 (2002): 45-47.

35. Smith D. "1800-MWe supercritical coal power for Shanghai," *Modern Power Systems* 19, special issue - Germany supplement (1999): 33-38.

36. Kuznetsov V.A., V.M. Skarkat, and M.F. Demenin. "Constructing China's largest supercritical power station," *Modern Power Systems* 20, no. 5 (2000): 43-47.

37. Kuznetsov V.A., V.M. Skarkat, and M.F. Demenin. "Constructing China's power plant Suijun" [in Russian], *Elektricheskie Stantsii*, no. 7 (2000): 59-63.

38. "China's 1st large supercritical units go commercial at Waigaoqiao site," *Modern Power Systems* 25, no. 4 (2005): 43-44.

39. Logan T.M. and Un-Hak Nah. "Tangjin 5 and 6: Korea's first ultrasupecritical units," *Modern Power Systems* 22, no. 10 (2002): 23-25.

40. "China may prefer ultra supercritical," *Turbomachinery International* 46, no. 5 (2005): 13-14.

41. Leyzerovich A. "New Benchmarks for Steam Turbine Efficiency," *Power Engineering* 106, no. 8 (2002): 37-42.

42. Hoffstadt U. "Boxberg achieves world record for efficiency," *Modern Power Systems* 21, no. 10 (2001): 21-23.

Rise of Steam Turbine Output and Efficiency with Steam Parameters 41

43. Hoffstadt U. "Boxberg - a New Benchmark for Modern Power Plant Technologies" [in German], *BWK* 53, no. 3 (2002): 53-57.
44. Scarlin B. "Steam Turbines" [in German], *BWK* 49, no. 4 (1997): 91-96.
45. Schlessing J. "Determination of operating parameters for steam generation in coal-fired power plants," *VGB PowerTech* 77, no. 12 (1997): 660-666.
46. Leyzerovich A.Sh. "New Developments of ABB for Steam-Turbine Power Plants of Germany" [in Russian], *Elektricheskie Stantsii,* no. 12 (1999): 57-60.
47. Pruschek R. "Coal-Fired Power Plants of the Future" [in German], *BWK* 53, no. 12 (2001): 40-47.
48. "Schwarze Pumpe: A new era in lignite fired power generation," *Modern Power Systems* 17, no. 9 (1997): 27-36.
49. Dubslaff E. "Lignite-Based Power Generation by Modern Units—An Important Option for German Utilities," *VGB PowerTech* 80, no. 6 (2000): 1.
50. "Tachibana-wan unit 2 takes a supercritical step forward for Japan," *Modern Power Systems* 21, no. 11 (2001): 41-47.
51. Kishimoto M., Y. Minami, K. Takayanagi, and M. Umaya. "Operating Experience of Large Supercritical Steam Turbine with Latest Technology," *Advances in Steam Turbine Technology for the Power Generation Industry,* PWR-Vol. 26, ASME, New York: 1994: 43-47.
52. Matsukuma M., R. Magochi, T. Nakano, et al. "Design and Operating Experience of 1000-MW High-Temperature Steam Turbine," *Proc. of 1999 Joint Power Generation Conference,* PWR-Vol. 34, New York: ASME, 1999, Vol. 2, 107-112.
53. Wani M., H. Fukuda, M. Tsuchiya, et al. "Design and Operating Experience of a 1000 MW Steam Turbine for the Chugoku Electric Power Co., Inc. Misumi No. 1 Unit," *Mitsubishi Heavy Industries Technical Review* 36, no. 3 (1999): 66-74.
54. Obara I., T. Yamamoto, and Y. Tanaka "Design of 600°C Class 1000 MW Steam Turbine," *Mitsubishi Heavy Industries Technical Review* 32, no. 3 (1995): 103-107.
55. Nameki Y., T. Murohoshi, F. Hiyama, and K. Namura. "Development of Tandem-Compound 1,000-MW Steam Turbine and Generator," *Hitachi Review* 47, no. 5 (1998): 176-182.
56. Sakai K., S. Morita, T. Yamamoto, and T. Tsumura. "Design and Operating Experience of the Latest 1,000-MW Coal-Fired Boiler," *Hitachi Review* 47, no. 5 (1998): 183-187.
57. Sakai K., S. Morita, and T. Sato. "State-of-the-art Technologies for the 1,000-MW 24.5-MPa/600°C/600°C Coal-fired Boiler," *Hitachi Review* 48, no. 5 (1999): 273-276.
58. Matsumoto N., H. Iwamoto, T. Kaneda, and K. Shigihara. "Completion of Hitchinaka Thermal Power Station Unit No. 1—1,000-MW Power Generating Plant of the Tokyo Electric Power Co., Inc." *Hitachi Review* 53, no. 3 (2004): 104-108.
59. Nomoto H., Y. Kuroki, M. Fukuda, and S. Fujitsuka. "Recent Development of Steam Turbines with High Steam Temperatures," presented at the Electric Power 2005 Conference, Chicago, 2005.
60. "Is 700+°C steam temperature economically viable?" *Modern Power Systems* 18, no. 5 (1998): 73-77.
61. Smith D. "Ultra-supercritical chp: getting more competitive," *Modern Power Systems* 19, no. 1 (1999): 21-30.
62. Heitmüller R.J., H. Fischer, J. Sigg, et al. "Lignite-fired Niederaußem K aims for efficiency of 45 per cent and more," *Modern Power Systems* 19, no. 5 (1999): 53-66.
63. Tippkötter Th. and G. Scheffknecht. "Operating Experience with New BoA Unit and Outlook" [in German], *VGB PowerTech* 84, no. 4 (2004): 48-55.
64. Lamberttz J. and G. Gasteiger. "Concept and Experience Gained during Start-up of the Power Unit with Optimized Plant Engineering (BoA) at Niederaußem" [in German], *VGB PowerTech* 83, no. 5 (2003): 82-87.
65. Horstmann A. "Modern Power Plant Technology in North Rhine-Westphalia," *VGB PowerTech* 83, no. 7 (2003): 1.
66. Haasa R., H. Breuer, and U. Gade. "Power Plant Technology Based on Fossil

Fuels," *VGB PowerTech* 86, no. 1/2 (2006): 39-43.

67. Meier H.-J., M. Alf, M. Fischedick, et al. "Reference Power Plant North Rhine-Westphalia (RPP NRW)," *VGB PowerTech* 84, no. 5 (2004): 76-89.

68. Baumgartner R., J. Kern, and S. Whyley. "The 600 MW Advanced Ultra-Supercritical Reference Power Plant Development Program for North Rhine Westphalia—A Solid Basis for Future Coal-Fired Plants Worldwide," presented at the 2005 Electric Power Conference, Chicago, 2005.

69. Okura R., S. Shioshita, Y. Kawasato, and J. Shimizu. "Completion of High-Efficiency Coal-fired Power Plant," *Hitachi Review* 52, no. 2 (2003): 79-83.

70. Irie K., H. Suganuma, T. Momoo, et al. "Commencement of the Commercial Operation of World's Top Performing 900 MW Unit Maizuru No. 1 Thermal Power Station of the Kansai Electric Power Co., Inc." *Mitsubishi Heavy Industries Technical Review* 41, no. 5 (2004): 1-5.

71. Momma H., J. Ishiguro, T. Suto, et al. "Commencement of the Commercial Operation of 600 MW Unit, Hirono No. 5 Thermal Power Station of the Tokyo Electric Power Co., Inc." *Mitsubishi Heavy Industries Technical Review* 41, no. 5 (2004): 1-5.

72. Tremmel A. and D. Hartmann. "Efficient Steam Turbine Technology for Fossil Fuel Power Plants in Economically and Ecologically Driven Markets," *VGB PowerTech* 84, no. 11 (2004): 38-43.

73. Jäger G. and K.A. Theis. "Increase of Power Plant Efficiency," *VGB PowerTech* 81, no. 11 (2001): 21-25.

74. Rukes B. and R. Taud. "Perspectives of Fossil Power Technology," *VGB PowerTech* 82, no. 10 (2002): 71-76.

75. Watanabe S., T. Tani, M. Takahashi, and H. Fujii. "495-MW Capacity Genesee Power Generating Station Phase 3: First Supercritical Pressure Coal-fired Power Plant in Canada," *Hitachi Review* 53, no. 3 (2004): 109-114.

76. "Genesee 3—a supercritical first for Canada," *Modern Power Systems* 25, no. 4 (2005): 26-28.

77. Peltier R. "Genesee Phase 3, Edmonton, Alberta, Canada," *Power* 149, July/August 2005: 46-50.

78. Hansen T. "Projects of the Year," *Power Engineering* 110, no. 1 (2006): 22-26.

79. "First U.S. supercritical plant in 14 years," *Power* 149, November/December 2005: 7.

80. Tateishi A., H. Fujii, T. Koga, and H. Kimura. "Approach to Overseas EPC Project—Technical Features of MidAmerican Project in the U.S." *Hitachi Review* 54, no. 3 (2005): 98-104.

81. "Joint Venture to Develop Oklahoma Power Plant," *Power Engineering* 110, no. 8 (2006): 18.

82. Ando K., E. Asada, K. Yamamoto, et al. "Design, Construction, and Commissioning of the Nos. 1 and 2 Units of the Ratchaburi Thermal Power Plant for the EGAT of Thailand as a Full Turn-key Contract," *Mitsubishi Heavy Industries Technical Review* 39, no. 3 (2002): 95-100.

83. "Kogan Creek: a king-size supercritical project for Queensland," *Modern Power Systems* 24, no. 9 (2004): 15-17.

84. Appleyard D. "Australia joins the supercritical ranks," *Modern Power Systems* 25, no. 4 (2005): 39-41.

Chapter 3

Configuration of Modern Power Steam Turbines

LARGE POWER STEAM TURBINES

Since the early 1960s, the quest for raising a single capacity of steam turbines to the record-breaking figures frequently led to designing some turbines as cross-compound (CC), that is, double-shaft, to provide a necessarily great exhaust area of the LP cylinders with the limited length of the last stage blades (LSBs), without an excessive increase of the cylinder number and, as a result, the total turbine length. This was especially typical for U.S. and Japanese power plants working with the grid frequency of 60 Hz. In this case, the centrifugal stresses in the LSBs for a full-speed turbine would be 1.44 times greater than those if the frequency made up 50 Hz, all other things being equal. As a result, for example, with the available LSBs of those years, the above mentioned 1,300-MW turbines of ABB, installed at several U.S. power plants, needed four double-flow LP cylinders. Thus, settled on the same shaft-line together with the double-flow HP and IP cylinders, they would have shaped a caterpillar-like six-cylinder machine with many potential problems in service. In addition, at that time it would be rather problematic to manufacture an electric generator of such an output. So the turbine was made with two full-speed shafts, and all the six cylinders were "evenly" distributed between them: one high-temperature (HP or IP) cylinder and two LP ones on each shaft.[1] Such an approach does not look like optimal and is obviously inferior in effectiveness to the CC concept with all the LP cylinders located separately on the half-speed shaft, making possible to use much longer LSBs and reduce the number of LP cylinders. An example of such a CC turbine (a 1,000-MW-class turbine of MHI) can be seen in Figure 2-12, and all the modern large CC turbines, mostly for the grid frequency of 60 Hz, with the single capacity of about 900-1,050 MW, follow this concept and have two double-flow LP cylinders at the half-speed shaft and

HP and IP cylinders settled at the full-speed shaft. This scheme makes it possible to cope with large volumetric steam flow amounts through the LP exhausts, not excessively increasing their number, and simultaneously provide a high internal efficiency in the full-speed HP and IP cylinders. Nevertheless, it is understandable that the TC turbine configuration would be much more preferable in terms of capital expenditures and service convenience. Recent achievements in creating longer LSBs (this subject is considered in one of the next chapters of this book) now allow turbine designers to make full-speed TC turbines of a 1,000-MW-class even for the electric power utilities with the frequency grid of 60 Hz, that is, for the rotation speed of 3,600 rpm.

In particular, Mitsubishi Heavy Industries announced of completing a new standard series of integrally-shrouded LSBs, including the 45-in (1,143 mm) titanium LSB for 3,600 rpm and 48-in (1,220 mm) steel blade for 3,000 rpm. By applying them, MHI declared it is ready to produce a 1,000-MW-class 3,600-rpm four-cylinder turbine with four LP flows (Figure 3-1) as an alternative to the CC configuration (see Figure 2-12) having been used until the recent years.[2] The 48-in steel LSBs were already used at the 600-MW 3,000-rpm TC turbine, commissioned in 2004 at the Japanese power plant Hirono's Unit 5;[3] the turbine is made in two cylinders: with

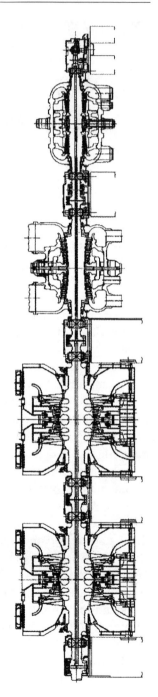

Figure 3-1. Longitudinal section of MHI's proposed TC-type steam turbine of a 1,000-MW class for advanced steam conditions. *Source: By courtesy of Mitsubishi Heavy Industries*

the integrated HP-IP cylinder and one double-flow LP one—Figure 3-2. This 600-MW turbine seems to be the largest two-cylinder machine in service and have the worldwide biggest load per LP exhaust—300 MW. According to MHI, in the future it is supposed to increase the single capacity of TC two-cylinder turbines (HP-IP + one double-flow LP cylinders) up to 750 MW; thus a three cylinder turbine (HP-IP + two LP cylinders) could provide an output of up to 1,000 MW.[2,4]

Nevertheless, until recently the largest TC steam turbines produced by Japanese manufactures were two 735-MW supercritical-pressure turbines of MHI installed at the Thailand power plant Ratchaburi's Units 1 and 2, put in operation in 2000.[5] The maximum gross output of such a turbine makes up 841 MW; the rotation speed is 3,000 rpm. The turbine repeats in its configuration the proposed 1,000-MW turbine shown in Figure 3-1: a four-cylinder machine with double-flow HP and IP cylinders and two double-flow LP ones, but in this case the LSBs are 900 mm (35.4 in) long.

The first 1,000-MW full-speed TC turbines for the grid frequency of 60 Hz (that is, with the rotation speed of 3,600 rpm) were manufactured by Toshiba and went on line at the Hekinan Units 4 and 5 in 2004.[6] The readiness to produce TC steam turbines of a 1,000-MW class has also been announced by Hitachi—instead of CC turbines like those recently

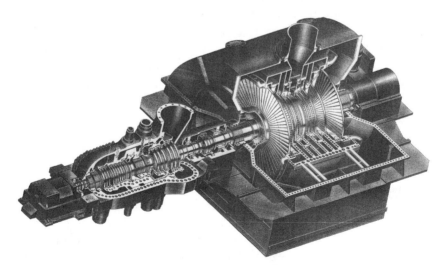

Figure 3-2. Outside view of a 600-MW TC turbine of MHI for advanced steam conditions. *Source: By courtesy of Mitsubishi Heavy Industries*

installed, for example, at the 1,000-MW 3,000 rpm power units Haramachi 2 and Hitachinaka 1.[7,8] Longitudinal sections of such proposed turbines of Hitachi for the rotation speed of 3,600 rpm and 3,000 rpm are presented in Figure 3-3. Even though these turbines were supposed to have the LSBs of a smaller length compared to the above mentioned LSBs of MHI, these turbines also comprise only two double-flow LP cylinders. They are to be furnished with the LSBs of a 1,016-mm (40-in) length for 3,600 rpm and 1,092-mm (43-in) length for 3,000 rpm. The same 40-in LSBs are applied to Hitachi's standard 700-MW turbines for Japanese power plants with the grid frequency of 60 Hz,[9] as well as are to be applied to the 495-MW turbine for the first Canadian supercritical-pressure unit of Genesee Phase 3 (with one double-flow LP cylinder—TC2F-40) and 870-MW turbine for the U.S. CBEC Unit 4 (with two double-flow LP cylinders—TC4F-40).[10] Their analogue for the grid frequency of 50 Hz—a 700-MW three-cylinder turbine for the steam conditions of 25 MPa, 600/600°C (3,625 psi, 1,112/1,112°F)—is shown in Figure 3-4a.[11] A similar 700-MW 3,600-rpm turbine (Figure 3-4b) with the main and reheat steam temperatures of 593°C (1,100°F) was produced by Toshiba for the power plant Nakano.[6] For a time being, these 700-MW turbines seem to be the worldwide largest machines using integrated HP-IP cylinders.

If MHI designs its turbines as reaction-type machines, Hitachi and Toshiba turbines are traditionally designed with impulse-type blading what considerably tells on the turbine configuration. In particular, the HP cylinders of the 1,000-MW turbines shown in Figure 3-3 are single-flow. The turbines are designed with nozzle-group steam admission control, and their control stages are made double-flow to reduce large alternating forces influencing their rotating blades at partial loads.

Until 2002, the largest West-European TC turbines for fossil-fuel power plants in service had the single gross capacity of 933 MW—two ABB turbines for the German lignite-fired power plant Lippendorf with the steam conditions of 25.9 MPa, 550/580°C (3,755 psi, 1,022/1,076°F).[12-14] The turbine's longitudinal section is presented in Figure 3-5a. The turbine is designed with the use of standardized modules for the main steam valve blocks (see Figure 3-5b), HP (see Figure 3-5c), IP, and LP (see Figure 3-5d) cylinders.[15] Both the HP and IP cylinders have horizontal inlets with single-seat angle valves, connected directly to the cylinders, without any crossover pipes. This makes the turbine more compact, improves the turbine's responsiveness, diminishing the steam volumes between the valves and steam path, reduces inlet energy losses, provides

Configuration of Modern Power Steam Turbines 47

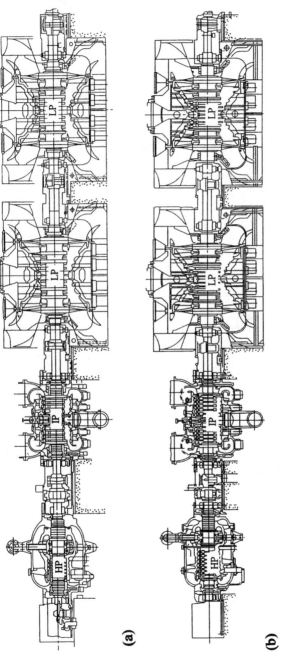

Figure 3-3. Longitudinal sections of Hitachi's proposed TC-type steam turbines of a 1,000-MW class for advanced steam conditions as applied to the grid frequencies of 60 Hz (a) and 50 Hz (b). Source: Y. Nameki et al.[7]

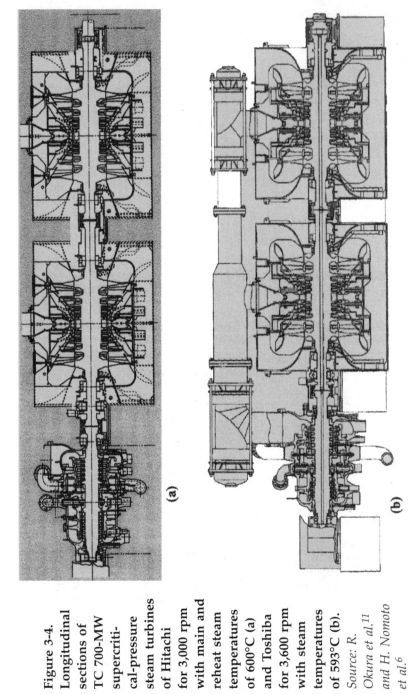

Figure 3-4. Longitudinal sections of TC 700-MW supercritical-pressure steam turbines of Hitachi for 3,000 rpm with main and reheat steam temperatures of 600°C (a) and Toshiba for 3,600 rpm with steam temperatures of 593°C (b). *Source*: R. Okura et al.[11] and H. Nomoto et al.[6]

an easy access to the turboset on the foundation, and makes dismantling and reassembling the turbine less laborious. The inlet passages of all the cylinders are designed as a scroll, or spiral, with a radial first blade row, ensuring lower energy losses in the inlet zones. With its three double-flow LP cylinders, this turbine has the total number of cylinders equal to five—the maximal admissible number. Earlier, the Russian record 1,200-MW turbine and standard 800-MW turbines were designed by LMZ with the same number of cylinders. The above mentioned 1,000-MW-class Siemens turbine presented in Figure 2-10 is also arranged in five cylinders. Such long and heavy machines need special countermeasures to provide the freedom of their thermal expansion. This subject is considered in one of the subsequent chapters.

With the net efficiency of 42.4%, the Lippendorf has been one of the most efficient solid-fuel-fired power plants worldwide. But the single capacity of its turbines was excelled by the steam turbine of Siemens PG installed at the power plant Niederaußem's Unit K with the gross output at the generator terminals equal to 1,012 MW (the net capacity of 965 MW).[16-18] The unit was put in commercial operation in fall 2002. Its turbine's configuration, with the single-flow HP cylinder, double-flow IP one, and three double-flow LP cylinders, is identical with that of the Boxberg Q's turbine shown in Figure 2-10. The main difference is that the 39-in (1,000-mm) LSBs used for the Boxberg Q turbine were replaced with the 45-in (1,146-mm) LSBs, increasing the annular exhaust area by 25%—from 10.0 m^2 per flow up to 12.5 m^2. Along with this, as applied to warmer cooling water, these longer LSBs allowed Siemens to make a 1,000-MW turbine with only two double-flow LP cylinders, as it is already suggested by Siemens for the Chinese Yuhan power units.[19] And, at last, transition to the titanium 55-in (1,400-mm) LSBs made it possible to design the above mentioned 600-MW turbine for the RPP NRW project consisting of three cylinders—that is, with only one double-flow LP cylinder, without sacrifice to the turbine efficiency.[20,21] The longitudinal section of this turbine is given in Figure 3-6.

Some cogeneration turbines, even of a smaller single capacity, can also be designed with a great number of cylinders, especially if it refers to double-reheat turbines. Thus, for example, the 410-MW double-reheat turbines of ALSTOM for the Nordjyllandsværket Unit 3 and Skærbæek Unit 3 (see Figure 2-8) and 570-MW USC-pressure turbine of Anasldo for the Avedøre Unit 2 have five cylinders with two double-flow LP cylinders and an additional asymmetrical double-flow IP cylinder, whose

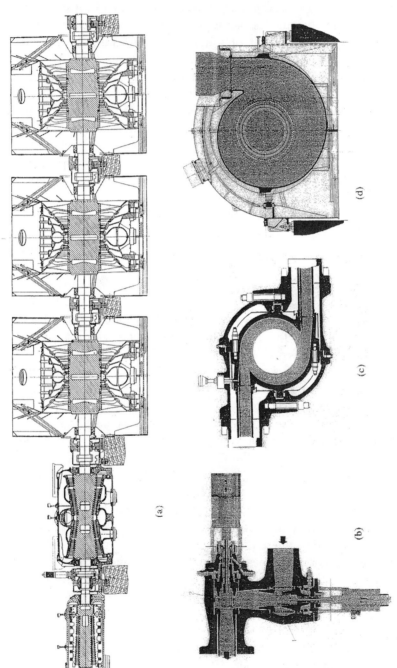

Figure 3-5. Longitudinal section of a TC 933-MW steam turbine of ABB for the Lippendorf power plant (a), main steam valve block with main stop (1) and control (2) valves (b), and scroll inlets for the HP (c) and LP (d) cylinders. *Source: L. Busse and K.-H. Soyk[12] and H. Lageder and P. Meylan[15]*

Configuration of Modern Power Steam Turbines 51

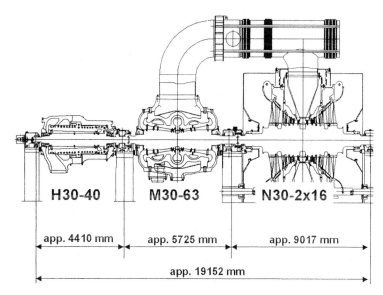

Figure 3-6. Longitudinal section of a proposed 600-MW steam turbine of Siemens for the RPP NRW project. *Source: H.-J. Meier et al.*[21]

right-hand and left-hand flows feed separate LP cylinders and different district network-water heaters. Besides, such a turbine comprises a separate VHP cylinder and an integrated HP-IP one, which cannot be united into a common structure. Along with this, the above mentioned 700-MW USC-pressure double-reheat turbine of Toshiba (see Figure 2-6) is arranged in four cylinders, with the integrated SHP-HP, double-flow IP, and two double-flow LP ones.

Of interest is to compare general design solutions of the considered Siemens and ABB 1,000-MW-class turbines (see Figures 2-10 and 3-5a) with those of the proposed TC 1,000-MW turbine of MHI shown in Figure 3-1. Such a comparison seems especially instructive since turbines of all these three companies have reaction-type steam paths. If MHI turbines are more traditional and conservative in their design, ABB and Siemens use some original and non-trivial design solutions. Along with this, as we can see, MHI, as well as the other Japanese turbine producers, is more courageous in applying advanced steam temperatures.

The most obvious difference is that both the ABB and Siemens turbines have single-flow HP cylinders, whereas all the cylinders of the MHI turbine are double-flow. The single-flow version potentially brings lesser secondary energy losses in the steam path due to taller stage blad-

52 *Steam Turbines for Modern Fossil-Fuel Power Plants*

ing, whereas the double-flow design provides fully counterbalanced axial thrust and less steam leakage through the end gland seals. The high-temperature (HP and IP) cylinders of all the considered turbines have a two-casing design. However, both ABB and Siemens turbines, as distinct from the considered ALSTOM, Hitachi, LMZ, MHI, and Toshiba turbines, are designed with throttle steam admission control, without nozzle boxes in the HP cylinders. As a result, without the control stage of a greater diameter relating to the rest HP stages, the inner HP casing can acquire a simpler, almost cylindrical shape, without a "hump," characteristic for HP cylinders with nozzle boxes (see Figures 2-2, 2-4a, 2-12, 3-2—3-4). On the other hand, with the nozzle boxes, the inner HP casing undergoes less steam pressure difference and lower steam temperatures what makes it less prone to distortion under action of thermal stresses.

As to steam turbines of ABB, in order to decrease these stresses in the HP inner casing, it is commonly made flangeless with *shrinking rings*. These rings are fitted onto the inner casing when it is fully assembled together with the rotor, before it is finally laid into the bottom of the outer casing. The outer casing acquires a simpler (almost cylindrical) geometric shape and is heated more even. Such a design solution was also applied by ABB even if the turbine was made with an integrated HP-IP cylinder and nozzle-group steam control (with the nozzle boxes and control stage), as for the 600-MW supercritical-pressure turbine installed at the U.S. power plant Cardinal Unit 3, or with a double-flow HP cylinder, as for the above mentioned 1,300-MW supercritical-pressure turbines installed at several U.S. power plants—Figure 3-7.

The HP cylinders of Siemens turbines (see Figures 2-10 and 3-6) are traditionally designed with a so-called "barrel-type" outer casing without a horizontal joint, although the inner casing traditionally consists of two halves with a horizontal flange joint. The highly symmetrical design of the HP inner and outer casings allowed them to avoid high thermal stresses even under conditions of very high steam pressure and temperature. The compact design of the HP outer casing joined together with the integrated main stop and control valve units can be seen in Figure 3-8. The inner casing is of a practically cylindrical shape with small longitudinal flanges not exposed to a great pressure difference. The set of the assembled inner casing and rotor is vertically loaded into the outer casing, and the cylinder is delivered to the site as a whole. Even though Siemens PG designs its large steam turbines with full-arc steam admission (that is, without a control stage) and reaction-type blading, the first

Configuration of Modern Power Steam Turbines

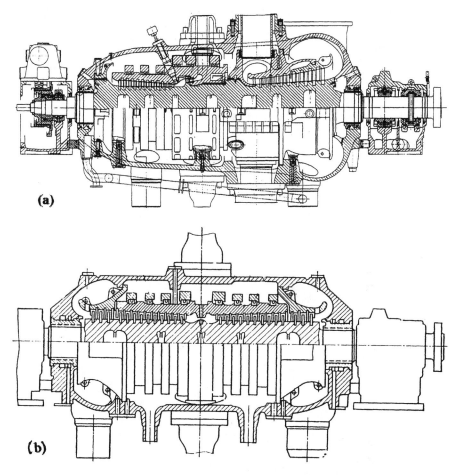

Figure 3-7. Longitudinal sections of an integrated HP-IP cylinder (a) and double-flow HP cylinder (b) for 600-MW and 1,300-MW supercritical-pressure steam turbines of ABB, respectively. *Source: F. Hard[22] and K. Reinhard et al.[23]*

HP stage with tilted stationary blading is of an impulse type. It not only provides optimal conversion of radial to axial flow patterns, but also decreases the leakage losses and rotor metal temperatures owing to the greater enthalpy drop across the impulse-type stationary blade row. This advantage is illustrated by the schematic of local temperature reductions shown in Figure 3-9a. Similar conditions take place for the double-flow IP turbine admission. The IP rotor surface in the steam admission zone is shielded against the steam entering the cylinder and is mainly swept by steam after the first stage's tilted stationary blading, passing through the vortex bores in the shield; the resulted effect is shown in Figure 3-9b.

More often than not, the main and reheat valves for most of modern large steam turbines are settled immediately close to the turbine—see Figures 2-10 and 3-5b, whereas for turbines of the past vintages the valve steam-chests are connected to the turbine cylinders with relatively long

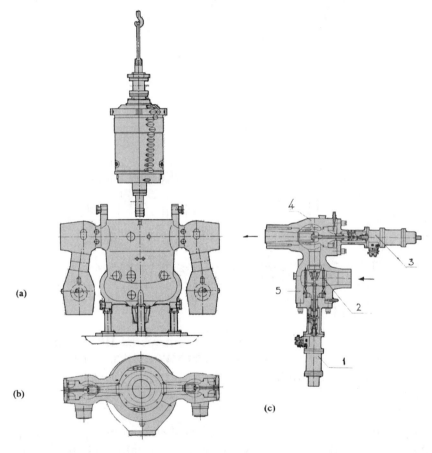

Figure 3-8. Assembling a barrel-type HP cylinder (a) of a Siemens turbine, its cross section through the steam admission plane (b), and the unit of main stop and control valves (c)
1 and 3: actuators, 2: steam strainer, 4: control valve steam chest, 5: stop valve steam chest. *Source: H. Oeynhausen et al.*[24] *and U. Sill and W. Zörner*[25]

Configuration of Modern Power Steam Turbines 55

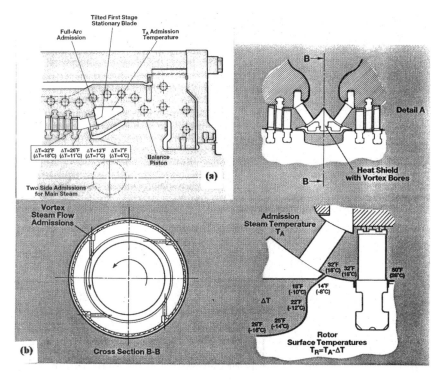

Figure 3-9. Steam temperature reduction in the HP (a) and IP (b) steam admission zones of a 1,000-MW-class turbine of Siemens. *Source:* H. Oeynhausen et al.[24]

crossover pipes—see Figures 2-8b and 2-12. In the case of turbine trips, when the turbine valves are rapidly closed, the amount of steam rested in these pipes causes an additional increase of the rotation speed. Besides, these crossover pipes require special heating procedures at start-ups to avoid steam cooling and even water induction when steam is given to the turbine. The problem is aggravated by the fact that during the turbine outages the crossover pipes because of their much lesser masses and relatively great areas of the outer surfaces cool down much sooner than the turbine cylinders. If the steam-chests of the stop and control main or intercept valves are settled separately, the same problem arises with the crossover pipes between these steam-chests. Fortunately, most modern large steam turbines are furnished with stop and control valves united in blocks. Both ABB and Siemens PG use single-seat angle valves (see Figures 3-5b and 3-8c) well advanced in terms of energy losses and sensitivity to vibration.

An important peculiarity of the ABB turbines is the arrangement of their steam admission parts as applied to both the HP, IP, and LP cylinders. The inlet passage of all these sections is designed as a scroll, or spiral—see Figures 3-5c and 3-5d. This ensures low energy losses in the inlet zones and an optimal steam flow to the first radial-axial blading row. Presently, scroll-type steam inlet is also used by some other turbine producers—in particular, it is applied by MHI at single-cylinder steam turbines for CC power units.[26] The diffusers after the main steam and intercept control valves also act as the inlet ducts to the corresponding inner casings of the HP or IP cylinder.

Both the ABB and Siemens turbines have the journal bearings mutual for the adjacent rotors, whereas each rotor of steam turbines designed by the other producers commonly rests on two "own" bearings, and, for example, a four-cylinder turbine has eight journal bearings instead of five for ABB and Siemens machines. The greater number of bearings increases the turbine length. This makes thermal expansion of the turbine somewhat more difficult, as well as somewhat increases the capital expenditures, but simplifies the turbine assembling and disassembling. The bearing systems of the ABB and Siemens turbines also have some additional important features that make thermal expansion of the turbine easier which are to be considered below.

All the HP, IP, and LP rotors of the ABB turbine (see Figure 3-5a) are of a welded type, that is, they are welded from individual forged parts. The main and undisputable advantage of such rotors is a high controllability of the welded parts and a possibility to use forgings of relatively small sizes and check them thoroughly before welding. This allows manufacturing the rotor (or its constituents) without a central bore. The maximum centrifugal stresses on the rotor axis are as much as approximately twice lesser than those in a forged rotor with a central bore, all other things being equal. Noteworthy is also that stress fields in relatively thin discs are practically two-dimensional, that somewhat additionally decreases the centrifugal and thermal stresses in them. The welded rotors are more rigid and have smaller weight sag, so the turbine is less prone to vibration, and its critical rotation speed increases. Nevertheless, most of modern large steam turbines, like those shown in Figures 2-2, 2-4, 2-6, 2-8, 2-10, 2-12, and 3-1—3-4, are traditionally furnished with forged HP and IP rotors.

Configuration of Modern Power Steam Turbines 57

By contrast, the LP rotors used to be mainly made combined, with shrunk-on wheel discs. Some modern large steam turbines (as, for example, standard Russian 800-MW turbines of LMZ—see Figure 2-4a) continue to use these rotors, but for most of modern turbines combined LP rotors have been replaced with welded (see, for example, Figure 2-4b) or forged ones. In particular, this takes away the problem of stress corrosion cracks (SCC) arising on the shrunk-on wheel discs' fit surfaces, being very topical for many types of turbines in the 1980s. Combined rotors with shrunk-on discs feature high stresses at the shrunk-on discs' fit surfaces and stress concentration near the keyways. In addition, steam wetness occurs in direct contact with these areas what makes these rotors more vulnerable for SCC especially in the Wilson (phase-transition) region. For forged and welded LP rotors, the highest centrifugal stresses take place in zones completely isolated from the steam flow, and these rotors are free from stress concentrators of the shrunk-on discs. The forged LP rotors can be made with or without a central bore. Solid (or monoblock) forged rotors (without a central bore), as well as welded rotors, allow using the longest rotating blades not fearing great centrifugal stresses caused by them. On the other hand, the forged bored rotors provide a helpful access to their central parts for quality control during manufacturing what is very important for large forges. Nevertheless, some turbine producers, including Siemens (see Figures 2-10 and 3-6), frequently use forged rotors without a central bore in the HP and IP cylinders, too, reducing in this way the maximal stresses in the metal depth.

It is known that high-temperature rotors of modern large steam turbines are their most thermally stressed design elements. Creep of the rotor steel under action of centrifugal stresses is one of the main factors determining the turbine lifetime, and unsteady thermal stresses at the turbine transients are the main factor limiting the rate of these transients, that is, the turbine's flexibility.[1] For these reasons, it is very desirable to reduce wherever possible the highest heating steam temperatures and, hence, maximal metal temperatures in the HP and IP rotors. Forced cooling of the rotor surfaces with steam is widely used by Japanese turbine producers. In order to reduce the temperature of steam sweeping the HP rotor surface against the nozzle boxes, the control stage discs are made with inclined holes that pump the steam after the control stage into the chamber between the nozzle box and rotor surface—Figure 3-10a. Similarly, the IP rotor surface in the steam admission zone is cooled by steam from the HP exhaust given into the chamber between the flow guide ring of the first IP stages and the rotor surface—Figure 3-10b. The same approach is used at the USC-pressure double-reheat 700-MW Toshiba turbines (see Figure 2-6) for steam cooling of the rotor surface in the HP steam admission zone of the integrated SHP-HP cylinder—Figure 3-10c, even though this turbine operates with relatively conservative steam temperatures of 566°C (1050°F). Similar design solutions for steam cooling of the high-temperature rotor surfaces were also widely applied before by Westinghouse (see below—Figure 3-13) and some European turbine producers—Figure 3-10d.

The IP cylinders of steam turbines with the single capacity of 800 MW and more are practically inevitably made double-flow (see Figures 2-4, 2-10, 2-12, 3-1, 3-3, 3-5) because of large volumetric steam flow amounts, whereas the HP cylinder of such a turbine can be either double-flow, just like the IP one (Figures 2-12, 3-1, 3-5), or single-flow (Figures 2-4, 2-10, 3-3). For turbines of a lesser output, their HP and IP sections can be settled in separate cylinders or enveloped in an integrated HP-IP cylinder. Along with this, such turbine producers, as, for example, ABB, LMZ, Siemens PG and some others, traditionally prefer using separate cylinders. Single-flow HP cylinders (as well as the SHP or VHP cylinders for USC and/or double-reheat turbines) are commonly made once-through (see Figures 2-2, 2-8, 2-10, 3-3, 3-5, 3-6). Alongside, for example, LMZ has traditionally used a loop-flow design of the HP cylinder for supercritical-pressure turbines with the output of from 300 MW up to 1,200 MW—see Figure 2-4a, as well as Figure 12-15. A main advantage of such a design scheme is a more even and intense heating of both the inner and outer casings swept by the full steam flow amount going through the cylinder's steam path. As a result, variations of the cylinder's relative rotor expansion at the transients are remarkably less as compared to the once-through scheme which makes it possible to set somewhat reduced axial clearances in the steam path. In addition, such a scheme is characterized with a counterbalanced axial thrust and somewhat lesser steam flow leakages through the front end seal. Furthermore, a hot steam-admission zone of the cylinder turns out to be moved away as far as possible from the adjacent bearings contrary to the once-through scheme.

It should be noted that double-flow IP cylinders are frequently used even for steam turbines whose single capacity would technically allow applying a single-flow solution—see, for example, Figures 2-6, 2-8, and

Configuration of Modern Power Steam Turbines

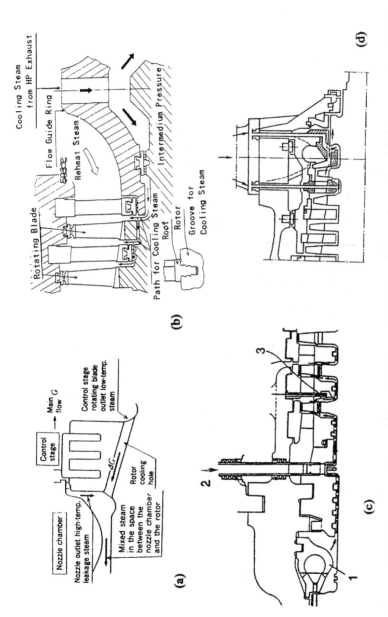

Figure 3-10. Steam cooling of the HP (a) and IP (b) rotor surfaces for high-temperature 1,000-MW-class turbines of MHI, HP steam admission zone of 700-MW USC double-reheat turbines of Toshiba (c), and steam admission zone for the single-flow IP cylinder of 600-MW turbines of GEC Alsthom (d)

1: SHP nozzle boxes, 2: cooling steam from the SHP exhaust, 3: steam temperature measurement at the inlet of the 2nd HP stage's diaphragm seal. Sources: I. Obara et al.[27], M. Kishimoto et al.[28], and A. Leyzerovich[1]

3-6. In this case, it is a good idea to make the cylinder and its steam path asymmetrical with different steam extractions settled in the cylinder's right-hand and left-hand halves. If the turbine has two LP cylinders, they can also be fed from the different IP flows. On the other hand, from the reasoning of efficiency, in order to reduce the secondary energy losses by using taller blades, it would be advisable to make the IP steam path single-flow as far as the volumetric steam flow amount allows. With the greatest consistency, this concept has been put into effect by Alsthom (presently ALSTOM Power), resulting in a so-called "optiflow" configuration of the LP cylinder.[29] In this case, not only the IP steam path (see, for example, Figure 2-2), but also the first LP stages, settled in the central part of the LP cylinder(s), are made single-flow, and only the few remaining LP stages are double-flow. Materialization of this concept is shown in Figure 3-11. In particular, subcritical-pressure steam turbines of GEC Alsthom with the output set in the range of 500-700 MW for the rotation speed of 3,600 rpm were designed with an integrated HP-IP cylinder and two "optiflow" LP ones. A similar approach was implemented at some wet-steam turbines of GEC Alsthom for nuclear power plants.[30]

Even though the IP cylinders are fed with reheat steam of a much lesser steam pressure as compared to main steam entering the HP cylinders, they are made of a two-casing design to decrease the maximal metal temperatures of the outer casing, as well as temperature differences and thermal stresses in both the outer and inner casings. These casings are made with common horizontal flange joints. Most of turbine producers use them also in the HP casings, and temperature differences in these flange joints (across the flange width, between the flange and bolts, between the casing wall and flanges, and so on) arising in the process of turbine transients and sometimes taking place under stationary operating conditions can cause plastic distortion of

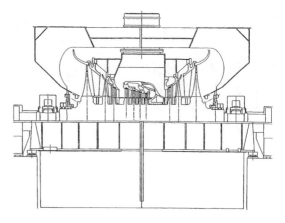

Figure 3-11. LP cylinder of an "optiflow concept." *Sources: By courtesy of ALSTOM Power*

the flanges.[1] To avoid this danger, the casing flanges of high-temperature cylinders (both HP and IP) should be done as thin as possible—Figure 3-12. For turbines with a great width of their HP and IP casing flanges (about 350-400 mm, or 14-16 in, and more), it makes sense to heat additionally the flange joints in areas of their highest metal temperatures by steam taken from the internal casing space and given into the recesses on the flange joint split surfaces, with subsequent discharging into the condenser or one of the LP feedwater heaters. Such steam heating systems were well worked up as applied to different steam turbines of LMZ, including those of supercritical steam pressure with the output of up to 1,200 MW, and have been widely used at many power plants. In addition, such heating systems, mainly used at cold and warm start-ups, reduce the changes of the cylinder's relative rotor expansions and allow reducing axial clearances in the cylinder's steam path.[1]

As said before, a popular design solution for turbines with the single capacity of 600-700 MW and less is the use of an integrated HP-IP cylinder with both main and reheat steam admissions in the central part of the cylinder—see Figures 2-6, 2-8, 3-2, and 3-4. Such a cylinder turns out to be pretty compact, with rather counterbalanced axial thrust and minimized steam flow leakages through the end seals.

A fortunate design concept for the integrated HP-IP cylinder was developed by Westinghouse—Figure 3-13. Such a "building block" designated BB44 was widely used in many Westinghouse turbines with steam conditions of 16.6 MPa, 538/538°C (2,400 psi, 1,000/1,000°F) ranging in their size between 350 MW and 680 MW. The inner casing of the cylinder contains both the HP and IP steam paths. The main steam flow comes into the cylinder through the nozzle boxes, then turns through 180° after the control stage and after going through the rest HP stages leaves the

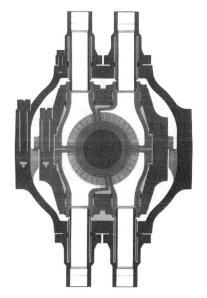

Figure 3-12. Cross section at the HP steam admission zone for a large steam turbine of ALSTOM. *Source: By courtesy of ALSTOM*

cylinder at its rear end. Steam in the IP section flows in the opposite direction, that is, toward the front bearing. Leaving the inner casing, it also turns through 180° and, making its way to the cross-over pipes connecting the HP-IP cylinder to the LP ones, goes between the outer and inner casings, heating the former and cooling the latter with pretty favorable heat transfer conditions.

There also exist some other turbine types with the integrated HP-IP cylinders where only the first IP stages are settled within the inner casing, common with the HP part, whereas the rest IP stages are located in the cylinder ring of the outer casing, and steam comes there, turning through 180° and passing between the inner and outer casings.[1]

Despite obvious advantages of these schemes, they might look too complicated. In addition, the steam flow turns are accompanied with remarkable energy losses. So, more often than not the integrated HP-IP cylinders, if exist, are made with once-trough HP and IP steam paths. In doing so, both the HP and IP sections can be enveloped within the common inner casing (see, for example, Figure 2-8b) or have separate inner casings. These two approaches can be illustrated by examples of two turbines of ALSTOM—Figure 3-14. If the inner casing contains both the HP and IP parts (see Figure 3-14b), they are separated from one another with a special diaphragm built into this casing. Anyway, the chambers within the outer casing against the HP and IP steam paths should also be separated

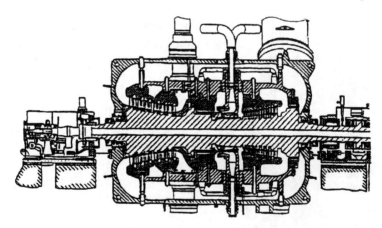

Figure 3-13. Integrated HP-IP cylinder for subcritical-pressure steam turbines of Westinghouse with the single capacity ranged between 350 and 680 MW. *Source: By courtesy of Siemens Westinghouse Power Corporation*

Configuration of Modern Power Steam Turbines 63

one from another. In the case of individual HP and IP inner casings (see Figure 3-14a), the steam flow within the intercasing space goes into the IP steam path through the front seal of the IP inner casing.

With the absence of organized steam flow in the intercasing chambers, they are characterized with rather uncertain heat transfer conditions from steam to the casing surfaces. These conditions can diverge considerably at diverse surface areas, resulting in significant metal temperature unevenness along the outer casing outline, especially at the transients. This threat is aggravated if the IP stages are settled in the cylinder rings of the outer casing without any "own" inner casing (for

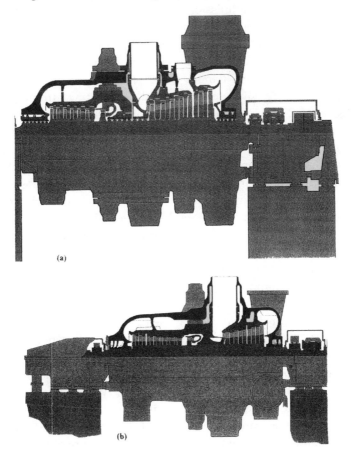

Figure 3-14. Integrated HP-IP cylinders of ALSTOM turbines with separate (a) and common (b) inner casings for the HP and IP sections
Source: By courtesy of ALSTOM

64 *Steam Turbines for Modern Fossil-Fuel Power Plants*

reasons of a moderate steam pressure), and the outer casing in the IP steam admission part is swept directly by reheat steam, as it takes place, for example, for 500-MW supercritical-pressure turbines of GE and Doosan Heavy Industries and Construction—Figure 3-15a. This is especially serious for turbines with elevated reheat steam temperatures.

Different temperatures of the heating steam and different heat transfer conditions in the neighboring chambers of the outer casing can cause sharp changes in the axial metal temperature gradients along the casing's length. The greatest "breaks" of the metal temperature distribution curve lengthwise of the casing outline take place in the vicinity of the HP and IP steam admission zones (that is, in the casing's central part) and near the transition to the IP exhaust hood. The resulted bending thermal stresses, acting together with unsteady thermal stresses caused by radial temperature differences in the outer casing, can result in distortion of the casing flanges and cracks in the casing walls. Shielding the outer casing with special shells, done by some turbine producers (see Figures 3-2 and 3-4), somewhat improve the situation, but cannot completely remedy it. In these cases, the metal temperature fields of the outer casing require special attention, including experimental measurements and investigation at turbines in service, wherever possible.

These factors were taken into consideration by Doosan Heavy Industries at the development of a new HP-IP cylinder with the inner casing enveloping both the HP and IP stages—Figure 3-15b—intended for newly projected steam-turbine power units with the elevated steam temperatures.[31]

It might be well also to point out that turbines with the integrated HP-IP cylinders commonly have a somewhat reduced number of stages in the IP steam path (for reasons of reducing the total length of the cylinder and its rotor and decreasing their sag) with a resulted increased temperature of steam leaving the IP steam part and entering the LP cylinder(s). In this case, especially for turbines with the elevated reheat steam temperatures, the LP rotors can experience high thermal stress at start-ups combined with centrifugal stresses. As a result, the LP rotors can limit the rate of starting up such turbines as well as their HP-IP rotors and require operating monitoring of their thermally stressed states at start-ups relying on the heating steam temperature at the IP section's outlet to be measured.

Raising the main and reheat steam temperatures up to 593-600°C (1,100-1,112°F) and even 610°C (1,130°F) for the new generation of large

Configuration of Modern Power Steam Turbines 65

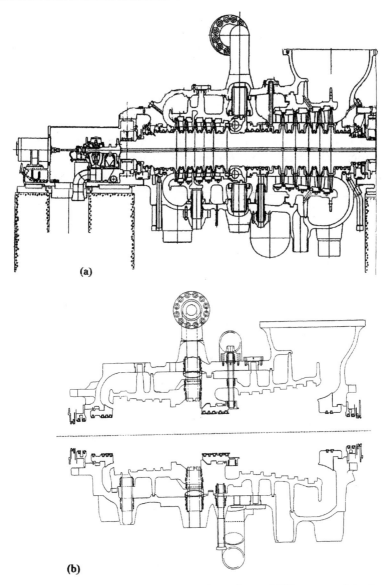

Figure 3-15. Integrated HP-IP cylinder of 500-MW supercritical-pressure turbines of General Electric in cooperation with Doosan Heavy Industries and Construction (a) and an alternative design of its casings for newly projected steam-turbine power units with elevated steam temperatures (b). *Source: By courtesy of Doosan Heavy Industries and Construction*

66 *Steam Turbines for Modern Fossil-Fuel Power Plants*

steam turbines became possible due to the use of chrome-molybdenum-vanadium steel materials with 9% to 12% contents of chrome (Cr) for the high-temperature components instead of common chrome-molybdenum-vanadium (Cr-Mo-V) steel.[2,6,7,11,27,28,32-35] Table 3-1, mainly composed on the basis of Japanese publications, gives a general idea of these replacements.

Table 3-1. Design materials used for main high-temperature design components of large steam turbines with the main and reheat steam temperatures of 538-566°C (1,000-1,050°F) and 593-600°C (1,100-1,112°F)

Major Turbine Components	*Turbines with the Steam Temperatures of 538-566°C*	*Turbines with the Steam Temperatures of 593-600°C*
Steam-chests of the main and intercept stop and control valves	2-1/4%Cr-1%Mo-V forged (or cast) steel	9%Cr forged steel
HP and IP inlet pipes	Cr-Mo-V forged steel	9%Cr forged steel
HP nozzle boxes	2-1/4%Cr-1%Mo-V forged (or cast) steel	12%Cr forged steel
HP, IP, or HP-IP rotors	Cr-Mo-V forged steel	New (advanced) 12%Cr forged steel
HP, IP, or HP-IP inner casings	2-1/4%Cr-1%Mo-V cast steel	12%Cr cast steel
HP, IP, or HP-IP outer casings	1-1/4%Cr-1/2%Mo-V cast steel	2-1/4%Cr-1%Mo cast steel
HP and IP first blade rings	1-1/4%Cr-1/2%Mo-V cast steel	12%Cr cast steel
HP and IP second blade rings	1/2%Cr-1/2%Mo-V cast steel	2-1/4%Cr-1%Mo cast steel
HP and IP rotating blades	Cr-Mo-Nb-V or 12%Cr forged steel	Austenic refractory alloy (R-26) or new 12%Cr forged steel
LP rotors	Ni-Cr-Mo-V forged steel	3-1/2%Ni-Cr-Mo-V superclean forged steel

Based on materials of "Tachibana-van 2..."[2]; H. Nomoto et al.[6];Y. Nameki et al.[7]; R. Okura et al.[11]; I. Obara et al.[27]; M. Kishimoto et al.[28], and M. Wani et al.[32]

All the materials newly used for main high-temperature design components (new 12%Cr forged steel, 12%Cr cast steel, and 9%Cr forged steel) are ferritic heat-resistant steels. In high-temperature rotating blades, austenitic refractory alloy replaces the originally used ferritic materials, whereas 12%Cr forged steel, with sufficient creep rupture strength for operation at 600°C class temperatures, is used as a high-temperature rotor material. Some disadvantage of this steel is its high hardness. For this reason, in the journal and thrust collar sections of high-temperature 12%Cr rotors overlay welds are built up with a low Cr weld material to reduce wear of the bearings. As an alternative decision, Mitsubishi Heavy Industries proposes the use of "hetero-material" welded high-temperature rotors with the central (high-temperature) part made of 12%Cr steel and the rotor ends made of 2-1/4%Cr-Mo-V steel.[4,36] A longitudinal section of a 1,000-MW-class 3,600-rpm three-cylinder turbine with 45-in titanium LSBs, integrated HP-IP cylinder, and welded rotors of all the three cylinders is shown in Figure 3-16. A further development of this idea led to the design of a single-cylinder single-exhaust 105-MW turbine with a welded rotor, whose front end is made of 2-1/4%Cr-Mo-V steel; the central (high-temperature) part is manufactured of 12%Cr steel, and the LP portion is made of 3.5%Ni-Cr-Mo-V steel.

"Hetero-material" rotors with high- and low-temperature forged parts welded together are also used by other turbine manufacturers, as, for example, Siemens PG and Toshiba, for their steam turbines of the moderate capacity with integrated HP-IP-LP and IP-LP cylinders (see below Figures 3-19 and 3-31a).

Even for the very industrially developed countries with a rich experience in operating large supercritical-pressure steam-turbine power units, along with the most advanced, most efficient ones of the highest (for current possibilities) steam parameters and utmost single capac-

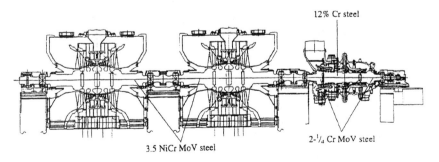

Figure 3-16. Longitudinal section of a proposed 1,000-MW-class 3,600 rpm turbine of MHI with welded rotors. *Source: R. Magoshi et al.*[4]

ity as a basis for the further progress in the power industry and steam turbomachinery, it seems reasonable to consider an alternative concept of ordinary, mass, widespread steam turbines.[37] They should be characterized with reduced capital expenditures and enhanced reliability and efficiency and designed on the basis of the whole previous experience in design and operation of the best steam-turbine units. It can be said that nowadays such turbines also have to match requirements of operation for deregulated power systems; so they should be capable of operating in both base-load and cycling modes. It seems now obvious that such turbines should be designed as TC single-reheat machines for a supercritical steam pressure and steam temperatures acceptable with the use of well worked out steel materials—that is, up to 593-600°C (1,100-1,112°F). Since the raise of the main steam pressure up to the USC level almost inevitably leads to the use of a second reheat, making the unit's scheme extremely intricate, it seems reasonable to set the main steam pressure at a certain moderate level of 24-28 MPa (3,480-4,060 psi). The turbine's single capacity should be chosen keeping in mind a possibility of minimizing the number of cylinders and main and reheat steam-lines without sacrifice in the turbine efficiency caused by the increased pressure drops in steam-lines and energy losses with the exit steam velocity. On this basis, it seems also desirable to use the integrated HP-IP cylinder instead of the separated high-temperature ones to provide the minimum specific metal amount and length of the turbine. In the late 1990s, when this approach was firstly developed and proposed, the single turbine capacity of up to 500 MW seemed to be maximal for the use of an integrated HP-IP cylinder and one double-flow LP cylinder. Yet now the progress in the turbine design made it possible to raise this limit up to 600-700 MW. With this output, the unit can be made with two main steam-lines and two hot and cold reheat steam-lines. By way of example of applying a similar approach, it could be possible to address to the above mentioned 700-MW turbines of Japanese producers with their integrated HP-IP cylinder and two double-flow LP ones (see Figure 3-4) and steam parameters equal to 25 MPa, 600/600°C (3,625 psi, 1,112/1,112°F).[6,11] For special cases (for example, as applied to the use by an independent power producer), the steam temperatures can be lowered to 566/566°C (1,050/1,050°F) or 538/566°C (1,000/1,050°F).[9] Lowering the rated output to 600 MW allows settling the turbine in two cylinders—see Figure 3-2. With separated HP and IP cylinders, such a turbine, with one double-flow LP cylinder, is settled in three cylinders—see Figure 3-6.

Configuration of Modern Power Steam Turbines 69

With a moderate output of about 300 MW, the turbine can be made in two cylinders (HP-IP + one double-flow LP) with single main and reheat steam-lines.

Along with minimizing the number of the turbine cylinders, main and reheat steam-lines, the capital expenditures can be additionally reduced by simplifying the turboset scheme and the unit's start-up system. This approach was applied to the project of a Russian supercritical-pressure power unit of a "new-generation" with the single capacity of 525 MW and steam conditions at the turbine entrance of 29 MPa, 595/597°C (4205 psi, 1,103/1,107°F).[38-40] The turbine for this unit is supposed to consist of three cylinders: a loop-flow HP cylinder, single-flow IP one, and one double-flow LP cylinder. It might be well to point out that it would seem more reasonable to use an integrated HP-IP cylinder instead of two separated ones. The proposed simplified steam/water flowchart and start-up system for such a unit are shown in Figure 3-17. Compared with the flowchart of a standard 800-MW unit shown in Figure 2-16, it looks much simpler. This is mainly reached thanks to the use of so-called "deaeratorless scheme"—without a deaerator whose role is mainly played by the mixing LP feed-water heaters. This scheme was well worked out and checked up in long-term operating service at a series of supercritical-pressure power units with the output of 300 MW and 800 MW.[41,42] All the proposed alternative solutions should be verified for their influence on the unit's efficiency, as it is shown by example in Figure 3-18. To simplify the start-up system and make the unit's control at the transients easier for the personnel, the start-up system could be taken of a one-bypass type, which was well worked up at standard supercritical-pressure power units with the single capacity of 300, 500, and 800 MW, in particular, under conditions of cycling operating conditions.

PROFILES FOR MODERN POWER STEAM
TURBINES OF A MODERATE CAPACITY

Along with large steam turbines with the power output of over 300-400 MW, many important novelties have appeared in the recent years in the design of modern power steam turbines of a moderate capacity—from 100 MW up to about 300 MW. As a rule, such turbines are designed for rather conservative steam parameters: subcritical steam pressure and main and reheat steam temperatures at the level of 538-

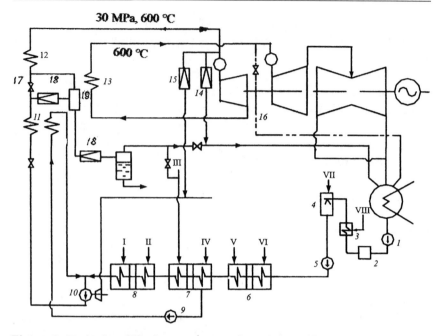

Figure 3-17. A simplified steam/water flowchart and start-up system for a proposed Russian supercritical-pressure 525-MW power unit.
1: condensate pumps of the first stage, 2: the unit's water-treatment facility, 3: surface-type LP feedwater heater (FWH) no. 1, 4: mixing LP FWH no. 2, 5: condensate pumps of the second stage, 6: integrated surface-type FWHs nos. 3 and 4, 7: integrated FWHs nos. 5 and 6, 8: integrated FWHs nos. 7 and 8, 9: FWH drain pump, 10: turbine driven boiler feed pump, 11: steam/water boiler path before the embedded valve, 12: boiler's superheater, 13: boiler's reheater, 14: turbine bypass, 15: start-up feeding line for the plant auxiliary header, 16: discharge line from the hot reheat steam-line, 17: embedded valve, 18: controllable throttle valves, 19: start-up separator, I-VIII: steam extractions of the turbine.
Source: G.D. Avrutskii et al.[39]

566°C (1,000-1,050°F). A modest capacity and, as a result, relatively low steam flow amount together with recent achievements in creating longer blades open possibilities for minimizing the installed cost of such turbines by means of reducing their size and number of flows and cylinders without a sacrifice in efficiency.

A single-cylinder, single-flow reheat turbine of Toshiba with the rated output set in the range from 100 to 250 MW (Figure 3-19) can be

Configuration of Modern Power Steam Turbines

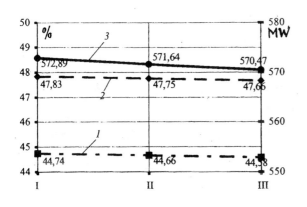

Figure 3-18. Change of the unit's output and efficiency depending on the used LP feedwater heaters.
1: the unit's efficiency, 2: the turbine's efficiency, 3: the unit's output I – mixing LP FWH nos. 1 and 2, II – mixing LP FWH no. 1 and surface-type LP FWH no. 2, III – surface-type LP FWHs nos. 1 and 2. *Source: G.D. Avrutskii and I.A. Savenkova*[43]

taken as an example of a characteristic machine of this class. The turbine is arranged in one cylinder with opposite flow directions in the HP and IP-LP sections. Besides the horizontal flange joint, the turbine casing has a vertical flange joint of the high-temperature HP-IP portion with the LP part and exhaust hood. Similarly, the welded rotor consists of two (HP-IP and LP) parts made of different steel materials. The front (high-temperature) part's steel has sufficient creep rupture strength to withstand relatively high main and reheat steam temperatures, while the LP part's steel should withstand high centrifugal force caused by the LSBs under conditions of relatively low metal temperatures.

A 220-MW turbine of Hitachi (Figure 3-20) gives another example of

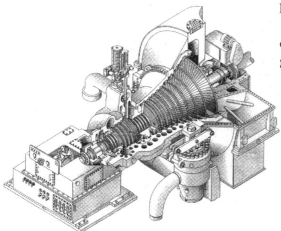

Figure 3-19. Outside view of Toshiba's single-cylinder, single-flow reheat turbine with the rated output of between 100 and 250 MW. *Source: By courtesy of Toshiba*

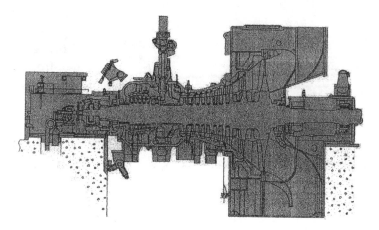

Figure 3-20. Longitudinal section of Hitachi's single-cylinder, single-flow 220-MW turbine. *Source: T. Umezawa et al.*[44]

a single-cylinder, single-flow reheat turbine with the steam conditions of 16.6 MPa, 566/566°C (2,400 psi, 1,050/1,050°F).[44] It was brought in service in 2002 at the coal-fired power plant Kin operated by Okinawa Electric Power Co. and becomes its largest generating capacity. Relatively warm cooling water determines low vacuum in the condenser of 6.67 kPa (0.97 psi), making it possible to use a single flow LP exhaust with titanium 1,016 mm (40 in) long LSBs at the rotation speed of 3,600 rpm. It might be well also to note that the preceding turbine of this type produced by Hitachi, with the output of 156 MW for approximately the same steam parameters and vacuum, was arranged in two cylinders with two LP flows and the LSBs of 660-mm (26 in) length.

It is understandable that the possibility of arranging the turbine in one or two cylinders mainly depends on the turbine output (speaking more accurately, the steam flow amount through the turbine into the condenser), vacuum at the turbine exhaust, and the available outlet annular area of the LSBs. The diagram of Figure 3-21 gives a rough idea of such a subdivision as applied to a new standard LSB series of MHI. It is supposed that a reheat condensing turbine can be arranged in one cylinder up to the single capacity of about 250 MW, and the further increase of the output or the vacuum depth requires the use of a two-flow LP steam path located in a separate cylinder.

However, in some instances the number of cylinders occurs to be not directly connected with the number of LP flows. Thus, in the late

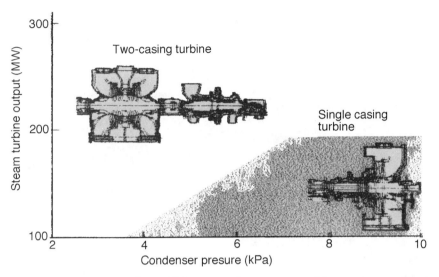

Figure 3-21. Areas of possible arranging steam turbines of a moderate capacity as a single-cylinder or two-cylinder machine (with one or two LP flows) as applied to a new LSB series of MHI. *Source: T. Nakano et al.*[26]

1980s GEC Alsthom developed a project titled the *EUREKA-Turbine* of a "compact" two-cylinder 315-MW turbine with the steam conditions of 18 MPa, 540/540°C (2,610 psi, 1,004/1,004°F)—Figure 3-22. The turbine was designed consisting of an integrated HP-IP cylinder and a single-flow LP cylinder with titanium LSBs of a 1,360-mm (53.5 in) length and axial exhaust.[45] In this case, transition to the two-cylinder configuration was determined by the reasoning of convenience in design of the integrated HP-IP cylinder. Direct arrangement of the HP, IP, and LP steam paths within a common cylinder would make the span between the journal bearings too great.

In modern practice, steam turbines of a moderate capacity are primarily needed for separate electric power utilities with a limited total capacity (like, for example, the above mentioned Okinawa Electric Power Company) or for independent power suppliers—for example, large industrial works. In the last case, more often than not it is supposed that the turbine will provide the consumer with both electric power and steam for technological needs. Such turbines are often made single-cylinder non-reheat back-pressure, based on the HP cylinder of larger condensing or cogeneration turbines. An example of such a 100-MW back-pressure non-

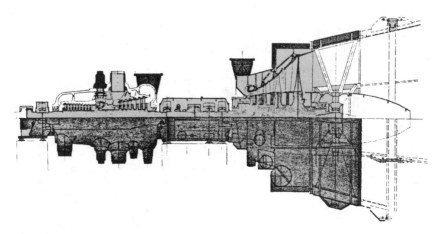

Figure 3-22. The *EUREKA-Turbine* project of a "compact" 315-MW turbine of GEC Alsthom with a single-flow axial exhaust. *Source: D. Tremmel et al.*[45]

reheat Russian turbine R-100-12.8/1.5, produced by Turbine-Engine Works (TMZ), is shown in Figure 3-23. The main steam parameters make up 12.8 MPa, 540°C (1,855 psi, 1,004°F); the rated back-pressure is 1.5 MPa (213 psi), and the turbine practically repeats the HP cylinder of a 210-MW non-reheat cogeneration turbine for district heating.[46] Back-pressure turbines utilize the entire turbine steam flow as a heat source after expanding in the turbine. Their advantage is that they provide the highest fuel utilization. Such turbines are widely accepted in diverse industry branches. The world's largest back-pressure steam turbine with the electric output of 150 MW was announced to be manufactured by Siemens for Finnish pulp and paper works.[47]

In the recent years, moderate-capacity steam turbines have found the very wide application for large CC units. In the mid-1990s, CC facilities accounted to approximately 35% of the total amount of all the newly constructed and commissioned fossil-fuel electric power generating capacities throughout the world.[48] In its classical form, a CC unit consists of one or several gas turbines, each one connected to its "own" heat recovery steam generator (HRSG), and one steam turbine, fed by steam generated in the HRSG(s) due to the heat of the gas turbine exit gas. High gas temperature at the gas turbine inlet and low temperature at the "cold end" of the steam turbine allow attaining a high efficiency of the binary cycle (the Brayton cycle topping the Rankine one), which cannot be achieved for either the

Configuration of Modern Power Steam Turbines 75

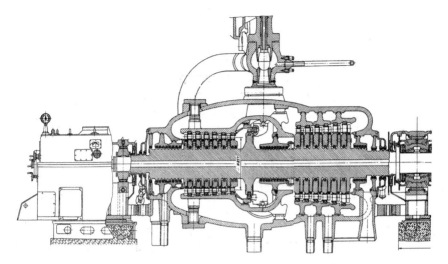

Figure 3-23. Longitudinal section of a single-cylinder back-pressure 100-MW turbine of TMZ. *Source: By courtesy of TMZ*

gas or steam turbine taken separately. Combined-cycle units are rather compact, require relatively small space, and can be constructed in short time intervals. They can be easily standardized which allows cutting down the project expenses and the concept-to-connection time period. Just these and some other merits make CC units so popular for the contemporary power industry. However, the most attractive advantage of CC units is their lower capital cost with a high operating efficiency and flexibility. So in the mid-1990s, the specific price level for a CC unit with a single capacity of about 350 MW amounts to approximately $300 per kW against $1,000 per kW and more for other power generating capacities.[49] Under these conditions, reducing the installed cost of steam turbines for CC units becomes especially topical. It is also worth noting that CC units are quite environmentally friendly since modern combustors of gas turbines feature low level of harmful emissions into the atmosphere. In addition, generation of a large portion of power by the gas turbines in CC units leads to lower demands for cooling water and smaller thermal contamination of the environment as compared to traditional steam-turbine power units of the same capacity.

A typical arrangement of a CC unit of a double-pressure non-reheat cycle with two gas turbines, their HRSGs, and one steam turbine is shown in Figure 3-24. Two equally sized gas-turbine-and-HRSG sets provide main

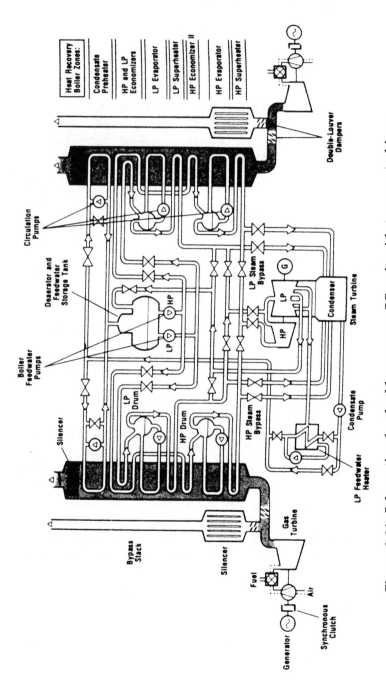

Figure 3-24. Schematic of a double-pressure CC unit with two gas turbines and one steam turbine. *Source: H. Brückner et al.*[50]

Configuration of Modern Power Steam Turbines

(HP) and additional (LP) steam to the steam turbine. Correspondingly, the HRSGs feature HP and LP drums followed by superheating sections. The steam condensate is preheated in the HRSGs with low-temperature flue gas before it is discharged through the stacks. In the considered scheme, the non-reheat steam turbine consists of a single-flow HP and double-flow LP cylinders. The LP steam is added to the LP crossover pipes. For more advanced CC units, the scheme becomes somewhat more complicated—triple-pressure by means of arranging an additional IP circuit with steam reheat. Each gas turbine can be shut down, and the unit will work generating approximately a half output with the remaining gas turbine and the steam turbine. Along with this, one or both gas turbines can be operated in a simple cycle mode without their HRSG(s) and the steam turbine; in this case, the exit gas is discharged besides the HRSG(s) through bypass stacks.[51]

U.S. and Japanese manufacturers mainly prefer HRSGs with a horizontal gas path. In this case, tubes of the heating surfaces are settled vertically, that enables the use of a natural circulation in the evaporator. By contrast, European designers rather prefer a vertical configuration with the heating surfaces arranged lengthwise of the gas pass in vertical direction from the bottom upwards. In this case, the tubes of the heating surfaces go horizontally, and the HRSGs are designed on the assisted circulation principle, even though once-through boilers are also possible. The first once-through Benson HRSG, developed by Siemens PG as applied to future CC units with an upgraded gas turbine model V94.3A, has been successfully operated at the power plant Cottam, the UK, since 1999. The absence of the HP drum in the Benson HRSG and its capability of handling better the steam temperatures at the transients should allow operating such a unit in a cycling, daily stop/start, manner without a risk of thermal-fatigue cracks in the HRSG's most thick-wall elements.[52-55] The use of a once-through HRSG also opens a possibility of transition to supercritical steam pressure in the steam circuit.[56] With supercritical steam conditions for the steam turbine and with the inlet gas temperature of over 1,427°C (2,600°F), the net efficiency of CC units is believed to reach 60%.[57] Nevertheless, until now all the commercial CC units have continued to be designed with subcritical steam pressure, and their efficiency is mainly raised by means of increasing the inlet gas temperature up to 1,500°C, or 2,732°F (a so-called "H" technology for gas turbines with steam cooling of their combustion chambers and nozzles of the first stages), as well as improving the internal equipment efficiency, including steam turbines.

The output ratio for the gas turbine(s) and steam turbine for the most widespread, typical CC schemes usually lies between 1.4 and 3.3. While the single capacity of gas turbines was rather small, it was reasonable to join one steam turbine with a few gas turbines. In addition, even in the recent past gas turbines featured a considerably lower operating reliability as compared to steam turbines. So, according to data of Strategic Power Systems Inc., for the CC units operated in 1991-95, their unavailability was caused by outages of the gas turbosets in 55% (mostly, because of the gas turbines themselves), whereas the share of the steam turbine outages was only 23%.[58] On this basis, it was reasonable to provide a possibility to keep a CC unit in service even if one of the gas turbines was out of order. The usage of a few gas turbines associated with one steam turbine also made a CC unit more flexible and allowed using it as a peak stand-by capacity. As single capacities of modern gas turbines rise, as well as their operating reliability does, more CC units of a greater capacity have been designed with one gas turbine and, correspondingly, one HRSG, associated with one steam turbine. In this case, it becomes possible to make the CC unit even more compact and decrease further the capital expenditures by settling both the gas and steam turbines on one shaft with a common generator, that is, by going over to a so-called single-shaft (SS) CC scheme.[59,60] In this case, both turbines can have common systems of lubrication and governing, and, as a result, the amount of auxiliaries also decreases. In addition, this scheme cuts down the required number of step-up transformers and high-voltage feeders and declines the scope of civil engineering work.

A characteristic steam turbine of GEC Alsthom for large CC power units is presented in Figure 3-25. It does not basically differ from "usual"

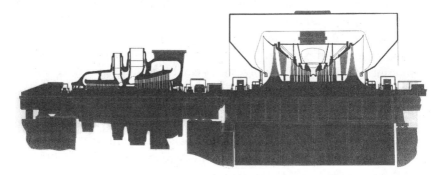

Figure 3-25. Longitudinal section of a triple-pressure reheat turbine of GEC Alsthom for CC units. *Source: By courtesy of ALSTOM Power*

Configuration of Modern Power Steam Turbines 79

large power steam turbines and consists of an integrated HP-IP cylinder and one double-flow LP cylinder.[61] This triple-pressure reheat turbine was mainly intended for operation with a 226-MW 9FA gas turbine, having been jointly developed by GE and GEC Alsthom. With the main steam pressure of 11-13 MPa (1,595-1,885 psi) and steam temperatures of 540/540°C (1,004/1,004°F), the steam turbine can give about 36% of the unit's total capacity. An arrangement of one steam turbine associated with two 9FA gas turbines and their HRSGs provides a capacity of about 700-720 MW. Along with this, such a turbine coupled with one gas turbine 9FA provides a total capacity of about 350 MW. They can be arranged on one shaft with a common generator, shaping a SS CC turbine unit, as it is shown in Figure 3-26, and the outside view of such a CC SS unit with a vertical HRSG, as applied to the power plant Eems in the Netherland (3×355 MW), is presented in Figure 3-27.[62]

A similar configuration has a steam turbine of Mitsubishi Heavy Industries—Figure 3-28. It is intended for operation with more advanced gas turbines of the "G" technology. For gas turbines of this family, owing to steam cooling of the combustor walls, the inlet gas temperature reaches 1,450°C (2,640°F). With two gas turbines M701G and one 265-MW steam turbine with the steam conditions of 13.7 MPa, 566/566°C (1,986 psi, 1,050/1,050°F), the CC unit provides the rated total capacity of 805 MW. The Higashi Niigata Unit 4 of this type was put in commercial operation in 1999, and field tests completely confirmed its calculated performances.[63]

Perhaps the first SS CC units were developed by GE under the trademark STAG.[64] There, it was initially deemed advisable to design them on the basis of single-cylinder non-reheat steam turbines. Both (steam and gas) turbines were located on both sides of the generator; each turbine had its own thrust bearing and was connected to the generator with a flexible coupling. Preference was given to steam turbines with axial exhaust and axial arrangement of the condenser. In this case, the machine hall can be made without a basement what additionally cuts down the civil engineering work and capital expenditures. The steam turbine was supposed to be taken off the foundation whenever the generator rotor has to be withdrawn from the stator. Therefore, all the steam turbine's external joints were made flanged, not welded. Along with this, GE also suggested SS CC turboset versions with two-cylinder steam turbines. In this case, the generator is driven from its front end connected to the steam turbine shaft's rear end; all the rotors are connected to each other with rigid couplings, and both turbines had a common thrust bearing—Figure 3-29a.

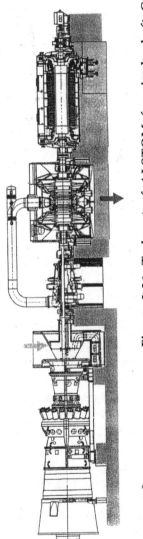

Figure 3-26. Turboset of ALSTOM for single-shaft CC units of the power plant Eems, 5x355 MW, the Netherlands.
1: exhauster-gas diffuser, 2: gas turbine, 3: air-inlet plenum, 4: steam turbine, 5: generator, 6: excitation brushes, 7: to condenser. *Source: By courtesy of ALSTOM Power*

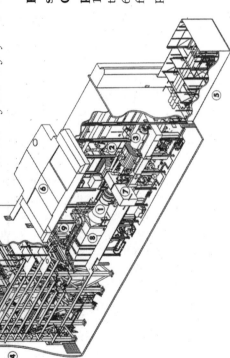

Figure 3-27. General three-dimensional CAD view of a single-shaft CC unit with vertical HRSG for the power plant Eems, the Netherlands. 1: gas turbine, 2: steam turbine, 3: generator, 4: vertical HRSG, 5: transformer area, 6 and 7: process air inlet filter and duct, 8: flue gas ducts and silencer, 9: main steam pipes. *Source: By courtesy of ALSTOM*

Configuration of Modern Power Steam Turbines   81

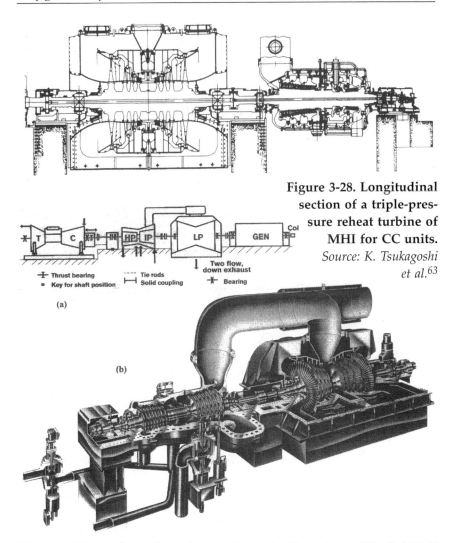

Figure 3-28. Longitudinal section of a triple-pressure reheat turbine of MHI for CC units. *Source: K. Tsukagoshi et al.*[63]

Figure 3-29. Configuration of a single-shaft CC turboset STAG S109H of GE (a) and general view of a steam turbine of GE for CC units with 7FA and 9FA gas turbines. *Source: "Structured steam turbines..."*[66] *and R. Swanekamp*[67]

On the basis of its gas turbine series "G" with air cooling (MS7001G and MS9001G for 3,600 and 3,000 rpm with the rated capacity of 240 and 282 MW, respectively), GE Energy has designed SS CC units STAG with the output of 350 and 420 MW, respectively. With the inlet gas temperature of 1,427 °C (2,600 °F) and the gas turbine efficiency in the simple-cycle of 39.5%, the net efficiency of such a CC unit amounts to 58%. With transition to the "H" technology and raising the inlet gas temperature up to 1,500 °C (2,732 °F), the efficiency amounts to 60%, and the output increases to 400 and 480 MW.[65-67] As in the case of the above mentioned CC SS turbosets of ALSTOM (see Figure 3-17), the generator is connected to the "cold" end of the steam turbine shaft. The first SS CC units of this type with a capacity of 480 MW on the basis of the "H"-technology, developed by GE together with Toshiba, were put in operation in 2001. The steam turbine is a dual-pressure two-cylinder machine with single-flow HP and integrated IP-LP cylinders. In another version, the steam turbine of the D11 family, specifically configured for CC units with 7FA and 9FA gas turbines (see Figure 3-29a), is made consisting of the integrated HP-IP cylinder and double-flow LP cylinder—Figure 3-29b.

The same scheme of locating the gas turbine, steam turbine, and generator on the common shaft and connecting them with rigid couplings was also used at many other SS CC units, including, for example, characteristic Japanese ones of the "third generation." Seven such units of Hitachi, with a single capacity of 243 MW each, were commissioned at the Kawagoe power plant's Phase 3.[60,68,69] Along with this, most of subsequent SS CC units have been designed with the use of a somewhat different approach.

Among the most widespread worldwide types of large gas turbines are V84.3A and V94.3A of Siemens Power Generation with the rotation speed of 3,600 and 3,000 rpm and the rated output of 170 and 240 MW, respectively. In the process of bench tests, V84.3A showed the efficiency of 38%. On the basis of these turbines, Siemens offers on a turnkey basis its SS CC units with the guaranteed gross efficiency of 58%. Schematic of such a unit for the grid frequency of 60 Hz is shown in Figure 3-30a. The first CC units of this type were installed at the power plant Portland in the USA; the first unit for 50 Hz was commissioned at the power plant Buggenum in the Netherlands,[59] and then similar units were constructed and put in operation in many industrially developed and developing countries: from the USA to the Philippines. The main feature of the turboset is a centrally mounted generator located between the gas and steam turbines and connected with the steam turbine by means of a self-shifting and synchronizing (SSS) clutch.

An SSS clutch is a fully automatic tooth-type "freewheel" clutch with a high torque transmitting capacity. The largest SSS clutch cur-

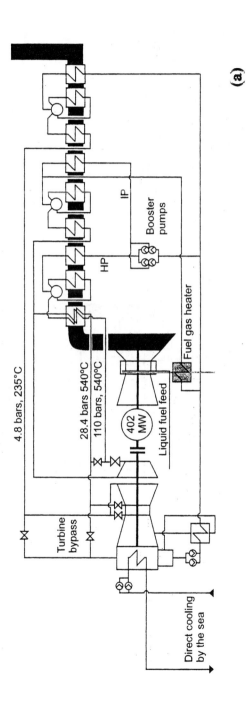

Figure 3-30. Schematics of characteristic single-shaft CC units of Siemens PG (a) and ALSTOM Power (b). *Source: E. Jeffs[70] and "GT 26 Kicks off..."[71]*

(Continued)

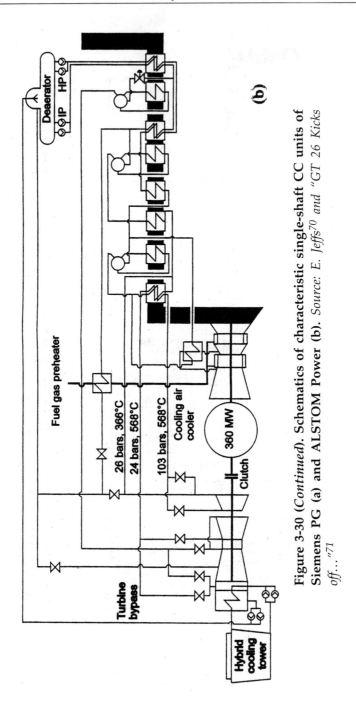

Figure 3-30 (*Continued*). Schematics of characteristic single-shaft CC units of Siemens PG (a) and ALSTOM Power (b). *Source: E. Jeffs[70] and "GT 26 Kicks off..."[71]*

rently in service for a turbine drive is capable of transmitting 300 MW at 3,000 rpm. The clutch teeth automatically engage at the instant when the input rotation speed tends to overtake that of the output. Conversely, the clutch teeth disengage the joined rotors when the input speed slows down relative to the output. First of all, at the operator's request, the SSS clutch allows operating the gas turbine independently of the steam turbine—in a simple-cycle mode. This makes the unit more flexible and allows using it as a peak stand-by capacity. The ability to de-clutch the steam turbine from the generator permits the gas turbine and generator to be commissioned in advance of the steam turbine to ease the commissioning process and get earlier power generation. But the most important advantage is that the clutch simplifies the start-up technology. In this case, the steam turbine remains shut down while the gas turbine is started up, and the steam turbine begins running up only when the HRSG produces steam of due conditions and in proper amounts, avoiding the use of a special boiler needed if the steam turbine were solidly coupled to the gas turbine. In addition, because the disengaged clutch allows the gas turbine and generator to be accelerated without the steam turbine, the starting power demand reduces, and a smaller variable-frequency start-up converter can be used. The start-up technologies for CC units, including SS ones with the use of the SSS clutch, are to be considered in the next part of the book.

The SSS clutches are also widely used at SS CC units of ALSTOM. The first SS CC unit K24 with the centrally mounted generator, GT26 gas turbine of ABB (presently ALSTOM), and steam turbine connected to the generator via the SSS clutch was commissioned in 1997 at New Zealand's Taranaki power plant. Its schematic is shown in Figure 3-30b. Then, similar SS CC units of ALSTOM were commissioned at the power plants Enfield and Shoreham in the UK, Agawam and Midlothian (four SS CC units with GT24 gas turbines) in the USA, and others.

The considered SS CC units of Siemens and ALSTOM (Figure 3-30) also feature floor-mounted steam turbines with axial exhaust and axially arranged condensers. The Siemens turbine (Figure 3-31a) is made in two cylinders with a separate HP cylinder and integrated once-through, single-flow IP-LP one. This turbine was especially designed for SS CC turboset arrangements with the generator driven at the turbine's front end through the SSS clutch, which can be seen in the drawing. The HP cylinder has a service-proven two-casing design with a barrel-type outer casing and the inner casing and rotor made of 10%Cr steel. These measures would allow the turbine to work with the main steam parameters of up to 16 MPa, 600°C (2,320 psi, 1,112°F),[72] even though the existing gas turbines and HRSGs do not provide so high steam conditions, and the IP steam admission part of the same turbine, with its single-casing design, is obviously intended for a lower reheat steam temperature. True, there exists a project of an advanced IP-LP cylinder for this turbine with an inner IP casing permitting the use of elevated reheat temperatures— Figure 3-31c. In order to provide the necessary annular exhaust area, if the power unit is operated with the grid frequency of 60 Hz, the turbine can be furnished with titanium integrally shrouded LSBs of 1,070-mm (42 in) length. For 50 Hz applications, these LSBs are replaced with 1390-mm (55 in) or 1,540-mm (61 in) ones. To withstand both the reheat steam temperature in the front part and huge centrifugal forces caused by the LSBs in the rear part, the IP-LP rotor is of a welded "hetero-material" type consisting of two forgings made of Cr-Mo-V steel with high-creep strength and a Ni-Cr-Mo material with high-fracture toughness. An axial exhaust and axially arranged condenser make it possible to built the turbine hall without any basement and minimize the turbine hall's sizes to the length of 45 m (150 ft) and height of 20m (65 ft). The general length of the unit with a horizontal HRSG, from the transformer to the stack, makes up to 122 m (400 ft).[73,74]

For CC units of a greater capacity with deep vacuum, Siemens suggests two-cylinder steam turbines with an integrated HP-IP cylinder and a double-flow LP cylinder with two side condensers on either side of the turbine—see Figure 3-31b. In this case, the turbine island can also be made without a basement. The LSBs length is chosen to provide an optimal annular exhaust area for each specific application. Both steam turbine versions of Siemens PG for SS CC units with the net power output of about 400 MW as applied to the grid frequency of 50 Hz are standardized as SCC5-4000F Reference Power Plants.[55] The first standardized SS CC units of this class were constructed at the Spanish power plant Arrubal (2×380 MW). Thanks to standardization, the total project and construction time period from the beginning to grid connection was cut down by two months and took 22 months, which, according to Siemens, is the market record but can be shortened further.[75]

At the same time, Siemens has especially tailored newly developed components of a so-called "HE" steam turbine product line for future advanced CC power plants with a SS turboset configuration. The emphasis is placed on fulfillment of the key market requirements: high

Configuration of Modern Power Steam Turbines 87

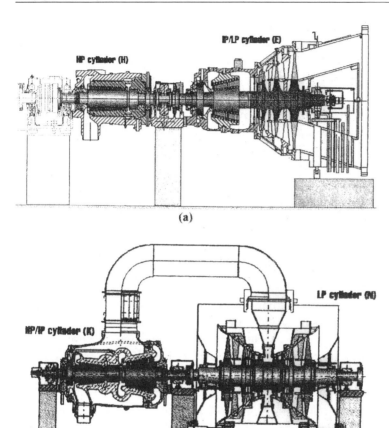

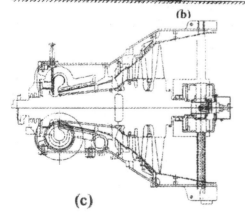

Figure 3-31. Longitudinal sections of Siemens steam turbines for single-shaft CC units: with an axial (a) and side (b) exhausts, and an advanced IP-LP cylinder with an axial exhaust (c). *Source:* "Rationalized product lines ..."[74]

availability, operational flexibility, low life-cycle costs, and short production and installation times. As well as the former turbine shown in Figure 3-31, the HE steam turbine consists of single-flow HP and integrated IP-LP cylinders with axial exhaust. The product line is founded on a modular concept for main turbine design elements as applied to 50 Hz and 60 Hz applications. In addition, the modular design of the "E" (IP-LP) portion enables a wide range of turbine applications owing to the use of different standard LSBs with the exhaust annular area of up to 16 m^2.[76]

Floor-mounted turbosets with the single-flow LP steam paths, axial exhausts, and axially arranged condensers are also used in SS CC units of ALSTOM—see Figure 3-30b. The footprint of a 360-MW SS CC unit with a horizontal HRSG measures 37.5 m (123 ft) wide and 96.3 m (316 ft) long to the center line of the stack, and the turbine table is positioned 3.8 m (12.5 ft) above the turbine hall floor. Along with this, ALSTOM has developed another arrangement concept for SS CC units based on the use of a side condenser that enables the steam turbine to be designed with a separate double-flow LP cylinder and have an output of up to 500 MW. The condenser stands on its own foundation, supported by sliding mounts, or it is welded to the turbine and placed on a foundation shared with the turbine. Locating the condenser in the side makes it possible to route the cooling water pipes above the floor in the building, thus reducing the lengths involved. The plan view of such a SS CC unit is given in Figure 3-32. The turbine table is located at the level of 4.5 m. This solution is incorporated in ALSTOM's standardized concept for CC power plants built on a turn-key basis. As compared to a conventional steam turbine arrangement with a basement condenser, the ground floor arrangement of the turbine in the case of axial or side condensers reduces the turbine foundation cost by approximately 20-25%.[77]

In conclusion, it might be well to point out that in many cases steam turbines of a moderate capacity have been successfully used for verification of some design novelties mainly intended for large power steam turbines, as well as for special experimental investigations of aerodynamic and thermal processes in steam turbines. Such an approach can be demonstrated by the example of a reheat steam turbine of MHI installed at the "T-point" CC power plant of the Takasago Machinery Works of MHI.[78] This CC unit with the total capacity of about 330 MW, on the basis of a 230-MW gas turbine 501G with inlet

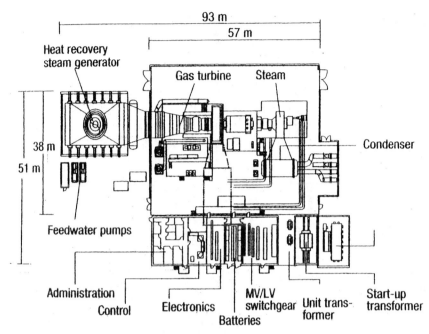

Figure 3-32. Plan view of a 250-MW SS CC unit of ALSTOM with a one-side condenser and vertical HRSG. *Source: H. Klotz et al.*[77]

gas temperature of 1,427°C (2,600°F) developed by Westinghouse together with MHI, was launched in 1997. Later, the newly developed 105-MW steam turbine was installed instead of the existing one. The new turbine was made one-cylinder single-flow with axial exhaust and a welded "hetero-material" rotor. In addition, the turbine was designed and manufactured with the use of some advanced technologies, including fully three-dimensionally designed HP and LP blading, leaf and active clearance control (ACC) seals, scroll-type steam inlets and outlets, highly efficient axial exhaust hood, and some others—Figure 3-33a. A bulk of special measurements for field tests intended to verify the effectiveness of these novelties and assess their effect is shown in Figure 3-33b. The gotten field test results facilitate MHI in design of a larger single-cylinder steam turbine for the single capacity of about 160 MW with a scroll-type steam inlet, hetero-material welded rotor, LSBs of the 45-in (for 60 Hz) and 48-in (for 50 Hz) length, axial exhaust, and other features proved in service at the "T-point" plant.[26]

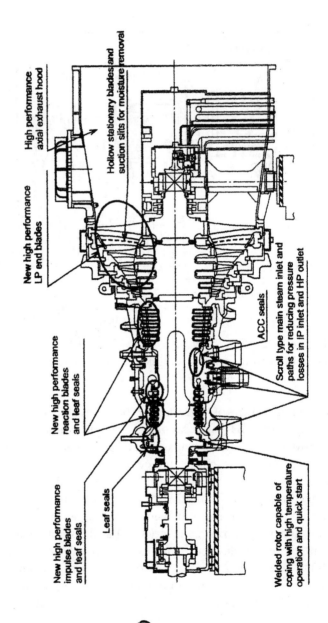

(a)

Configuration of Modern Power Steam Turbines 91

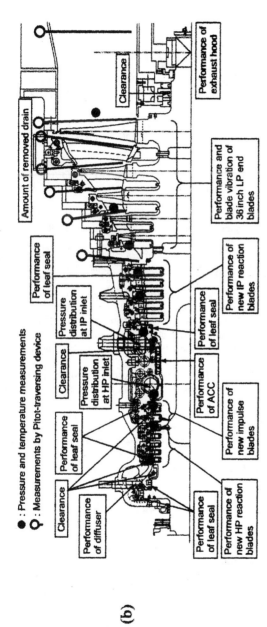

Figure 3-33. New technologies applied at the "T-point" CC power plant's 105-MW steam turbine of MHI (a) and special measurements (b) installed at the turbine. *Source: E. Watanabe et al.*[78]

References

1. Leyzerovich A. *Large Power Steam Turbines: Design & Operation*, in 2 vols., Tulsa (OK): PennWell Books, 1997.
2. "Tachibana-wan unit 2 takes a supercritical step forward for Japan," *Modern Power Systems* 21, no. 11 (2001): 41-47.
3. Momma H., J. Ishiguro, T. Suto, et al. "Commencement of the Commercial Operation of 600 MW Unit, Hirono No. 5 Thermal Power Station of the Tokyo Electric Power Co., Inc." *Mitsubishi Heavy Industries Technical Review* 41, no. 5 (2004): 1-5.
4. Magoshi R., T. Nakano, T. Konishi, et al. "Development and Operating Experience of Welded Rotors for High-Temperature Steam Turbines," presented at the 2005 Electric Power Conference, Chicago, 2005.
5. Ando K., E. Asada, K. Yamamoto, et al. "Design, Construction, and Commissioning of the Nos. 1 and 2 Units of the Ratchaburi Thermal Power Plant for the EGAT of Thailand as a Full Turn-key Contract," *Mitsubishi Heavy Industries Technical Review* 39, no. 3 (2002): 95-100.
6. Nomoto H., Y. Kuroki, M. Fukuda, and S. Fujitsuka. "Recent Development of Steam Turbines with High Steam Temperatures," presented at the 2005 Electric Power Conference, Chicago, 2005.
7. Nameki Y., T. Murohoshi, F. Hiyama, and K. Namura. "Development of Tandem-Compound 1,000-MW Steam Turbine and Generator," *Hitachi Review* 47, no. 5 (1998): 176-182.
8. Matsumoto N., H. Iwamoto, T. Kaneda, and K. Shigihara. "Completion of Hitchinaka Thermal Power Station Unit No. 1—1,000-MW Power Generating Plant of the Tokyo Electric Power Co., Inc." *Hitachi Review* 53, no. 3 (2004): 104-108.
9. Umezawa T., R. Okura, M. Sato, and Y. Nagashima. "Shinko Kobe No. 2 700-MW Power Station of Kobe Steel Ltd. as Largest IPP Facility in Japan," *Hitachi Review* 54, no. 3 (2005): 105-109.
10. Tateishi A., H. Fujii, T. Koga, and H. Kimura. "Approach to Overseas EPC Project—Technical Features of MidAmerican Project in the U.S." *Hitachi Review* 54, no. 3 (2005): 98-104.
11. Okura R., S. Shioshita, Y. Kawasato, and J. Shimizu. "Completion of High-Efficiency Coal-fired Power Plant," *Hitachi Review* 52, no. 2 (2003): 79-83.
12. Busse L. and K.-H. Soyk. "World's highest capacity steam turbosets for the lignite-fired Lippendorf power station," *ABB Review*, no. 6 (1997): 13-6.
13. "Lippendorf sets new standards in efficiency," *Modern Power Systems* 17, no. 4 (1997): 57-64.
14. Leyzerovich A.Sh. "New Developments of ABB for Steam-Turbine Power Plants of Germany" [in Russian], *Elektricheskie Stantsii*, no. 12 (1999): 57-60.
15. Lageder H. and P. Meylan "ABB modular reheat steam turbines," *ABB Review*, no. 5 (1990), 3-10.
16. Heitmüller R.J., H. Fischer, J. Sigg, et al. "Lignite-fired Niederaußem K aims for efficiency of 45 per cent and more," *Modern Power Systems* 19, no. 5 (1999): 53-66.
17. Lamberttz J. and G. Gasteiger. "Concept and Experience Gained during Start-up of the Power Unit with Optimized Plant Engineering (BoA) at Niederaußem" [in German], *VGB PowerTech* 83, no. 5 (2003): 82-87.
18. Tippkötter Th. and G. Scheffknecht. "Operating Experience with New BoA Unit and Outlook" [in German], *VGB PowerTech* 84, no. 4 (2004): 48-55.
19. "Yuhuan: a Chinese milestone," *Modern Power Systems* 25, no. 6 (2005): 27-31.
20. Baumgartner R., J. Kern, and S. Whyley. "The 600 MW Advanced Ultra-Supercritical Reference Power Plant Development Program for North Rhine

Westphalia—A Solid Basis for Future Coal-Fired Plants Worldwide," presented at the 2005 Electric Power Conference, Chicago, 2005.

21. Meier H.-J., M. Alf, M. Fischedick, et al. "Reference Power Plant North Rhine-Westphalia (RPP NRW)," *VGB PowerTech* 84, no. 5 (2004): 76-89.

22. Hard F. "75 Years of Brown Boveri Steam Turbines," *Brown Boveri Review* 63, no. 2 (1976): 85-93.

23. Reinhard K., F. Vest, and E. Vogelzang. "Experience with the World's Largest Steam Turbines," *Brown Boveri Review* 63, no. 2 (1976): 106-114.

24. Oeynhausen H., A. Drosdziok, W. Ulm, and H. Termuehlen. "Advanced 1000 MW Tandem-Compound Reheat Steam Turbine," presented at the American Power Conference, Chicago, 1996.

25. Sill U. and Zörner W. *Steam Turbine Generators Process Control and Diagnostics*, Erlangen: Publics MCD Verlag, 1996.

26. Nakano T., K. Tamaka, T. Nakazawa, et al. "Development of Large-Capacity Single-Casing Reheat Steam Turbines for Single-Shaft Combined Cycle Plant," *Mitsubishi Heavy Industries Technical Review* 42, no. 3 (2005): 1-5.

27. Obara I., T. Yamamoto, and Y. Tanaka "Design of 600°C Class 1000 MW Steam Turbine," *Mitsubishi Heavy Industries Technical Review* 32, no. 3 (1995): 103-107.

28. Kishimoto M., Y. Minami, K. Takayanagi, and M. Umaya. "Operating Experience of Large Supercritical Steam Turbine with Latest Technology," *Advances in Steam Turbine Technology for the Power Generation Industry*, PWR-Vol. 26, ASME, New York: 1994: 43-31.

29. De Paul M.V., M. Wallon, and A. Anis. "Twenty years' Progress in Steam Turbine Aerodynamics," *Proc. of the American Power Conference* 51, Chicago, 1989: 166-173.

30. Leyzerovich A. *Wet-Steam Turbines for Nuclear Power Plants*, Tulsa (OK): PennWell Corporation, 2005.

31. Nah U.-H., J. Powers, et al. "Recent Advances in Ultra Super Critical Steam Turbine Technology," presented at the POWER-GEN Asia Conference, Singapore, 2005.

32. Wani M., H. Fukuda, M. Tsuchiya, et al. "Design and Operating Experience of a 1000 MW Steam Turbine for the Chugoku Electric Power Co., Inc. Misumi No. 1 Unit," *Mitsubishi Heavy Industries Technical Review* 36, no. 3 (1999): 66-74.

33. Rukes B. and R. Taud. "Perspectives of Fossil Power Technology," *VGB PowerTech* 82, no. 10 (2002): 71-76.

34. Tremmel A. and D. Hartmann. "Highly Efficient Steam Turbine Technology for Fossil Fuel Power Plants in Economically and Ecologically Driven Markets," *VGB PowerTech* 84, no. 11 (2004): 38-43.

35. Wichtmann A., M. Deckers, and W. Ulm. "Ultra-Supercritical Steam Turbine Turbosets Best Efficiency Solution for Conventional Steam Power Plants," *VGB PowerTech* 85, no. 11 (2005): 44-49.

36. Shige T., R. Magoshi, S. Itou, et al. "Development of Large-Capacity, Highly Efficient Welded Rotors for Steam Turbines," *Mitsubishi Heavy Industries Technical Review* 38, no. 1 (2001): 6-11.

37. Leyzerovich A. "A Hypothetical Profile of Ordinary Steam Turbines with Reduced Cost and Enhanced Efficiency and Reliability for Contemporary Conditions," *Proc. of the American Power Conference* 60, Chicago, 1998, Part 2: 1021-1026.

38. Lysko V.V., G.I. Moseev, A.L. Shvarts, et al. "New-Generation Coal-Fired Steam-Turbine Power Units," *Thermal Engineering* 43, no. 7 (1996): 545-552.

39. Avrutskii G.D., V.V. Lysko, A.V. Shvarz, and B.I. Shmukler. "About creating coal-fired power units with USC steam conditions" [in Russian], *Elektricheskie Stantsii*, no. 5 (1999): 22-31.

40. Avrutskii G.D., I.A. Savenkova, M.V. Lazarev, et al. "Development of Technical Solutions for Creating a Supercritical Steam-Turbine Power Unit" [in Russian], *Elektricheskie Stantsii*, no. 10 (2005): 36-40.
41. Efimochkin G.I. "A Two-Cascade Deaeratorless Thermal Scheme for Steam Turbines of Thermal and Cogeneration Power Plants," *Thermal Engineering* 44, no. 1 (1997): 20-21.
42. Efimochkin G.I. "Auxiliary Equipment of Turbine Units: Achievements and Problems," *Thermal Engineering* 46, no. 12 (1999): 987-993.
43. Avrutskii G.D. and I.A. Savenkova "Improvement of Turboset's Thermal Schemes," *Elektricheskie Stantsii*, no. 10 (2003): 33-21.
44. Umezawa T., K. Ura, and Y. Hanawa. "Kin No. 1 220-MW Thermal Power Station of the Okinawa Electric Power Company, Incorporated," *Hitachi Review* 52, no. 2 (2003): 84-88.
45. Tremmel D., W. Kachler, and P. Bourcier. "Development of a compact 300-MW steam turbine with single-flow LP section and axial exhaust" [in German], *VGB Kraftwerkstechnik* 72, no. 1 (1992), 33-27.
46. Trukhnii A.D. *Stationary Steam Turbines* [in Russian], 2nd edition, Energoatomizdat, Moscow: 1990.
47. "Finnish utility plans largest backpressure steam turbine," *Power* 146, August 2002: 8.
48. Hennagir T. "Combined cycles meet the market need," *Power Engineering International* 5, no. 5 (1997): 25-32.
49. Makansi J. "It's time to calibrate financial models for life-cycle cost," *Power* 142, no. 4 (1998): 4-5.
50. Brückner H., D. Bergmann, and H. Termuehlen. "Various Concepts for Topping Steam Plants with Gas Turbines," *Proc. of the American Power Conference* 54, Chicago, 1992: 569-581.
51. Kehlhofer R., R. Bachmann, H. Nielsen, and J. Warner. *Combined-Cycle Gas & Steam Turbine Power Plants*, 2nd edition, Tulsa (OK): PennWell, 1999.
52. Franke J., U. Lenk, R. Taud, and F. Klauke. "Advanced Benson HRSG makes a successful debut," *Modern Power Systems* 20, no. 7 (2000): 33-19.
53. "Fast work for the Benson once-through HRSG," *Modern Power Systems* 24, no. 2 (2004): 23-9.
54. Jeffs E. "New combined cycle tackles cycling," *Turbomachinery International* 45, no. 5 (2004): 30-31.
55. Emberger H.-M., P. Mürau, and L. Beckman. "Siemens Reference Power Plants—Translating Customer Needs into Plants," *VGB PowerTech* 85, no. 9 (2005): 78-83.
56. Galopin J.F. "Going supercritical: once-through is the key," *Modern Power Systems* 18, no. 12 (1998): 39-43.
57. Stambler I. "Advanced steam turbines pushing combined cycle efficiency past 60%," *Gas Turbine World*, no. 3 (1996): 29-34.
58. Götz W. and S.A. DellaWilla. "Plant operation databases improve system reliability," *Modern Power Systems* 18, no. 4 (1998): 19-22.
59. Swanekamp R. "Single-shaft combined-cycle packs power in at low cost," *Power* 140, no. 1 (1996): 24-28.
60. Leizerovich A.Sh. "Single-Shaft Combined-Cycle Installations," *Thermal Engineering* 47, no. 12 (2000): 1123-1128.
61. Hassan A., P.D. Hemsley, and F. Lamarque. "A Complete Range of Steam Turbines for Combined Cycles," presented at the Symposium on Steam Turbines and Generators of GEC Alsthom, Monaco: 1994: 8/1-8/13.
62. Swanekamp R. "Europe welcomes single-shaft CC units—in a big way," *Electric*

Configuration of Modern Power Steam Turbines

Power International, no. 3 (1998): 34-36.

63. Tsukagoshi K., N. Yoshida, S. Takahashi, et al. "Design and Operation Experience of Most Advanced Combined Cycle Plant—Construction of Unit 4-1 Higashi Niigata Thermal Power Station of Tohoku Electric Power Co., Inc." *Mitsubishi Heavy Industries Technical Review* 37, no. 3 (2000): 69-73.

64. Baily F.G., A.P. Rendine, and K.E. Robbins. "Steam Turbines for STAG Combined-Cycle Power Systems," *Proc. of the American Power Conference* 51, 1989: 389-393.

65. "GE breaks 60% net efficiency barrier," *Power Engineering* 99, no. 6 (1995): 34-36.

66. "Structure steam turbines for advanced combined cycles," *Modern Power Systems* 20, no. 12 (2000): 41-46.

67. Swanekamp R. "Merchant power projects push for competitive edge," *Power* 144, January/February 2000: 32-39.

68. Kawauchi A., F. Hirose, and M. Musashi. "High-Efficiency Advanced Combined-Cycle Power Plants," *Hitachi Review* 46, no. 3 (1997): 121-128.

69. Leizerovich A.Sh. "Several Current Aspects of the Development of Thermal Engineering in Japan," *Thermal Engineering* 46, no. 10 (1999): 885-892.

70. Jeffs E. "Siemens on Fast Track to Commission Santa Rita," *Turbomachinery International* 41, no. 2 (2000): 31-33.

71. "GT26 Kicks off New Zealand Combined Cycle Program," *Turbomachinery International* 38, no. 2 (1997): 31-34.

72. "Preparing for full output testing at Cottam," *Modern Power Systems* 19, no. 6 (1999): 19-20.

73. Termuehlen, H. *100 Years of Power Plant Development. Focus on Steam and Gas Turbines as Prime Movers*, New York: ASME Press, 2001.

74. "Rationalized product lines aim to anticipate future markets," *Modern Power Systems* 23, no. 10 (2003): 38-40.

75. "Going for the record," *Modern Power Systems* 24, no. 5 (2004): 26-28.

76. Pfitzinger E.-W., M.-A. Schwarz, and F. Hiss. "The New HE Product Line. A Compact Steam Turbine for Highest Efficiency Levels in Combined-Cycle Power Plants" [in German], *VGB PowerTech* 84, no. 11 (2004): 32-37.

77. Klotz H., L. Rausch, P.-H. Weirich, and P. Zhang. "Innovative Arrangement Concepts for Modern Steam Turbosets and New Trends in Power Houses," *VGB PowerTech* 80, no. 7 (2000): 19-22.

78. Watanabe E., Y. Tanaka, T. Nakano, et al. "Development of New High Efficiency Steam Turbine," *Mitsubishi Heavy Industries Technical Review* 40, no. 4 (2003): 1-6.

Chapter 4

Design of
Steam Path, Blading,
Gland Seals, and Valves

FEATURES OF MODERN THREE-DIMENSIONAL
STEAM PATH DESIGN

According to Siemens Power Generation, implementation of the advanced steam path design for large steam turbines of the mid-1990s brought approximately the same gain in the turbine efficiency (about 5%) as transition from subcritical-pressure steam conditions of 16.6 MPa, 538/538°C (2,400 psi, 1,000/1,000°F) to supercritical-pressure ones with elevated steam temperatures of 24.1 MPa, 593/593°C (3,500 psi, 1,100/1,100°F).[1,2] Acceptance field tests of the Boxberg Q's 907-MW turbine of Siemens gave the internal efficiency data of 94.2% for the HP cylinder and 96.1% for the IP one.[3] These efficiency figures, even being record-breaking, are quite characteristic for steam turbines with a modern steam path designed with the use of advanced computation technologies. Thus, for example, when ALSTOM Power retrofitted the steam turbine's HP-IP cylinder at the U.S. power plant J.K. Spruce, the subsequent acceptance tests showed the internal efficiency figures of 93% for the HP section and 95.7% for the IP section,[4] whereas even for the best turbines of the 1980s these figures were at the level of about 90% and 93%, respectively.[5] By another example, for the HP and LP sections of the retrofitted steam turbine of Siemens at the German power plant Mehrum, the internal efficiency values were found equal to 93.6% and 89.9%, respectively, compared to the original values of 86.5% and 87.2%.[6]

The major losses in turbine stages, causing a decrease in their internal efficiency, are the profile energy losses, secondary losses (or the losses at the end-walls, or merely end losses), and the losses with leakages of steam passing beyond the stage's rotating blades. In a general way—as

97

applied to an impulse-type turbine stage, these losses are conditionally shown in Figure 4-1; the nature of secondary flows in a blade passage is explained by a sketch of Figure 4-2.

Reduction of Profile Energy Losses.

The profile energy losses in the blade rows are conditioned by three main factors: 1) friction losses on the profile surface and the losses with vortices if the boundary layer separates from the surface, 2) losses with vortices downstream from the profile's trailing edge, and 3) wave losses under conditions of a supersonic flow through the row channel. Until the 1980s, turbine designers had applied their main efforts just to decrease the profile losses, and experimental investigations and calculations with the use of the newest, more advanced calculation methodologies and refine

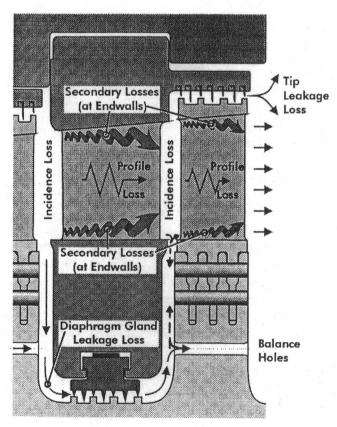

Figure 4-1. Major losses in an impulse-type stage

Design of Steam Path, Blading, Gland Seals, and Valves 99

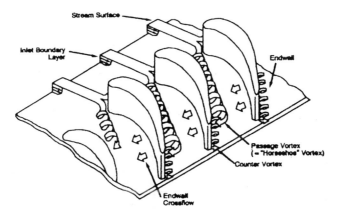

Figure 4-2. Secondary flows in a blade passage.

computer programs were aimed at developing and implementing more advanced, lower-loss blading profiles—see, for example, Figure 4-3.

This process of improving and polishing the blade profiles has not stopped in the subsequent years. By example, the development of reaction-type blade profiles of Siemens is illustrated by Figure 4-4a: beginning with a so-called T2 profile in 1970, via a T4 profile in 1980, and toward an improved TX profile in 1995.[1] According to Siemens, the TX profile affords a higher overall stage efficiency compared to the T4, being at the same time less prone to deposits on the suction side. While the earlier T4 profile, with its pretty flat optimum efficiency curve, was well suitable for a wide range of applications, the TX profile yields advantages for part-load operation over a pre-defined load range, corresponding to the current operational practice for large steam turbines.

For turbine stages with a relatively great height-to-mean-diameter ratio ($l/d_m > 0.1$), it is necessary to take into consideration an uneven radial pressure distribution lengthwise in the stage height counterbalancing centrifugal forces caused by a circular constituent of the steam velocity with resulted changes of the reaction degree along the rotating blade length, as well as the change in the circular rotation speed and, therefore, the velocity triangles. Hence such stages should be designed with profiles varying lengthwise of the stage height and twisted fixed and rotating blades. This requirement is partially pertinent to IP stages of large steam turbines with great volumetric steam flow amounts, but primarily refers to LP stages, especially the LSBs. The velocity diagrams, steam flow character, and Mach number values for these stages essential-

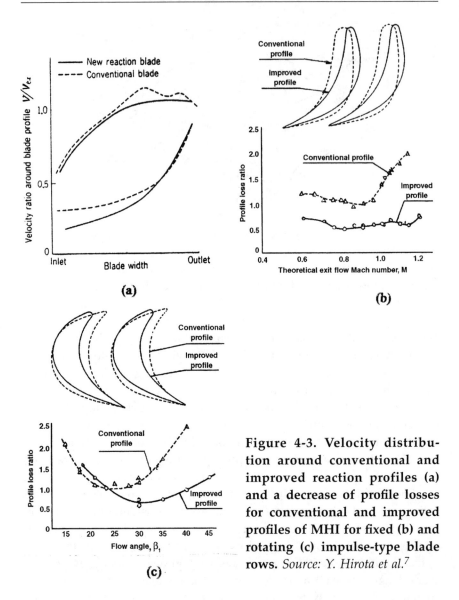

Figure 4-3. Velocity distribution around conventional and improved reaction profiles (a) and a decrease of profile losses for conventional and improved profiles of MHI for fixed (b) and rotating (c) impulse-type blade rows. *Source: Y. Hirota et al.*[7]

ly vary lengthwise in the stage height, and the applied blading profiles are set substantially different for different radial sections. Of importance is that LP stages with large blade lengths and great steam expansion and speed-up of the steam flow, primarily the LSBs, require supersonic profiles for the tip sections. Characteristic steam velocity fields at the tip,

Design of Steam Path, Blading, Gland Seals, and Valves 101

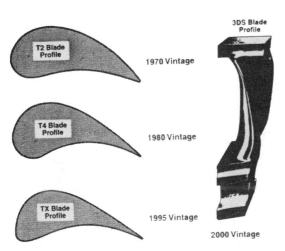

Figure 4-4. Evolution of reaction-type profiles (a) and fully three-dimensional 3DS blades (b) of Siemens. Source: H. Oeynhausen et al.[1]

median, and root sections of a characteristic LP LSB can be seen in Figure 4-5, and the diagrams of Figure 4-6 demonstrate possible variations of the profile types for the root, mean, and tip sections of a LSB to obtain the minimum profile losses in the wide range of Mach number values.

Reduction of Secondary (End) Losses

In parallel with combating the profile losses, in the 1990s the main efforts were redirected to the secondary losses, generated by the interaction of complicated three-dimensional flows near the channel end-walls—see Figure 4-2. These flows are caused by cross pressure gradients ensued from the channel curvature. Due to the increase in pressure along the concave profile face, the flow stream deviates to the back surface where the pressure is less. As a consequence, the boundary layer at the profile back becomes vitally thicker. The pressure certainly decreases towards the end-walls on the concave surface and increases on the convex one. These secondary flows within the inter-blade channels result in two vortex regions at the root and tip end-walls near the convex blade surfaces. In these vortex regions, there appear spiral flows of the opposite rotation directions. The secondary losses are especially significant for relatively short blades, when the height of the row channel is comparable to the profile width or chord length and the vortex regions at root and tip end-walls can interact and even merge.

Adequate mathematical modeling and detailed Computational Fluid Dynamics (CFD) investigations of the secondary flows becomes

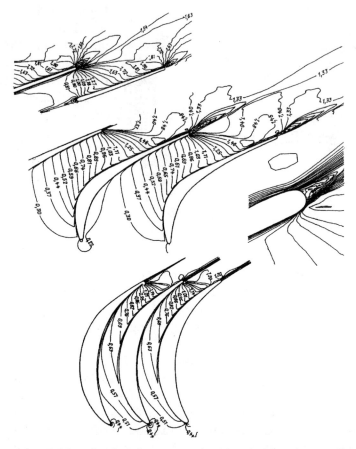

Figure 4-5. Fields of related steam velocities in the tip, median, and root sections of a characteristic LP LSB. *Source: F.P. Borisov et al.[8]*

possible only in the 1990s with developing fully three-dimensional (3D) computational models and corresponding software running on powerful computers. In the early 1990s, to decrease the secondary losses, GEC Alsthom proposed to extend to steam turbine blading the use of a so-called compound lean geometry design approach earlier developed for gas turbines.[10,11] According to such an approach and its variety having been called a "controlled flow" principle, the root and tip sections are displaced tangentially relative to the mid-height section in such a manner that the blade is curved in the direction of rotation. As a result, the concave blade surface intersects with the root and tip end-walls at

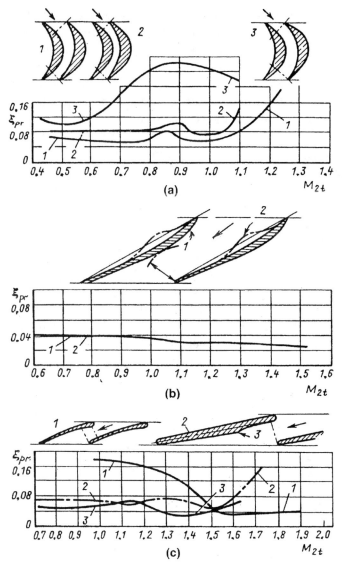

Figure 4-6. Characteristic blade rows and dependence of profile losses on the Mach number for different sections of an LSB.
A root section: 1-row with convergent channels, 2-pure impulse-type blading, 3-row with divergent channels, b-mean section: 1-common row, 2-row with double-convex profiles, c-tip section: 1-row with divergent channels and common profiles, 2-row with convergent channels and common profiles, 3-row with convergent channels and a ridge at the profile back. *Source: A.V. Shcheglyaev*[9]

an acute angle. This causes higher statistic pressure and hence lower velocities. In addition, the blade sections are skewed in such a manner that the channel throat areas are reduced at the root and tip sections, but increased at the mid-height. This results in reduced mass steam flows in the end-wall regions, where the secondary losses are generated, and increased mass flow in the most efficient mid-height blade zone. The resulted improvement in the stage efficiency, attained due to the use of newly profiled blades and controlled flow and verified by experiments at a model air turbine, is illustrated in Figure 4-7 as a function of stage loading (velocity ratio).

Transition to a fully three-dimensional design of the steam path with twisted, leant, and bowed (curved) blading was the next step. It might be well to remark that this conception was earlier developed and experimentally investigated and verified as long ago as in the early 1960s in the Moscow Power Engineering Institute (MEI) under guidance of Prof. M.E. Deich.[12,13] In particular, G.A. Filippov and Huang Chungchi of MEI investigated the blades, called saber-like ones, with a co-rotational lean varying along the stage height.[14] To diminish the secondary flows, the angle of inclination of such blades is reduced lengthwise in the blade height and can become negative at the tip, resulting in diminishing

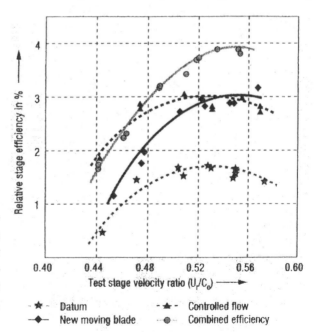

Figure 4-7. Efficiency gain with controlled-flow blades of ALSTOM. Source: A. Nowi and P.J. Walker[11]

of the secondary losses—Figure 4-8a. Much later, these experiments were repeated by specialists of Mitsubishi Heavy Industries (Figure 4-8b), and such a method of suppressing the secondary losses has begun applying by MHI (Figure 4-9), Siemens PG (Figure 4-4b), ALSTOM Power and other leading turbine producers of the world.

Apart from featuring consistently refined blades with decreased profile losses, the aerodynamic burden on the blade sections in the tip and root areas is eased by inclining the blades in the direction of rotation and bowing them, thus reducing the secondary losses. If initially the "saber-like" (leant and curved, or bowed) fixed and rotating blades were mainly intended to find application primarily for the LP LSBs (and, they truly bring there the greatest gain), then they has become to be used extensively throughout the entire steam path, in all the HP, IP, and LP stages.

In particular, this refers to Siemens steam turbines of the 2000 vintage with three-dimensionally designed (3DS) blading. Test results for the first HP stages' blading, shown in Figure 4-10, reveal the obvious

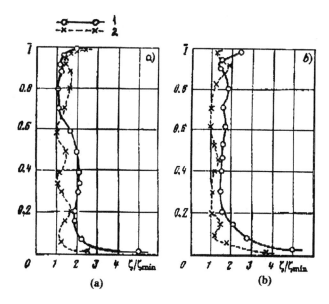

Figure 4-8. Local-to-minimum loss ratio variation over the height of a fixed blade rows with radial, straight (1) and "saber-like", bowed (2) blades according to experimental data of MEI of the 1960s (a) and MHI of the 1980s (b). *Source: M.E. Deich et al.*[14]

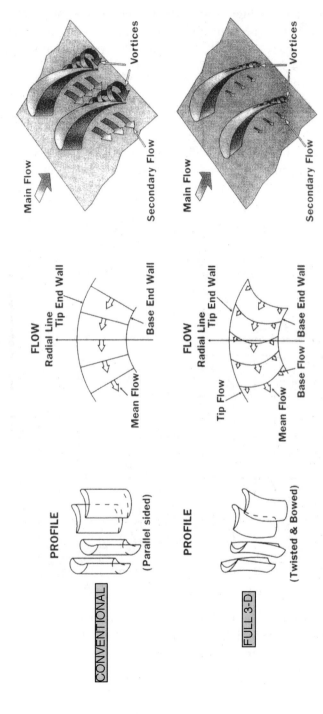

Figure 4-9. Formation of secondary losses in a blade row with conventional, cylindrical (a) and fully three-dimensionally designed, twisted and bowed (b) blades. *Source: By courtesy of Mitsubishi Heavy Industries.*

reduction of the secondary losses for a 3DS, leant and curved, blade over a conventional, cylindrical blade. At the same time, because the blade is pretty short, suppression of the secondary losses not only tells on the losses in the nearest vicinity of the blade platform (hub end-wall), but also considerably raises the internal efficiency along the whole blade height. Detailed experimental investigations conducted on a model turbine under representative Mach and Reynolds number values showed that 3DS blading gives the stage efficiency improvement of up to 2% points as compared to purely cylindrical blades.[16,17] In full measure, the advanced three-dimensional CFD design of the turbine steam path, including the HP and IP sections, Siemens Power Generation carried into effect at the above mentioned 907-MW Boxberg turbine (see Figure 2-10), resulting in its record efficiency. Major pre-requisite for the development of the 3DS blading have been the introduction of advanced CFD methods, computational optimization, Finite Element Analysis (FEA), experimental verification, linked Computer-Aided Design, Engineering, and Manufacturing (CAD/CAE/CAM) tools, and modern blade manufacturing facilities.[16]

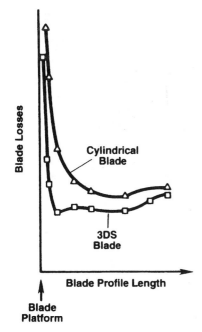

Figure 4-10. A decrease of losses along the blade height for a three-dimensionally designed (3DS) short blade compared to a conventional, cylindrical blade. *Source: H. Oeynhausen et al.*[15]

For stages with a small relative height (the blade-height-to-chord ratio $l/b < 1\text{-}1.5$), in 1960 Prof. M.E. Deich also suggested to suppress the secondary losses by means of special meridional profiling of the nozzle channel, later called "countered walls."[18] In this way, the steam flow is artificially driven to the root section, counterbalancing the centrifugal force, which presses out the flow to the outlying end-wall. Later, this approach was accepted and implemented by General Electric and Mitsubishi Heavy Industries for low aspect ratio (short) impulse-type stages with the resulted efficiency gain of up to about 1.5%—Figure 4-11.[18,19] Simultaneously, the flow patterns in the IP steam path were also optimized on the meridional plane. A similar approach has also been practiced by Siemens to increase the IP stage height and make the hub and outlying contours more smooth.[20] However, the meridional profiling of the steam path is most important for the LP sections taking into account the greatest steam expansion in this area and hence the considerable rise of the stage height.

According to a classic approach to the LP cylinder design, more often than not the root (hub) diameters of the stages are set constant (see Figures 2-2, 2-5, 2-6, 2-10, 3-3, 3-4, 3-5, 3-6, 3-11, 3-25, 3-28, 3-31, and 4-12). Sometimes, the root diameters are made slightly decreasing toward the last stage (see Figures 2-12, 3-1, 3-16, and some others, especially in LP cylinders of wet-steam turbines[21]). In these cases and with high values of the length-to-mean-diameter ratio for the LSB, the peripheral streamline pitch angles reach about 60°. Under such circumstances, transition from a broken, "piece-linear" peripheral outline of the steam path to a conical meridional outline gives a remarkable gain in the local stage efficiency for the tip sections.[22] In addition, under conditions of a classical design, the steam flow tends to separate from the hub, and the mass flow is mainly driven to the outlying contour. The steam streamlines cut the radially stacked vane at acute angles and hence experience a significant sweep angle—Figure 4-12c. In the case of leant and curved vanes of the last stage (Figure 4-12b), the streamline

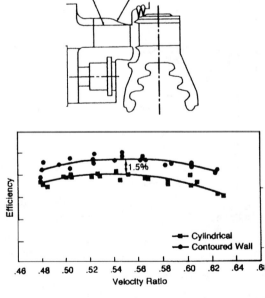

Figure 4-11. Test data for a control stage with conventional (1) and contoured (2) sidewalls. Source: J.I. Coffer, IV 18 and Y. Hirota et al.[19]

Design of Steam Path, Blading, Gland Seals, and Valves

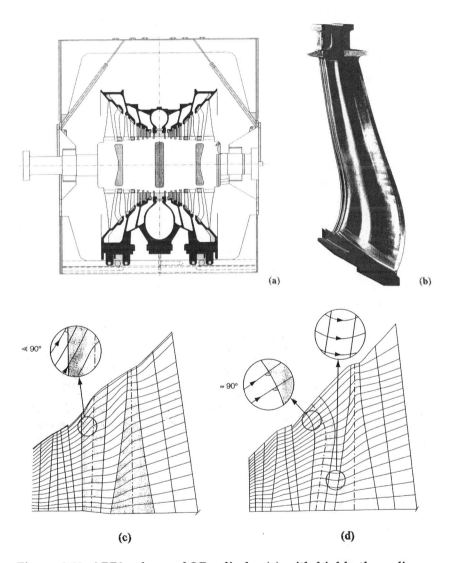

Figure 4-12. ABB's advanced LP cylinder (a) with highly three-dimensional geometry of the last stage vane (b) and meridional flow patterns in the LP steam path with a "classical" (c) and leant and curved (d) last stage vane. *Source: A.P. Weiss*[22]

inclination angles tend to be perpendicular to the blade edges—see Figure 4-12d compared to Figure 4-12c. The more uniform streamline fanning provides an evener mass flow distribution in the radial direction, along the stage height. The leant and curved vane (see Figure 4-12b) also produces an additional radial force which influences on the flow, increasing the root reaction of the stage and preventing flow separation. These principles of the LP steam path design were carried into effect, in particular, in the 933-MW turbines of ABB for the Lippendorf power plant—see Figure 3-5a, as well as employed for refurbishing some aging turbines—see Figure 18-5a.

The influence of saber-like vanes used in the LP last stage was also investigated in calculations and experiments of All-Russia Thermal Engineering Institute (VTI) together with LMZ.[23,24] Under consideration was the standard LP steam path of LMZ with the LSB of 960 mm long (see Figure 2-4a). Its conventional cylindrical vanes were replaced with saber-like ones with the angle of inclination near the root varying beginning from 10-to-30°, and the bowed portion of the vane length was varied in the range between 0.3 and 0.5 (the upper portion of the vane was remained radial), as it is proposed for retrofitting steam turbines in service. (Earlier it was shown that for stages with the highest values of height-to-the-mean-diameter ratio, that is, for LP last stages, it is not necessary to lean the vanes at their periphery—the peripheral zones can remain radial.) The resulted changes in the stage efficiency are shown in Figure 4-13 depending on the volumetric steam flow amount. For newly designed turbines with fully three-dimensionally bowed vanes and buckets through the entire steam path, the effect should be much more noticeable.

The peripheral pitch angle of the LP steam path can be reduced by lowering sharply the stage hub diameter toward the last stage, as shown, for example, in Figure 4-14 as applied to the LP steam path of Mitsubishi Heavy Industries with the titanium LSB 1,143 mm (45 in) long. A similar approach was also applied by GE Energy for the LP steam paths with the LSBs of 1,106 mm (40 in) length for 3,600 rpm and 1,218 mm (48 in) for 3,000 rpm.[25] However, in these cases the designers have to deal with the length-to-mean-diameter ratio for the LSBs up to the highest available values of 0.4-41 and even more. (The problems of LSBs and their design are to be considered in more details in the next section.)

As seen from Figure 4-2, the secondary losses are substantially tied with conditions by which the steam flow enters the blade row. So

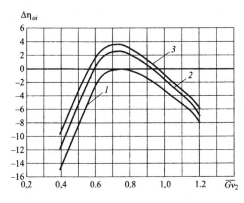

Figure 4-13. Changes in efficiency with the volumetric steam flow amount for the LP last stage of LMZ with the 960-mm LSB and original conventional vanes (1) and saber-like ones with the inclination angle at the root of 20° and the bowed length portion of 0.3 (2) and 0.5 (3). Source: L.L. Simoyu et al.[24]

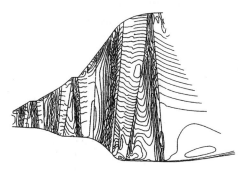

Figure 4-14. Calculated isotach field for the steam path of the three last LP stages with the 1,143 mm long titanium LSB of MHI's 3,600-rpm turbine. Source: By courtesy of Mitsubishi Heavy Industries

in order to reduce the secondary flow it would be advisable to change the nature of the flow at the channel entrance to suppress as much as possible so-called horseshoe-shaped vortices at the channel endwalls. In particular, it would be possible by changing the leading edge form. The traditional way of reducing the intensity of the considered vortex cords comes to decreasing in a reasonable manner the profile's leading edge thickness. However, in this case the channel becomes more sensitive to the angle of incidence. An original solution of this problem was developed in the 1960s by Prof. Deich, who proposed an attachment of a profiled, dolphin-head-like nose to the profile's leading edge.[12,13] Subsequent tests confirmed its high aerodynamic quality and its advisability, in particular, to use for wet-steam turbines. Along with this, in order to reduce the intensity of the horseshoe vortices, it was proposed a fundamentally different solution—to provide a common, smooth leading edge with a V-shaped cut, as shown in the inset of Figure 4-15.[26] This cut creates a steam pillow on the leading edge that adapts itself to the inlet flow angle and, as experiments showed, considerably reduces the intensity of the inlet horseshoe vortices. The diagram of Figure 4-15 just demonstrate the resulted decrease of the energy losses for a blade row profile with V-cutting the leading edge compared with a conventional profile. It is understandable that this solution is primarily intended for short blades (in this specific case with the related height l/b of 0.8) especially sensitive to the end losses.

Another proposed solution of the problem with the secondary losses can be a special suction of the boundary layer from the endwall surfaces.[13] In particular, it can be done by means of perforating the shrouds forming these surfaces—Figure 4-16.

Reduction of Parasitic Steam Flows within the Stage

Along with the profile and secondary (end) energy losses, parasitic steam leakages through the gland seals are a serious source of losses in

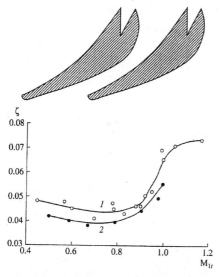
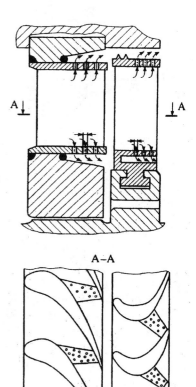

Figure 4-15. Energy losses in the blade rows with a relative height of 0.8 and smooth (1) and V-cut (2) profiles depending on the dimensionless outlet velocity. *Source: A.E. Zaryankin et al.*[26]

Figure 4-16. A turbine stage with perforated endwall surfaces. *Source: A.E. Zaryankin et al.*[26]

the turbine steam path and hence a reserve for raising the turbine efficiency. First of all, this concerns the steam leakage losses within the stage (see Figure 4-1). In order to diminish the tip leakage losses, as well as the windage resistance losses, rotating blades of all the HP, IP, and LP stages of modern steam turbines are made shrouded. The only exception is the LSBs of some turbine producers who prefer to make them free standing to reduce the centrifugal forces acting on the blade root and the rotor, whereas the tip steam leakages for the last stages are rather less important. Practically all the rotating blades (buckets) of modern turbines are manufactured integrally shrouded. With improvement of the blade airfoils and the use of twisted, leant, and curved blades, the tip gland seals improve their design as well—Figure 4-17. In order to heighten the aerodynamic resistance of the seals and diminish in this way the steam leakage amounts, the number of sealing ridges is increased; the ridges on the opposite gland surfaces are made with different pitches; the ridges are bent against the steam flow direction, and so on. The shrouds of IP and LP stages with a conical (on the meridional plane) peripheral outline are made of a special shape to supply them with tip seals—Figure 4-18.

For impulse-type stages, a general pattern of steam flows depends on the aerodynamic resistance ratios for the blade row, diaphragm and overshroud (tip) seals, pressure-balance holes in the wheel-discs, and axial clearance between the disc and diaphragm near the root seal. If the flow capacity of the pressure-balance holes is less than that of the diaphragm seal, there take place some suction of steam leaking through the

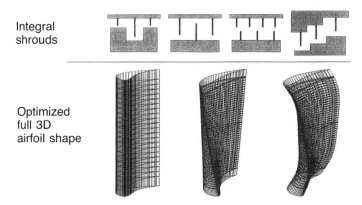

Figure 4-17. Modular concept of blade design of Siemens with improvement of both the blade airfoils and shroud seals. *Source: M. Deckers and E.W. Pfitzinger*[20]

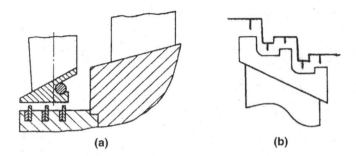

Figure 4-18. Two different types of shrouds and tip seals for LP stages with a conical meridional profile.

latter, additionally decreasing the stage efficiency. This occurs because the injected steam does not possess the necessary energy to work in the rotating blades, and in addition, it disturbs the major steam stream and retards it. By increasing the balance holes, it is possible to bring the steam flow amount through the root seal to zero. What is more, some steam leakage through the root seal can be even useful, withdrawing a portion of the disturbed steam layer with its secondary losses from the steam path. Experiments of Mitsubishi Heavy Industries showed that in this case, with optimizing the root leakage flow amount, the stage efficiency could be even increased by up to 0.4-0.5 %.[7] This effect was also corroborated by special calculations of ALSTOM Power intended for prediction of the internal turbine efficiency gain due to potential retrofits.[27]

To increase the accuracy of such predictions, analyze the general pattern of the steam flow through the stage rows, and estimate with more confidence their efficiency, it is desirable to use an overall mathematical model of the steam path with all features of its geometry. In particular, such a development was carried out by Mitsubishi Heavy Industries as applied to the experimental single-cylinder 105-MW turbine installed at the "T-point" power plant in Takasago Mechanical Works of MHI (see Figure 3-33a). The calculated results are supposed to be compared with the data of experimental measurements (see Figure 3-33b). One of the aids of this model is to optimize the shape of the flow path with minimizing the parasitic steam flows.[28]

The relative values of the profile, secondary, and leakage losses in the stages of the inlet and outlet zones of the HP and IP sections of modern Siemens turbines are shown in Figure 4-19. The blade profile

Design of Steam Path, Blading, Gland Seals, and Valves 115

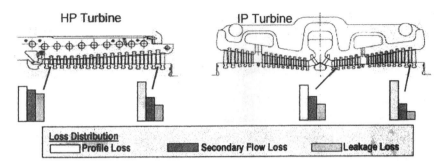

Figure 4-19. Relative values of the stage losses in the inlet and outlet zones of the HP and IP sections for modern steam turbines of Siemens. *Source: M. Deckers and E.W. Pfitzinger[20]*

losses are the largest single source of losses in the whole turbine. This clearly indicates that the overall losses should be primarily reduced by diminishing the profile losses throughout the turbine steam path. The secondary losses remain significant for those stages that are characterized by a low aspect ratio (the blade's height-to-chord ratio), that is, the front stages of the HP and double-flow IP sections. The leakage losses are most significant in both the HP and IP steam admission regions, and a leakage-free admission would be desirable. Such a comparison analysis[1,2,20,29] helps to reveal the most promising directions of developments to increase the turbine stage efficiency.

The circle diagram of Figure 4-20 demonstrates shares of different energy losses for a characteristic impulse-type HP stage. In the specific

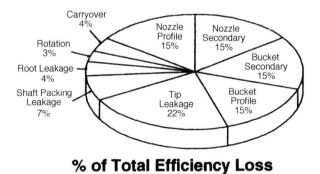

Figure 4-20. Shares of different energy losses for a characteristic impulse-type HP stage. *Source: J.J. Coffer, IV[18]*

considered case the total profile losses (for both the nozzle vane and rotating blade) make up 30%, as well as the total secondary losses, and the remaining 40% mainly relate to the leakage losses, with the largest share of 22% accounting for the tip leakage and the shaft packing leakage loss (loss with the leakage through the diaphragm seal) amounting to 7%.[18] It should be understandable that the cited figures of the loss shares are not absolute, and for each specific case the correlation of different energy losses occurs different, depending on the type of blading, blade profiles, and other specific design solutions.

Influence of a Blading Type on the Cylinder Efficiency

All the above considered effects relate to the efficiency and energy losses as applied to an individual turbine stage, impulse-type or reaction-type. Both of them have certain advantages and shortages depending on the specific operating and boundary conditions, and since the beginning of industrial production of steam turbines, for more than a hundred years, different turbine manufacturers have traditionally produced their turbines with either impulse-type blading (GE, GEC Alsthom, Hitachi, LMZ, Skoda, Toshiba, Turboatom, and some others) or reaction steam path (ABB, MHI, Parsons, Siemens, Westinghouse). True, this paradigm of employing either impulse or reaction blading and applying the same reaction degree to all the stages concerns only the HP and IP steam paths. For the LP stages with their great height-to-mean-diameter ratio values, the steam pressure in the gap between the fixed and rotating blades considerably rises longwise of the stage height, and the velocity diagrams substantially change while the circular rotation speed increases with the radius. Under these conditions, to avoid negative reaction in the root section and excessively high reaction near the stage tip, the adherents of both the impulse-type and reaction blading equally have to use blading with the variable reaction degree, increasing from 0.3-0.4 in the root section up to 0.6-0.7 to the blade tip. However, presently with the use of an individual fully three-dimensional blading design for all the HP, IP, and LP stages and each particular application, it becomes reasonable to have the stage reaction degree also set individually for each stage not only in the LP steam paths, but in the HP and IP sections as well.

In brief, the major advantages of a reaction-type steam path over an impulse type can be brought to the following items. First, steam flow channels in reaction-type stages, owing to their confuserness, usually have somewhat less profile and end energy losses. Second, all other

Design of Steam Path, Blading, Gland Seals, and Valves 117

things being equal, a reaction-type stage is characterized by a lesser enthalpy drop; this requires a greater number of stages to cover the taken total enthalpy drop for a cylinder/section and results in a greater heat recovery factor. On the other hand, steam leakages through the under-shroud seals of reaction stages, or the "tip" leakages through the seals between the fixed blades and the rotor surface, are greater than steam leakages through the diaphragm seals of impulse-type stages because of greater body diameters for rotors of reaction-type turbines, even though such rotors are more rigid and their seals are less prone to wearing out because of vibration. In addition, single-flow cylinders of reaction-type turbines need remarkable dummy-pistons to counterbalance the axial thrust caused by considerable pressure differences across the rotating blades which also results in greater steam leakages and an efficiency decrease. Special calculations were conducted to compare the efficiency of HP steam paths for some specific turbines, if they had been designed with different reaction degrees, based on the same design approaches.[30] According to their results, with the use of reaction-type blading, the cylinder's internal efficiency would be about 1.8% higher compared to that of the existing cylinders with impulse-type blading, even though this efficiency could decrease more rapidly in the course of operation, being more sensitive to the operation quality.

In the last decade, such a dichotomous attitude to the turbine blading choice has given way to a more flexible approach based on the use of merits of both blading types. This conception, shaped as "Impulse Blading? Reaction Blading?—Variable Reaction Blading!" was first declared and proposed by Siemens PG and materialized in the so-called 3DV blading technology.[31,32] The 3DV blades of the HP and IP steam paths feature the fully three-dimensional airfoil shapes of the 3DS family (see Figure 4-4b) with the ability to vary the stage reaction in the range between 10 and 60% to obtain the maximum efficiency—Figure 4-21. This allows the cylinder efficiency to be increased by up to 1% compared to conventional blading with constant 50% reaction. The 3DV blading conception was first applied to the 1,000-MW-class steam turbine of the German Niederaußem K Unit, the world's largest supercritical lignite-fired one, put in operation in 2002. Since then, Siemens declares the 3DV blading employed in all the newly developed steam turbines and the majority of those in service to be refurbished.[16]

If Siemens was always known as a consistent advocate of reaction-type blading, GE was traditionally considered to be the main representa-

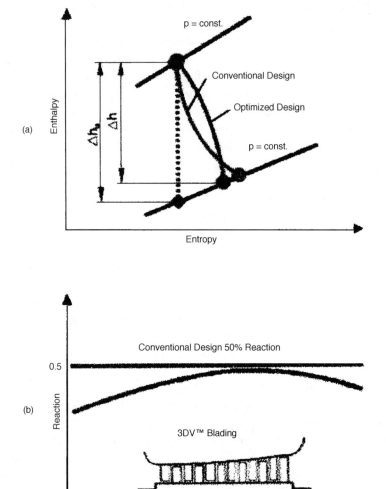

Figure 4-21. Changes in the steam expansion line (a) and reaction degree (b) of an optimized 3DV steam path design compared with conventional reaction-type blading . *Source: P. Hurd et al.*[16]

tive of impulse-type turbines. So it is of special interest that presently GE came to a similar idea "from the other end," developing the conception of the "Dense Pack" steam turbine design.[33] According to this conception, the turbine's HP and IP stages are designed with higher reaction, lower mean diameters, longer blades, and increased annular area. As a result, the stage number increases without increasing the overall cylinder length, what explains the taken name of Dense Pack. The mentioned changes lead to a reduction in flow-through velocities with a corresponding drop in the profile losses, while the improved aspect ratio reduces the secondary losses. The root reaction degree and number of stages are optimized to gain the maximum efficiency. To validate these design principles, with the use of special Steam Turbine Test Vehicle (STTV) facility, GE Energy tested some model steam paths designed on their basis. These tests indicated a noticeable gain in efficiency for the Dense Pack steam paths over conventional impulse-type turbines. Furthermore, to fully utilize the available energy, GE Energy developed the Advanced Dense Pack technology. In addition, to optimizing the rotor diameter, root reaction, and number of stages, the three-dimensional flow field is also optimized by applying compound tangential lean and modified vortex distribution for the nozzles and buckets. This technology is declared to be applied to the HP and IP sections for all the newly designed steam turbines, as well as the most of machines to be retrofitted.[25]

IMPROVEMENT OF TURBINE GLAND SEALS AND NON-BLADED AREAS

For the efficiency of individual turbine cylinders or sections, of significance are energy losses with the steam leakages through the end seals, as well as central seals between sections for the integrated cylinders. These (shaft) seals are mainly designed with the use of the same approaches as those applied to other types of gland seals—Figure 4-22. Along with this, even these very traditional components of steam turbine design are presently considerably modernized.

Speaking of labyrinth and once-trough gland seals, very attractive are so-called the seals with retractable packings, or 0, or "controlled" radial clearances. For this type of seals, their radial clearances increase when the turbine is stopped or operated with low steam flow amounts and decrease under normal operating conditions under load. If for conventional seals with fixed packings their flat springs press down the packing segments toward the shaft, with the retractable packings their springs push the packing segments outward, that is, toward the stator component (the diaphragm or casing ring). A possible scheme of such a seal is shown in Figure 4-23a, even though most often than not the

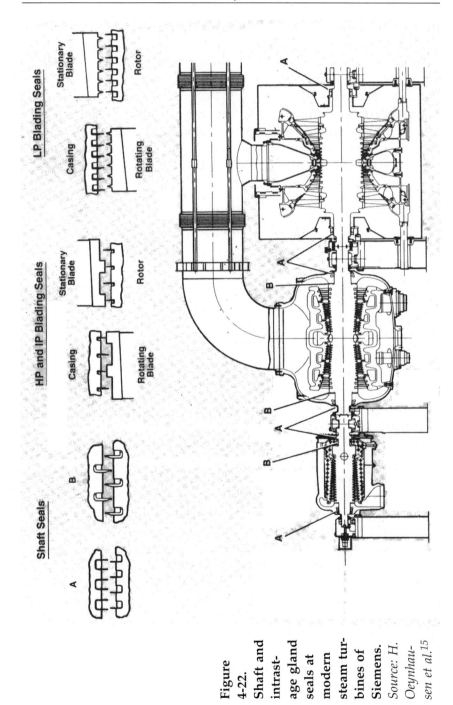

Figure 4-22. Shaft and intrastage gland seals at modern steam turbines of Siemens. Source: H. Oeynhausen et al.[15]

Design of Steam Path, Blading, Gland Seals, and Valves 121

springs pressing the packing segments outward are settled circumferentially, as in Figure 4-23b. The chamber over the packing segments is connected to the steam path space upstream of the seal, and the steam pressure within this chamber pushes the packing segments inward, toward the shaft. As a result, at the beginning stages of start-ups, when the probability of rubbing in the seal is maximum (because of vibration, thermal bowing of the shaft and/or casing, and other possible circumstances), the radial clearances in the seal are maximal, and only when the turbine begins operating under load the steam pressure difference

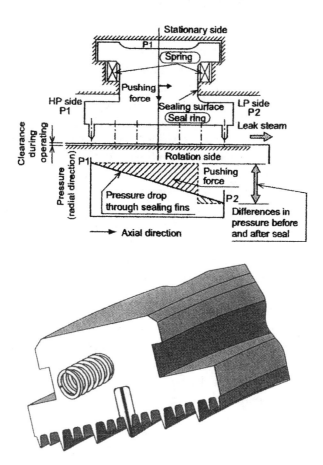

Figure 4-23. Schematic of a seal with retractable packings (a) and an actual retractable packing segment with a brush seal (b). *Sources: E. Watanabe et al.[28] and M.E. Foley[34]*

across the packing segments displaces them into the working, "closed" position, and the clearances decrease to the values recommended by the turbine producer—for example, about 0.6 mm (0.025 in), whereas when the turbine is stopped, the seal clearances are as much as about five-six times greater.

The first such seals were tested in operation at steam turbines in service in the 1980s. After nearly two decades of experience at hundreds turbines of ALSTOM, GE, Hitachi, Toshiba, Westinghouse, and others, the retractable packings have shown themselves to be a cost-effective and reliable measure for preventing the wear of seals and improving steam turbine efficiency.[35] So, for example, by 1999 TurboCare, Inc. had installed such seals within approximately 400 turbines with the single capacity of up to 1,100-1,300 MW with the longest operating experience over 11 years.[36] Replacement of conventional end tip and diaphragm seals with retractable ones at steam turbines in service in all the cases has brought a noticeable increase in the turbine efficiency. So, for example, for the 618-MW supercritical-pressure Gibson Unit 1 of Public Service Company of Indiana, the HP cylinder's post-overhaul efficiency was reported as equal to 84.4% after installing the retractable packing, that is, 2.9% over the pre-outage value and 0.6% better than the design efficiency.[37]

At many power plants, transition to the retractable packings completely solved a problem of rubbing in the steam path at start-ups without sacrificing the turbine efficiency. The U.S. Department of Energy (DOE) estimated the cumulative effect of retractable packings due to the steam turbine efficiency improvement from introduction through 1997 as equal to a fuel savings equivalent of some 55×10^{18} kJ (52.13 trillion Btu).[36]

Retractable seals in their simplest form, while being quite effective in the HP and IP sections, are not always fit for the LP stages because available pressure differences across the packing segments can be insufficient to retract them. For the LP stages it is proposed to use so-called "sensitized" seals, which, unlike retractable seals, does rely on steam pressure to displace. While the turbine is working under load, such seals are kept in the operating position as sensitized coil springs hold the packing segments in place.[35]

Retractable packing systems were additionally improved with the inclusion of a built-in brush bristle(s) sandwiched between two solid faceplates—Figures 1-23b and 1-24. The brush material is Haynes 25 cobalt-based super-alloy. Several thousand extremely fine diameter bristles, with the wire diameter in the range of 0.1 to 0.15 mm (4-6 mil), are packed together, forming a hedge against the leakage steam flow. The bristles are inclined radially in the direction of the shaft rotation, commonly by about 45°, to prevent them from picking up on the rotor. The back plate provides stiffness to the brush pack and prevents it from being deflected downstream by the steam pressure across the seal. During operation, the aerodynamic forces, due to the leakage flow and well known bristle "blow-down" effect, make the bristles move down and close up the bristle tip clearance further reducing the leakage flow. Because retractable brush seal packing operates at zero clearance when it is "closed," steam leakage is limited to the flow that can find its way through the tight maze created by the brush seal's bristle pack. In doing so, the steam leakage flow through the seal dramatically reduces and remains almost invariable in the operation process, even if the turbine works in a cycling manner—Figure 4-25.

Analysis of potential sources for efficiency degradation as applied to steam turbines with an integrated HP-IP cylinder showed that it can happen because of a wear of intermediate seals between the HP and IP sections, with a resultant increased leakage of steam after the control stage to the IP steam admission part. In particular, such a conclusion was made by TurboCare, Inc. for the Northport Power Station Unit 3 before its retrofit in the fall of 2004, its first since 1997.[38] This U.S. power plant comprises four G2 steam turbines of GE with the rated capacity of 375 MW each put in operation between 1967 and 1977. In order to decrease the mentioned leakage between the HP and IP sections, TurboCare proposed and carried out refurbishment of the existing intermediate seals based on the brush seal technology. On-line monitoring of the unit's

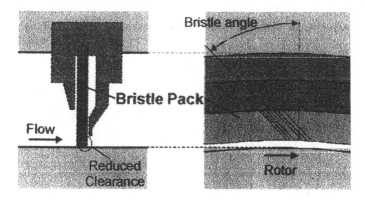

Figure 4-24. Schematic of a brush seal. *Source: P. Hurd et al.[16]*

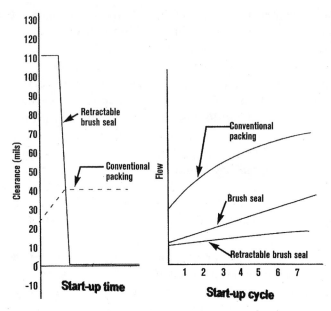

Figure 4-25. Changes of the radial clearance and steam leakage through the retractable brush seal compared to a conventional packing. *Source: M.E. Foley*[34]

heat-rate performances showed the resulted increase in the turbine output of 14.1 MW.

The use of brush seals is tied with a number of physical phenomena to be understood in order to apply these seals confidently and broadly in the turbine operation practice. In particular, these phenomena became the subject of special investigations of ALSTOM.[39] Presently, ALSTOM Power has had brush seals installed on the HP and IP tip seals of a number of retrofitted turbines and has accumulated several years of operating experience at these machines. Brush seal performance has also been monitored over many thousand operation hours in a test facility and on a trial application on the end gland seals of a boiler-feed pump's driving turbine. No significant degradation of leakage performance over time has been observed in the tests.

Another seal design concept, called "leaf seal," is under consideration by Mitsubishi Heavy Industries.[28] Such a seal comprises a number of thin metal plates ("leaves") inclined in the circumferential direction so that their tips are kept in a non-contact state with a negligibly small clearance when the rotor is rotating. This is provided by a lifting force produced due to a hydrodynamic pressure effect acting between the leaf and rotor—Figure 4-26. The tip of the leaf is lifted up by a balance of the pushing force due to pre-pressure of the setting, hydrodynamic lifting force, and additional lifting force due to the pressure difference across the seal. The result is that both the seal and rotor are prevented from wear and the durability of the seal is increased when the turbine is running. This distinguishes the leaf seal from the contact-type seals such as brush seals. In addition, since the seal itself is in the shape of a plate with axial width, it has a higher rigidity in the direction of the steam pressure difference and the sealing function can be kept up to a higher differential pressure values compared with brush seals. Figure 4-27 shows some results of special bench tests conducted by MHI to verify the seal performance, lifting and some other characteristics, including the electricity discontinuity, or potential difference, between the seal and rotor. The results confirmed that the flow rate through the seal is about one-third of that for a conventional labyrinth seals and the lifting-up force performs well even if the seal is eccentrically positioned against the rotor.

Leaf seals, as well as ACC seals (with retractable packings) were installed and investigated at the experimental single-cylinder 105-MW turbine at the "T-point" CC power plant in Takasago Machinery Works of MHI.[28]

For reaction-type turbines, the packing segments in the shaft glands, balance pistons, and low-height spring-backed under-shroud seals (between the fixed blades and rotor) can be coated with an abradable material, which decreases the effective clearance of the seal, but wears away if there appears rubbing against the opposed labyrinth teeth (ridges)—Figure 4-28. This design can effectively reduce the leakage flow in the seal by approximately 20% compared to uncoated seals without

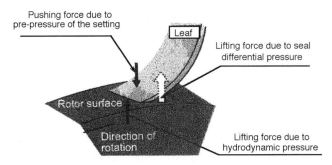

Figure 4-26. Leaf seal schematic. *Source: E. Watanabe et al.*[28]

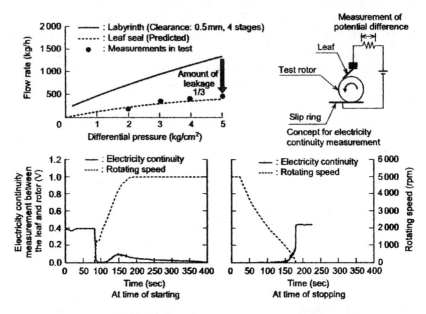

Figure 4-27. Some results of verification tests for leaf seal performance and lifting characteristic of the leaf. *Source: E. Watanabe et al.*[28]

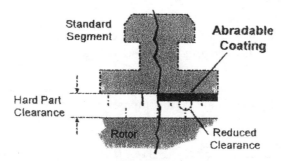

Figure 4-28. A gland seal segment with and without abradable coating. *Source: P. Hurd et al.*[16]

compromising operational safety. This technology was introduced by Siemens after several years of experience with abradable coatings for small-capacity turbines. An extensive test program, to optimize durability and wear behavior of the coating, has verified the effectiveness and contact properties. The coating wears locally and gives way to intruding labyrinth teeth, leaving the adjacent seal surfaces and teeth fully intact.

The considered advanced seal technologies, such as retractable, brush, and abradable seals, have already used in power plant practice, even though nowadays mainly at refurbished turbines, but with significant performance benefits. By example, Figure 4-29 demonstrates application of these technologies to the shaft seals at the integrated HP-IP cylinder of a BB44FA project of Siemens Westinghouse Power Corporation intended for retrofitting steam turbines in service of Westinghouse with a nominal rate between 350 to over 680 MW.[16] In particular such a retrofit was carried out in 2004 at the Madgett power plant (Florida) as applied to its subcritical-pressure 365-MW turbine # 1.[40] Presently, the considered sealing technologies are also offered for gland seal systems of large steam turbines, including those of advanced steam conditions.[17]

Besides reduction of the steam leakages through the seals, to increase the cylinder efficiency, of importance is to avoid or at least minimize the energy drops in the non-bladed steam path areas by reducing their aerodynamic resistance and diminishing pressure drops. This primarily concerns steam admission and outlet zones, bleeding chambers and channels connecting them to the steam mainstream, and all the areas where the steam flow turns. Upgrading and optimizing finally their sizes and shapes is commonly fulfilled with CFD methods and computer programs. The LP exhaust hoods also require special thorough analysis, but this subject is to be considered separately below.

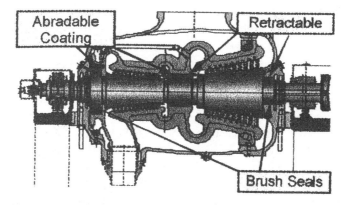

Figure 4-29. Application of advanced sealing technologies at the HP-IP cylinder of a BB44FA project of Siemens for retrofitted steam turbines. *Source: P. Hurd et al.[16]*

As applied to the 907-MW Boxberg Q steam turbine of Siemens (see Figure 2-10), optimizing the HP steam admission section led to the development of a helical configuration that yielded a more uniform flow distribution over the entire stationary blade ring area. New diffuser geometry was developed for the HP and IP exhaust hoods specifically to counteract backflow and vortex formation caused by flow separation, making it possible to reduce pressure drops and energy losses due to internal circulation flows. Special attention was paid to excluding any excessive aerodynamic resistances and vortex formation sources at all the steam flow turns, between the main steam flow and bleeding chambers, and so on.[3]

Attentive consideration and special analysis of many projects of steam turbines in service allow revealing numerous places with excessive steam leakages or energy drops which could be easily refurbished with a resulted gain in efficiency. It can be said that the increases in the turbine efficiency gained due to such specific individual measures are counted in fractions of a percent. However, implementation of all these design improvements taken together can raise the final turbine efficiency quite remarkably, and modern steam turbines demand paying attention to all these "small things" along with the "global" improvement of the steam path design.

REDUCTION OF LOSSES IN CONTROL VALVES

Existing stop and control valves of steam turbines for fossil-fuel power plants with substantially high main steam conditions are frequently characterized with very great steam velocities in the narrowest section of the diffuser-type seat (that is, practically at the turbine entrance) of approximately 100-140 m/s (330-460 ft/s). As a result, they have a pressure drop of about 4-7%, and hence the available enthalpy drop for the turbine decreases by about 7-10 kJ/kg.[26] Some special investigations of this problem[26,41-43] brought to the concept of profiled valves with a cup featuring a perforated surface as an alternative to plate-type valves. The proposed solutions can be illustrated by the example of the unit of the main stop and control valves for 360-MW steam turbine 18K-360 of Zamech (Poland) with subcritical steam pressure of 18 MPa (2,610 psi)—Figure 4-30. The unit comprises the stop valve 1 and control valve 2, arranged on both sides of the common saddle 3. The distinguishing

Design of Steam Path, Blading, Gland Seals, and Valves

features of the units are the following:

1) profiled valve heads, whose surfaces, together with the surfaces of the inlet and outlet parts of the saddle, forms an axially symmetrical annular duct;
2) perforation on the streamline valve head surfaces for damping possible flow pulsations;
3) a special system for additional loading of the valve under conditions of high lifting, when the tapered bushing (4 in Figure 4-30) obstructs the access of steam to the unloading valve, and
4) confuser saddle instead of diffuser saddle.

The last solution, being entirely unconventional for control valves, ensures their stable operation under any operating conditions. In order to reduce the aerodynamic resistance of the fully opened valves, the maximum steam velocity in the saddle throat was reduced from the initial value of 150 m/s (490 ft/s) for the old-type valves to 70 m/s (230 ft/s) by increasing the fit diameters of the valves. As a result, the total resistance of the considered unit did not exceed 1.5% of the inlet steam pressure. Such a solution became possible due to application of an unconventional system for counterbalancing the axial thrust on the control

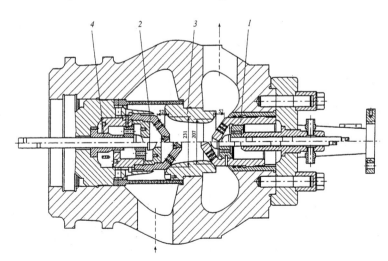

Figure 4-30. A unit of main stop and control valves for 360 MW turbine of Zamech (Poland) for subcritical steam pressure. *Source: A.E. Zaryankin et al.*[42]

valve stem when its internal cavity is connected to the confuser saddle duct, not through the central unloading orifice but through the holes perforated on the streamlined surface of the valve head. Thus, at small openings of the valve, its deep unloading from the acting steam pressure forces is ensured, and there arises a possibility to increase the fit valve diameters not increasing the servodrive power.

The single-seat angle valves of ABB (see Figure 3-5b) were also optimized in terms of power drops, flow losses, and sensitivity to vibration. The design of their spindle seals diminishes the steam leakage amount along the stem and reduces the required leakage piping to a minimum. The diffusers after the main and intercept control valves (see Figures 3-5b and 3-5c) also act as the inlet ducts to the inner casings. The turbine design in these HP and IP areas significantly reduces the cost and duration of overhauls by allowing easy dismantling and reassembling. The gland seal pipes of the valves are fitted directly to the valve body and do not need to be removed for the overhauls.[44,45]

The same to a great degree can be said about the combined stop-and-control-valves unit of Siemens—see Figure 3-8c. Their number and size are set according to the mass flow rate to diminish the maximum steam velocity and hence the pressure drop and energy losses. Sealing of the valve stems should minimize the steam leakages along the valve stems.

Steam leakages through the seals of the HP valves' stems are commonly quite remarkable, especially for supercritical-pressure steam turbines, and result in noticeable energy losses. In addition, the outside steam given to the seals and flowing along the valve stem causes an axial temperature unevenness in the valve steam-chests with a resulted increase of unsteady thermal stresses in them. The presence of these seals also makes the valve steam-chests cool faster when the turbine is stopped. The temperature differences, arising between the turbine's HP cylinder and valve steam-chests in the cooling-down process, produce certain difficulties at subsequent start-ups.

All these problems vanish if the valve stem's seal is replaced with a hermetic assembly without steam flows along the stem. This task was successfully solved by specialists of All-Russia Thermal Engineering Research Institute (VTI) together with LMZ with the use of a Liquid-Metallic Seal (LMS)—Figure 4-31.[46,47] Its behavior was experimentally tested even under super-elevated steam conditions.[48] Due to a low friction coefficient (less than 0.05), the LMS does not hamper the valve motion freedom. After thorough bench tests and long-term field tests at the HP control valves on an actual supercritical-pressure 300-MW steam turbine in service, since 1987 such LMSs were installed, in particular, at all the HP control valves of six 300-MW turbines of LMZ at the Konakovo power plant and four similar turbines of the Lukoml power plant. The LMSs have been successfully operated without additional inspections and maintenance between the overhauls (every six years). No forced outages have taken place because of these seals. According to the power plant data, the resulted increase in the turbine efficiency makes up about 0.2%. It is obvious that with the increase of the main steam conditions the effect of applying the LMSs rises. Altogether, the leakless LMSs were installed at 22 turbines of six types in the former Soviet Union and abroad, including 18 turbines of supercritical steam pressure. Presently, the LMSs are supposed to be used at new supercritical-pressure 300-MW turbines designed by LMZ for replacing the aging ones in service, as well as at newly designed steam turbines with the elevated steam conditions.[47,49-51]

References
1. Oeynhausen H., A. Drosdziok, and M. Deckers. "Steam Turbines for the New Generation of Power Plants," *VGB Kraftwerkstechnik* 76, no. 12 (1996): 890-895.
2. Drosdziok A. "Steam Turbines" [in German], *BWK* 50, no. 4 (1998), 120-124.

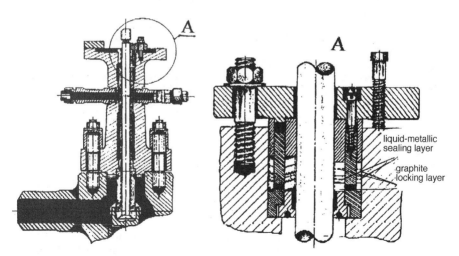

Figure 4-31. Hermetic sealing assembly for the HP control valve's stem. (The existent labyrinth seal of the valve stem are retained, but the pipelines of sealing steam are plugged and cut off).

3. Hoffstadt U. "Boxberg achieves world record for efficiency," *Modern Power Systems* 21, no. 10 (2001): 21-23.
4. Hogg S. and D. Stephen. "ALSTOM-CPS San Antonio Retrofit of JK Spruce Init 1 HP-IP Turbine—An Example of an Advanced Steam Turbine Upgrade for Improved Performance by a non-OEM Supplier," presented at the Electric Power 2005 Conference, PWR2005-50227.
5. West L.A., J.G. Neft, and M.H. Wallon. "Better Plant Efficiency Thanks to Aerodynamic Research on Turbines," *The 10th Interantional Conference on Modern Power Stations, Liege, Sept. 1989*, Liege, 1989: 47/1-47/8.
6. Varley J. "Modernising Mehrum: 40 green MW from a coal plant," *Modern Power Systems* 24, no. 7 (2004): #11-13.
7. Hirota Y., T. Mizutori, E. Watanabe, et al. "Recent Technology on Turbine Performance Improvement," *Mitsubishi Heavy Industries Technical Review* 22, no. 3 (1985): 234-240.
8. Borisov F.P., Ivanov M.Ya., Karelin A.M., et al. "Steam Turbine High-Efficiency Stage Design Using Ideal and Viscous Gas-Flow Calculations," *Thermal Engineering* 40, no. 5 (1993): 375-381.
9. Shcheglyaev A.V. *Steam Turbines* [in Russian], 2 vols., 6th ed., Revised and expanded by B.M. Troyanovskii, Moscow: Energoatomizdat, 1993.
10. Nowi A. and B.R. Haller. "Developments in Steam Turbine Efficiency," *VGB PowerTech* 77, no. 7 (1997): 551-555.
11. Nowi A. and P.J. Walker. "First Test Results for Steam Turbines with State-of-the-Art Blading," *VGB PowerTech* 79, no. 12 (1999): 48-52.
12. Deich M.E. and B.M. Troyanovskii. *Investigation and Design of Axial Turbine Stages* [in Russian], Moscow: Mashinostroenie, 1964.
13. Deich M.E. *Gas Dynamics of Turbine Blade Rows* [in Russian]. Edited by G.A. Filippov. Moscow: Energoatomizdat, 1996.
14. Deich M.E., B.M. Troyanovskii, and G.A. Filippov. "An Effective Way of Improving the Efficiency of Turbine Stages," *Thermal Engineering* 37, no. 10 (1990): 520-523.
15. Oeynhausen H., A. Drosdziok, W. Ulm, and H. Termuehlen. "Advanced 1000 MW Tandem-Compound Reheat Steam Turbine," presented at the American Power Conference, Chicago, 1996.
16. Hurd P., N. Thamm, M. Neef, et al. "Modern Reaction HP/IP Turbine Technology Advances & Experiences," presented at the Electric Power 2005 Conference, Chicago, 2005.
17. Wichtmann A., M. Deckers, and W. Ulm. "Ultra-Supercritical Steam Turbine Turbosets Best Efficiency Solution for Conventional Steam Power Plants," *VGB PowerTech* 85, no. 11 (2005): 44-49.
18. Coffer J.I., IV. "Advances in Steam Path Technology," *Transaction of the ASME. Journal of Engineering for Gas Turbines and Power* 118, April 1996: 337-352.
19. Hirota Y., T. Mizutori, E. Watanabe, et al. "Recent Technology on Turbine Performance Improvement," *Mitsubishi Heavy Industries Technical Review* 22, no. 3 (1985): 234-240.
20. Deckers M. and E.W. Pfitzinger. "The Exploitation of Advanced Blading Technologies for the Design of Highly Efficient Steam Turbines," presented at the 6th International Charles Parsons Turbine Conference, Dublin, 2003.
21. Leyzerovich A. *Wet-Steam Turbines for Nuclear Power Plants*, Tulsa (OK): PennWell Corporation, 2005.
22. Weiss A.P. "Aerodynamic design of advanced LP steam turbines," *ABB Review*, no. 5 (1998): 4-11.
23. Simoyu L.L., N.N. Gudkov, M.S. Indurskii, et al. "The Influence of the Saber Shape

Design of Steam Path, Blading, Gland Seals, and Valves 133

of the Nozzle Vanes on the Performance of the Last Stage in a Steam Turbine," *Thermal Engineering* 45, no. 8 (1998): 659-664.

24. Simoyu L.L., A.S. Lisyanskii, V.P. Lagun, et al. "Main Principles of Shaping the LP Steam Path of Steam Turbines" [in Russian], *Elektricheskie Stantsii*, no. 10 (2005): 51-55.

25. Boss M.J., M. Gradoja, and D. Hofer. "Steam Turbine Technology Advancements for High Efficiency, High Reliability and Low Cost of Electricity," presented at the POWER-GEN 2005 International Conference, Las Vegas, 2005.

26. Zaryankin A.E., V.A. Zaryankin, and B.P. Simonov. "Several Ways of Improving the Efficiency of the Flow Paths for Steam Turbines," *Thermal Engineering* 50, no. 6 (2003): 442-448.

27. Hesketh A., S. Hogg, and D. Stephen. "A Stage Efficiency Prediction Method and Related Performance Aspects of Retrofits on Disc/Diaphragm Steam Turbines," presented at the 2003 International Joint Power Generation Conference. 2003, Atlanta, PWR2003-40145.

28. Watanabe E., Y. Tanaka, T. Nakano, et al. "Development of New High Efficiency Steam Turbine," *Mitsubishi Heavy Industries Technical Review* 40, no. 4 (2003): 1-6.

29. Engelke W. "Specific Design Solutions for Steam Turbines of Siemens," *VGB Kraftwerkstechnik* 74, no. 4 (1994): 346-349.

30. Kostyuk A.G. and A.D. Trukhnii. "A Comparison of the Impulse and Reaction HP Cylinders of Steam Turbines," *Thermal Engineering* 52, no. 6 (2005): 439-450.

31. Simon V., H. Oeynhausen, R. Bürkner, and K.-J. Eich. "Impulse Blading? Reaction Blading? Variable Reaction Blading!" *VGB Kraftwerkstechnik* 77, no. 9 (1997): 648-652.

32. Simon V. and H. Oeynhausen. "3DV Three-Dimensional Blades—A New Generation of Steam Turbine Blading," *Proc. of 1998 Joint Power Generation Conference*, PWR-Vol. 33, New York: ASME, 1998, Vol. 2, 71-78.

33. Maughan J.R., L.D. Willey, J.M. Hill, and S. Goel. "Development of the Dense Pack Steam Turbine: A New Design Methodology for Increased Efficiency," presented at the International Joint Power Generation Conference & Exposition, Miami Beach, 2000.

34. Foley M.E. "Retractable brush seals allow sweeping improvements in steam turbine efficiency," *Modern Power Systems* 20, no. 7 (2000): 37-39.

35. Brandon D.R. "Retractable and Sensitized Packing Reduces Friction and Increases Turbine Efficiency," *Power Engineering* 108, no. 8 (2004): 58-62.

36. Sulda E. "Retractable Brush Seal Optimizes Efficiency and Availability for Cycling and Baseloaded Steam Turbines," *Power Engineering* 103, no. 11 (1999): 96-102.

37. Cerasoli D.J. "Retractable packings pass inspection," *Power Engineering* 96, no. 6 (1992): 30-32.

38. Blachley S.R. and Foley M.E. "Testing for Turbine Degradation and Improving Performance with Seal Optimization," presented at 2005 Power-Gen International Conference, Las Vegas, 2005.

39. Stephen D. and S.I. Hogg. "Development of Brush Seal Technology for Steam Turbine Retrofit Applications," presented at the 2003 International Joint Power Generation Conference, Atlanta, 2003, PWR2003-40103.

40. Cheski J.R., R. Patel, K. Rockaway, et al. "A Large Steam Turbine Retrofit Design and Operation History," presented at the POWER-GEN 2005 International Conference, Las Vegas, 2005.

41. Zaryankin A.E. and B.P. Simonov. "New Control Valves of Steam Turbines. Their Characteristics and Operating Experience," *Thermal Engineering* 43, no. 1 (1996): 18-23.

42. Zaryankin A.E. and V.I. Chernoshtan. "Aerodynamic Principles for Designing Steam Turbine Control Valves," *Thermal Engineering* 44, no. 1 (1997): 53 -57.
43. Zaryankin A.E., V.G. Gribin, and A.N. Paramonov. "Applying Nonconventional Solutions to Enhance the Efficiency and Reliability of Steam Turbines," *Thermal Engineering* 52, no. 4 (2005): 267-274.
44. Lageder H. and P. Meylan "ABB modular reheat steam turbines," *ABB Review*, no. 5 (1990), 3-10.
45. Busse L. and K.-H. Soyk. "World's highest capacity steam turbosets for the lignite-fired Lippendorf power station," *ABB Review*, no. 6 (1997): 13-22.
46. Levin A.Y., L.B. Izrailevskii, and A.S. Sozaev. "The Use of Liquid-Metallic Seals in Control Valves of Large Steam Turbines" [in Russian], *Elektricheskie Stantsii*, no. 1 (1985): 27-29.
47. Leyzerovich A. "Advanced Russian Technologies for Raising the Operating Performances of Large Power Steam Turbines," *Proc. of the 2000 International Joint Power Generation Conference*. Miami Beach, New York: ASME Press, 2000. IJPG2000-15016: 115-120.
48. Levin A.Y., L.B. Izrailevskii, A.S. Sozaev, and I.A. Dezhin. "Investigations of Closing Layer's Properties for Liquid-Metallic Seals under High Parameters" [in Russian], *Thermal Engineering* 33, no. 12 (1986): 55-58.
49. Kondrat'ev V.N., A.S. Lisyanskii, and Y.N. Nezhentsev. "A Project of Retrofitting the 300-MW Supercritical-Pressure Steam Turbines" [in Russian], *Elektricheskie Stantsii*, no. 7 (1999): 78-81.
50. Avrutskii G.D., V.V. Lysko, A.V. Shvarz, and B.I. Shmukler. "About creating coal-fired power units with USC steam conditions" [in Russian], *Elektricheskie Stantsii*, no. 5 (1999): 22-31.
51. Sozaev A.S., O.F. Fomchenko, Y.A. Lygin, and D.V. Remizov "About Installing Liquid-Metallic Seals in Steam Turlilne Valve" [in Russian], *Elektricheskie Stantsii*, no 1 (2007).

Chapter 5

Last Stages and Exhaust Hoods of LP Cylinders

LAST STAGE BLADES

Utmost Length

Last stage blades (LSBs), being the key element of steam turbines, in many instances dictate the turbine configuration, including the number of LP cylinders. They also to a great degree determinate the turbine's operating performances. At the taken rotation speed, n, the LSB length is mainly limited by the blade material strength under the action of centrifugal force. If the blade's bucket is not shrouded and not tied with the adjacent ones (that is, it is "free standing"), and its profile is invariable lengthwise of the blade height, the tensile radial stress in the bucket's root section is accounted as equal to

$$\sigma_r = 0.5 \; \rho\omega^2 l_2 d_2, \tag{5.1}$$

where ρ is the material density, $\omega = 2\pi n$ is the angular velocity, and l_2 and d_2 are the blade's length and mean diameter, respectively. In reality, the LSB profile essentially changes along the height (Figure 5-1), decreasing its area from the root value f_r to the tip one f_t. This can be approximated by entering the factor for unloading the root stress

$$k_{disl} \approx 0.35 + 0.65 \times (f_r/f_t)^{-1}. \tag{5.2}$$

With regard to this factor and taking the LSB annular exit area as $F = \pi d_2 l_2$, the tensile radial stress in the bucket's root section is:

$$\sigma_r = k_{disl} \times 2\pi F \times \rho n^2. \tag{5.3}$$

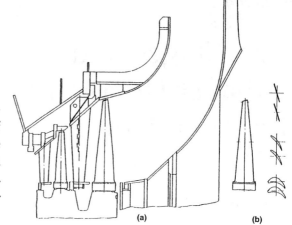

Figure 5-1. Characteristic last LP stages of a large steam turbine (a) and the LSB with its profiles at the tip, mean, and root diameters (b).

The area ratio f_r/f_t commonly lies in the range of 7-10, which corresponds to $k_{disl} \approx 0.4$. For the LSB buckets manufactured of stainless steel, $\rho \approx 8.0 \times 10^3$ kg/m³, and then:

$$F \approx \frac{\sigma_r}{2\pi k_{disl} n^2 \rho} \approx 0.48 \times 10^{-4} \frac{\sigma_r}{n^2}, \qquad (5.4)$$

where σ_r is measured in Pa, n in s⁻¹, and F in m².

Development of new blade steel materials with advanced strength properties has promoted significant progress in the development of new, longer blades. It follows that for a high-alloy steel, with the admissible stress value equal to about 620 MPa, the maximum accessible exit annular area of an LSB row can approach 11.9 m² (128 sq. ft.) for n = 50 Hz and 8.3 m² (89 sq. ft.) for n = 60 Hz. If $d/l \approx 2.7$, these annular area values can be achieved with the LSB length of 1,220 mm (48 in) for n = 3,000 rpm and 1,016 mm (40 in) for n = 3,600 rpm. These figures just coincide with specifications of the actual steel LSBs developed by General Electric in cooperation with Toshiba—Figure 5-2.[1,2]

The subsequent tables present main characteristics of some LSBs developed by various manufacturers for full-speed steam turbines with the rotation speed of 3,000 rpm (Table 5-1) and 3,600 rpm (Table 5-2).

Transition to the half-speed conception theoretically allows designers to quadruple the annular exit area by doubling the LSB length. But

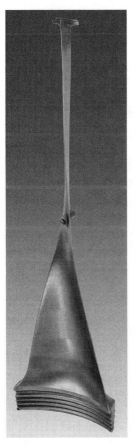

Figure 5-2. A steel LSB of GE/Toshiba with the length of 1,016/1,220 mm (40/48 in) for the rotation speed of 3,600/3,000 rpm, respectively. *Source: A. Mujezinovic[1]*

in practice, the maximal LSB length of half-speed turbines does not exceed one-and-a-half length of LSBs for full-speed turbines. The explanation of such prudence lies in the fact that the increased length unavoidably lowers the aerodynamic quality of the blades and makes their design more complicate because of the large length-to-mean-diameter ratio and an increased pitch of the meridional stage profile. Nowadays, the maximum length-to-mean-diameter ratio for the longest full-speed LSBs reaches 0.41-0.415, whereas for half-speed LSBs it remains at the level below 0.36. With an optimal circumferential-speed-to-steam-velocity ratio, the increased mean diameter means an increased enthalpy drop and, as a result, a greater difference in the specific steam volume values between the row entrance and exit. High, supersonic values of the steam velocities and their great variations along the row height hinder the achievement of optimal aerodynamic performances. In addition, the erosion impact of wet steam becomes more dangerous, the longer the LP stage blades and the greater their tip circumferential speed. In such a way, the closer the LSB is to its limiting length, the smaller the gain in efficiency and the higher the cost of these achievements. There still exists a substantial margin for increasing the size of half-speed steel LSBs, whereas for full-speed steel LSBs this margin has practically run out.

Presently, new, longer LSBs for the half rotation speed are mainly developed for large wet-steam turbines of nuclear power plants (mainly, as applied to the single capacity of 1,300 MW and more for 50 Hz and 1,000 MW and more for 60 Hz).[3] Along with this, these LSBs can also be employed for newly designed and retrofitted CC steam turbines of fossil-fuel power plants with the grid frequency of 60 Hz. As of now, such new turbines with the single capacity of 1,000-1,050 MW are produced and installed only in

Table 5-1. Main characteristics of some LSBs of various manufacturers for steam turbines with the rotation speed of 3,000 rpm.

Length, mm (inch)	Material	Developer	Annular Exit Area per Flow, m^2 (ft^2)	Length-to-Mean-Diameter Ratio	Tip Circumferential Speed, m/s (ft/s)	Notes
1,500 (59)	titanium	LMZ	17.9 (193)	0.40	832 (2,727)	Under development
1,423 (56)	titanium	Siemens	16.0 (172)	0.40	785 (2,574)	Under development
1,372 (54)	titanium	MHI	14.6 (157)	0.40	747 (2,449)	Under development
1,360 (53.5)	titanium	ALSTOM	14.7 (157)	0.40	754 (2,472)	Under development
1,219 (48)	steel, titanium	GE/Toshiba	11.9 (128)	0.39	680 (2,230)	Available since 2003
1,219 (48)	steel	MHI	11.3 (122)	0.41	655 (2,150)	-
1,200 (47)	steel, titanium	ABB	12.2 (132)	0.37	697 (2,285)	-
1,200 (47)	titanium	LMZ	11.3 (122)	0.40	660 (2,160)	In operation since 1979
1,146 (45)	steel	Siemens	12.5 (135)	0.33	720 (2,360)	Available since 2002 Free-standing
1,130 (44.5)	steel	ALSTOM	10.5 (113)	0.38	642 (2,105)	-
1,100 (43)	steel	Turboatom	10.4 (116)	0.36	645 (2,110)	Under development
1,092 (43)	steel	Hitachi	10.1 (109)	0.37	634 (2,079)	-
1,093 (43)	steel	Siemens	10.0 (108)	0.38	630 (2.060)	Free-standing
1,085 (43)	steel	Skoda	9.5 (102)	0.39	608 (1,993)	-
1,067 (42)	steel	Toshiba	9.5 (103)	0.38	613 (2,010)	-
1,050 (41.5)	steel	ABB	9.7 (104)	0.36	627 (2,056)	Free-standing
1,050 (41.5)	steel	Turboatom	8.4 (95)	0.39	576 (1,889)	-
1,030 (40.5)	steel	MHI	9.4 (102)	0.36	618 (2,026)	-
1,021 (40)	titanium	GE	8.8 (95)	0.37	591 (1,937)	-
1,016 (40)	titanium	Toshiba, Hitachi	8.76 (94)	0.37	590 (1,935)	-
1,000 (39.5)	steel	ABB	8.5 (91)	0.37	582 (1,908)	Free-standing
978 (38.5)	steel	Siemens	10.0 (108)	0.30	665 (2,180)	Free-standing
960 (38)	steel, titanium	LMZ	7.5 (81)	0.39	541 (1,773)	-

Last Stages and Exhaust Hoods of LP Cylinders

Table 5-2. Main characteristics of some LSBs of various manufacturers for steam turbines with the rotation speed of 3,600 rpm.

Length, mm (inch)	Material	Developer	Annular Exit Area per Flow, m^2 (ft^2)	Length-to-Mean-Diameter Ratio	Tip Circum-ferential Speed, m/s (ft/s)	Notes
1,194 (47)	titanium	Siemens	11.1 (120)	0.4	783 (2,567)	Under development
1,170 (46)	titanium	Hitachi	N/A	N/A	N/A	Under development
1,143 (45)	titanium	MHI	10.1 (109)	0.41	745 (2,445)	-
1,067 (42)	titanium	Siemens	10,3 (111)	0.35	780 (2,557)	-
1,016 (40)	titanium	GE/Toshiba	8.5 (92)	0.38	693 (2,272)	-
1,016 (40)	titanium	Hitachi	8.4 (90)	0.39	687 (2,252)	-
1,016 (40)	steel	GE/Toshiba	8.3 (89)	0.39	682 (2,236)	Available since 2003
1,016 (40)	steel, titanium	MHI	7.8 (84)	0.416	652 (2,137)	-
955 (38)	steel	Siemens	8.7 (94)	0.33	727 (2,384)	Free-standing
858 (34)	steel	Turboatom	5.68 (61)	0.41	559 (1,832)	-
852 (33.5)	steel	Hitachi, Toshiba	6.2 (67)	0.37	594 (1,948)	-
852 (33.5)	steel	GE	6.15 (66)	0.37	594 (1,948)	-
815 (32)	steel	Siemens	6.9 (74)	0.30	662 (2,170)	Free-standing

Japan (see, for example, Figure 2-12), but in the nearest future they will most likely give way to TC steam turbines of the same and even larger output (see, for example, Figures 3-1 and 3-3). Nevertheless, in the U.S. and Japan there exists a considerable fleet of aging CC turbines, and new LSBs can be needed for retrofitting them. Main characteristics of the longest LSBs of various developers for the rotation speed of 1,800 rpm are given in Table 5-3.

For full-speed LSBs, there are no in sight other real ways to increase radically their length and annular exit area than by transition to manufacturing the blades from titanium alloys (for example, Ti-5Al, Ti-6Al-4V, or Ti-6Al-6V-2Sn). As a rule, titanium-alloy blades are merely coined "titanium." The density of titanium alloys is approximately 1.8 times less that that of steel, with the same, or even greater, strength. Because of this, the length of titanium buckets can be extended appreciably. On the other hand, titanium alloys are considerably more expensive compared to steel and are much harder in machining. Nevertheless, even the most consistent former opponents of titanium LSBs have presently turned to developing and implementing them, and every major turbine producer

140 — *Steam Turbines for Modern Fossil-Fuel Power Plants*

Table 5-3. Main characteristics of the longest LSBs of various manufacturers for half-speed shafts of cross-compound steam turbines for 60 Hz

Length, mm (inch)	Material	Developer	Annular Exit Area per Flow, m^2 (ft^2)	Length-to-Mean-Diameter Ratio	Tip Circumferential Speed, m/s (ft/s)	Notes
1,525 (60)	steel	Siemens	20.3 (219)	0.36	543 (1,780)	Free-standing, under development
1,375 (54)	steel	MHI	17.8 (192)	0.33	518 (1,698)	-
1,370 (54)	steel	Siemens	17.3 (187)	0.34	512 (1,679)	Free-standing
1,320 (52)	steel	Hitachi, Toshiba,	16.7 (180)	0.33	504 (1,652)	-
1,320 (52)	steel	ABB	16.4 (176)	0.33	497 (1,630)	Free-standing
1,320 (52)	steel	GE	15.8 (170)	0.35	483 (1,584)	-
1,270 (50)	steel	Siemens	15.35 (165)	0.33	482 (1,580)	Free-standing
1,220 (48)	steel	Hitachi	NA	NA	NA	Under development
1,194 (47)	steel	ALSTOM, Westinghouse	13.4 (144)	0.33	449 (1,472)	-
1,170 (46)	steel	Siemens	13.4 (144)	0.32	453 (1,485)	Free-standing
1,170 (46)	Steel	MHI	12.5 (134)	0.34	431 (1,413)	-
1,170 (46)	steel	ALSTOM	12.2 (131)	0.35	424 (1,390)	-
1,143 (45)	steel	GE	12.3 (132)	0.33	431 (1,413)	-
1,118 (44)	steel	Westinghouse	11.8 (127)	0.33	422 (1,384)	-
1,092 (43)	steel	GE, Hitachi	11.5 (124)	0.33	419 (1,374)	-
1,041 (41)	steel	MHI	10.0 (108)	0.34	386 (1,266)	-
1,016 (40)	steel	Westinghouse	9.6 (103)	0.34	380 (1,246)	-

in the world employs or, at least, has at its disposal titanium LSBs commercially available. Effectiveness of titanium LSB has been well proved in the operational practice since the very early 1980s.[4-12] An appearance of a characteristic 1,093-mm (43-in) titanium LSB for the rotation speed of 3,000 rpm developed by Hitachi is shown in Figure 5-3.[13]

An important additional advantage of titanium alloys is their lesser susceptibility to erosion and corrosion compared to stainless steels. Because of this, some turbine designers propose to use titanium blades also for intermediate LP stages operating in the Wilson region (where steam goes over into the two-phase state, that is, becomes wet, and the danger of SCC is maximum), even though in practice, as applied to modern technologies, this looks too costly.

With the increase in the LSB length, not only does the tensile stress caused by centrifugal forces grow, but the danger of WDE increases, too, because of the increase in the tip circumferential speed. In addition, under operating conditions with low steam flows and/or high back-pres-

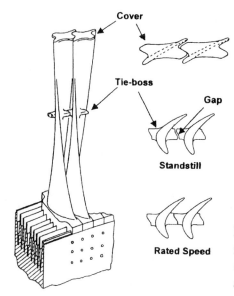

Figure 5-3. Hitachi's 43-inch titanium LSB for the rotation speed of 3,000 rpm. *Source: M. Machida et al.*[13]

sure, longer LSBs are also more intensely heated because of friction and fanning in the ambient steam. This also lowers the blades' strength and requires special attention to be paid.

Roots, Shrouds, and Snubbers

For many years, the most widespread type of attachment bases for LSBs were prong-and-finger (or fork-shaped) roots with a various number of prongs. For example, titanium LSBs of Hitachi with the length of 1,016 mm (40 in) and 1,092 mm (43 in), shown in Figure 5-3, were made with seven- and nine-prong roots, respectively. However, in recent years most LSBs have been designed with curved-entry fir-tree roots—see, for example, Figure 5-2. This attachment type is now employed by most of the world's major steam turbine manufacturers, even though some of them continue to use successfully fork-shaped roots.

For computational calculations of the blade stress state, the LSB is digitally modeled together with its attachment base and the adjacent steeples of the rotor with regard to possible clearances in the joint. An example of such a 3D finite-element model for an 1,143-mm (45-in) steel LSB of MHI for the rotation speed of 3,600 rpm with a fir-tree root is presented in Figure 5-4.[14] Another example of a similar model and the resulted relative stress field for an 1,067-mm (42-in) long steel LSB (for the rotation speed of 3,000 rpm) of Toshiba with the fork-shape root is presented in Figure 5-5. Sometimes, such models comprise several buckets connected by a shroud, tie-bosses, or arch bands.

The curved-entry fir-tree dovetail is currently supposed to be the most suitable attachment structure for the longest LSBs. The compactness of the dovetail enables a thinner wheel configuration, reducing the centrifugal stress in the rotor body. Furthermore, the fir-tree root is free from sharp edges or pin holes. This is especially important for blades made of titanium alloys, which are relatively brittle and sensitive to notches. In determining the dovetail shape and its machining tolerance, the difference in elasticity between the materials of the blade root and wheel disc or rotor should be considered. Of significance for the curved-entry fir-tree dovetail is a uniform distribution of load on all the blade root hooks. The stress contours and maximum stress values related to the tensile strength for the blade and wheel dovetails are shown in Figure 5-6a as applied to the 1,016-mm (40-in) LSB of Hitachi under conditions of the rated rotation speed of 3,600 rpm.[8] The maximum centrifugal stresses take place at the corner of the top hook for the blade root and at the corner of the bottom hook for the wheel dovetail. Their values are sufficiently lower than the material tensile strength. The load on each hook ranges from approximately 20% to 30% of the total load under normal conditions. However, these fractional loads can be considerably affected by errors while machining the dovetail, with resulted initial clearances on the individual hook surfaces. In Figure 5-6b the dotted line indicates the total

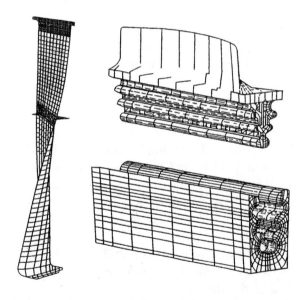

Figure 5-4. Computational 3D model for calculating the stress state and vibrational characteristics for MHI's 1,143-mm titanium LSB of MHI, with its root and adjacent steeple of the rotor. *Source: By courtesy of Mitsubishi Heavy Industries.*

Last Stages and Exhaust Hoods of LP Cylinders 143

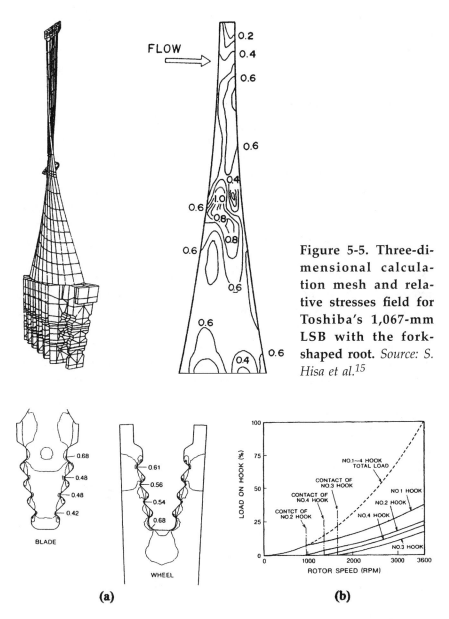

Figure 5-5. Three-dimensional calculation mesh and relative stresses field for Toshiba's 1,067-mm LSB with the fork-shaped root. Source: S. Hisa et al.[15]

Figure 5-6. Centrifugal stress contours and maximum stress values related to the tensile strength for a 1,016-mm titanium LSB with a fir-tree dovetail (a) and load distribution on hooks with machining error (b). Source: T. Suzuki et al.[8]

load on all the hooks, and the solid lines refer to the load on the individual ones. As the rotor speed increases, the lower hooks come into contact, and the load on the hooks is rather equalized, but the share for the top hook remains approximately 5% higher compared with normal conditions.

Modern blades, including LP ones, are mainly manufactured integrally shrouded, that is, made with the shrouding elements milled together with the bucket's airfoil (profiled body). The shrouding elements of the individual blades are connected together by means of special outside inserts and wedge-shaped grooves in the shrouds like a dovetail joint, or the shroud pieces are designed with special wedge-shaped edges that engage the blades in mesh under action of centrifugal forces. The second way is predominantly applied to modern LSBs—Figure 5-7. In addition, to increase the rigidity of the entire blade structure, the blades are supplementary coupled with a kind of *snubbers*—integrally formed tie-bosses at the mid-span of the blade height—see Figures 5-2—5-5. Their edges also engage under action of centrifugal forces. As a result, when the turbine rotates, all the LSBs are tied together, forming a continuous ring of blades. One of the major advantages of such a continuous annular blade structure, compared with blade groups (several units of several blades each connected with wire ties) more conventional in the past, is that it has fewer resonance points during rotation. The resulted structure with two contact supports (tie-bosses at the blade mid-height and integral shroud at the blade tip) provides well defined and easily controlled vibration modes and significantly reduces the buffeting stresses arising when the LSBs are subjected to low-steam-flow and high-back-pressure conditions.

Nevertheless, such leading turbine producers as Siemens and ABB for many years have successfully employed free-standing LSBs, not connected by shrouds, mid-span damping wire ties, or tie-bosses.[16-19] It is emphasized that modern CFD computation methods combined with extensive model trials, to-

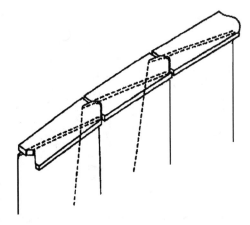

Figure 5-7. Connection of shrouding elements of LSBs with wedge-shaped edges.

gether with precise manufacturing technologies, make it possible to completely eliminate the need for any vibration damping elements, including shrouds. Even though shrouding the blades typically reduces tip leakage losses, for LSBs this is compensated by more effective peripheral water separation for unshrouded blades. In turn, the mid-span damping devices cause the increase of the airfoil thickness in their neighborhood, considerably increasing the profile losses. In addition, all the obstacles in the interblade channels (like tie-bosses or wire ties) disrupt the steam flow and lead to additional energy losses. Of importance also is that any local wetness concentration in the stage channels considerably contributes to the blade erosion. In particular, this concerns wire-ties and tie-bosses between the blades and brings another point in favor of using free-standing LSBs, as well as the shrouded blades without any additional ties in the preceding stages. Free-standing LSBs of ABB and Siemens can be seen in Figures 2-10, 3-5a, and 3-6. Along with this, it seems likely that free-standing LSBs can be acceptable only up to a certain threshold length. So the newest titanium LSBs of Siemens, with the annular exit area equal to 16.0 m^2 and 11.1 m^2 per flow for the rotation speed of 3,000 and 3,600 rpm, respectively, are characterized by an "interlocked" design and feature an integral shroud, as well as a mid-span snubber.[20,21]

Unshrouded blades allow more precise determination of their vibrational characteristics and thus a more reliable tuning out of the blades. In addition, for unshrouded LSBs, it is easier to arrange their non-contact continuous vibrational monitoring, in particular, for the operating purposes.[19,22] True, for fairness, it is worth noting that non-contact vibration measuring systems are developed and applied for the shrouded LSBs, too. In this case, the primary sensors are installed at the wheel side.[23]

Implementing Newer LSBs

The LSB geometry is derived from complex aerodynamic calculations. These aerodynamic calculations are confirmed by experiments at test cascades and model turbines with model and actual buckets. If the geometry of the developed blade is completely similar to that of the model bucket, and their dimensions are in inverse proportion to the rotation speed, all the aerodynamic, vibrational, and strength properties of both the model and actual blades are the same. This enables creating families of standard LSBs for different rotation speeds (3,600 rpm, 3,000 rpm, 1,800 rpm, and 1,500 rpm) based on a single model version to cover a wide range of the output and vacuum values.

The advent of newer, longer, more efficient LSBs allows replacement of older ones at steam turbines in service in the course of their refurbishment. So, for example, in the process of refurbishing the German power plant Mehrum's 712-MW steam turbine of Siemens the original LSBs were replaced with new ones with the resulted increase of the annular exhaust area from 6.3 m^2 to 8.0 m^2 per flow.[24] In the USA, Siemens successfully retrofitted LP cylinders at five power plants, employing its BB73-8.7m^2 project based on the use of 955-mm (38-in) LSBs with the annular exhaust area of 8.7 m^2 (94 ft^2) per flow.[25] At the U.S. power plant Labadie with two 570 MW of Westinghouse and two similar turbines of GE, the existent 30-in (762-mm) LSBs were replaced with more advanced 34-in (864-mm) LSBs, with necessary changes in the preceding stages; the resulted increase in the units capacity made up from 10 to 14 MW.[26]

For the last LP stages, it is especially important not only to obtain their high efficiency under the nominal (rated) operating conditions, minimizing the energy loss with the exit velocity. It is also necessary to ensure stable stage operation at reduced volumetric flow amounts, as well as maintain the stage efficiency as high as possible under these variable conditions. Significant changes in the calculated steam flow patterns, under low-flow operating conditions, with appearance of reverse vortex motion in the tip and root sections, capturing even the nozzle row and the preceding stage blades, are shown in Figure 5-8. Besides the streamline pattern, with the steam flow amount there also change the steam velocities and their ratio to the current acoustic velocity, that is, the Mach number.

Modern large turbines feature high length-to-mean-diameter ratios for the LSBs—see Tables 5-1 and 5-2. The value of this ratio can be considered to be an indicator of the three-dimensionality for the steam flow through the last stage, as well as through the LP steam path as a whole. The higher the chosen ratio value, the more complicated is the aerodynamic design of the steam path. Because the steam velocity diagrams, steam flow patterns, and values of the Mach number (the steam velocity related to the acoustic velocity corresponding to local steam conditions) essentially vary lengthwise of the stage height, it is understandable that the applied blading profiles also must be different. The inserts in Figure 4-6 demonstrate the profile types proposed to obtain the minimum profile losses in the wide range of Mach number values for the different LSB sections. Development of new, often quite nontraditional, profiles (es-

Last Stages and Exhaust Hoods of LP Cylinders

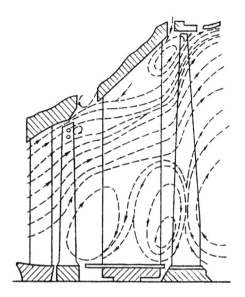

Figure 5-8. Appearance of reverse vortex motion in two last LP stages at low (14%) volumetric steam flow amount. *Source: A.V. Shcheglyaev[27]*

pecially for the longest LSBs operating with the highest values of the Mach number and their sharp variations along the stage height) brings considerable gains in the stage efficiency, as it is seen, for example, in Figure 5-9 with comparing the stage efficiency of conventional and newly developed LSB of Hitachi with the length of 660 mm (26 inch) for the rotation speed of 3,600 rpm.

For LSBs, the use of modern, inclined and bowed vanes is especially desirable and effective. Saber-like vanes can bring a noticeable increase in the stage efficiency (see Figure 4-13) and

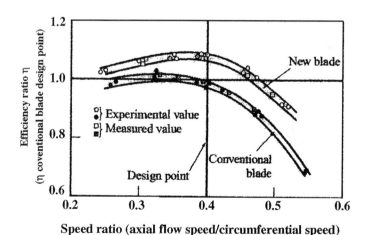

Figure 5-9. Comparison of stage efficiencies for newly developed, advanced and conventional 26-in LSBs of Hitachi. *Source: M.Machida et al.[13]*

a decrease of energy losses with the exit velocity, even without replacement of the existent LSBs.

In the early 1990s, special experimental investigations were conducted in the UK on a model turbine as applied to the planned refurbishment of the LP steam paths for British 500-MW and 600-MW steam turbines with replacement of the existing 914-mm (36-in) long LSBs with 945-mm (37-in) ones. Figure 5-10 presents some results of these experiments for three last LP stages with the 945-mm LSBs as a function of the steam flow amount characterized by the steam pressure ratio over the modeled stages. Under consideration are three factors: the use of saber-like vanes in the next-to-the-last and last stages, as well as the conical meridional profiling of the steam path outline. For the rated operating conditions with $p_0^I / p_2^{III} = 16$, corotational leaning and bowing of the next-to-the-last stage's vanes increased the efficiency by about 1.7%; the same measure for the last stage gave additional 1%, and peripheral meridional profiling of the steam path added 0.5%.

Turbine producers conduct 3D aerodynamic calculations of the LP steam paths with regard to the steam viscosity, taking into consideration the complex shape of the row profiles, appearance of the local supersonic velocities, and wave phenomena which accompany them. Some computer programs are also intended to solve the reverse problem—to obtain the row profiles on the basis of the set streamlines and distribution of the steam flow conditions. Because of the extreme complexity of the considered problems, all the developed and applied approaches and computer

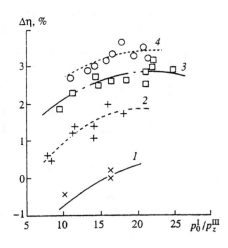

Figure 5-10. Changes of efficiency with the steam flow amount for a model of a three-stage LP steam path to be retrofitted with the use of 945-mm LSBs.

1: a conventional (initial) version, 2: with leant and bowed vanes of the next-to-the-last stage, 3: with leant and bowed vanes of the next-to-the-last and last stages, 4: the same with a conical meridional contour of the steam path's outline. *Source:* B.M. *Troyanovskii*[28]

Last Stages and Exhaust Hoods of LP Cylinders

programs unavoidably adopt some more or less serious assumptions. Nevertheless, they allow researchers and designers to obtain detailed space nets of meridional streamlines for the given boundary conditions, find lines of the constant relative velocities λ or M (isotaches) and pressure values (isobars) for different sections, distribution of energy losses along the stage height, and so on. Three-dimensional computations also result in the field of velocities in the exit edge trail, which opens a possibility to optimize the exit edge shape. Relative steam velocity fields for the tip, median, and root sections of a characteristic modern LSB were presented in Figure 4-5; another example of the steam velocity distribution, as applied to a 815-mm (32-in) LSB of Siemens, is shown in Figure 5-11, and Figure 4-14 shows the isotach field for the LP steam path of three stages developed by MHI with the titanium LSBs 1,143 mm (45 in) long for the rotation speed of 3,600 rpm.

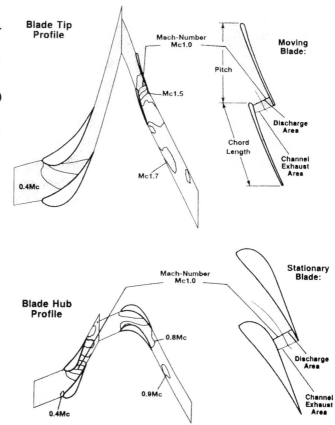

Figure 5-11. Velocity distribution for tip and root (hub) sections of an 815-mm (32-in) LSB of Siemens for 3,600 rpm. *Source: H. Oeynhausen et al.*[29]

LP EXHAUST PORTS

The LP exhaust ports (hoods) should feature sufficiently large axial sizes and smooth turns of the steam stream to minimize the exit energy losses. However, the exhaust ports are quite limited in their sizes; one of these limits is associated with the necessity of an easy access to the adjacent turbine bearings without an excessive increase in the turbine length. In addition, intricate spatial geometry of LP exhaust hoods makes the problem of their optimization fairy complicated. An example of a characteristic exhaust hood structure is presented in Figure 5-12. The problem of modeling mathematically and optimizing the LP exhaust hoods is also substantially complicated by a high level of the exit steam velocity after the LSBs, significant unevenness of this velocity field across the steam stream, with supersonic steam velocities near the tip and mainly subsonic ones near the hub (see Figures 4-5 and 5-11), and tangible changes of the steam flow pattern with the steam flow amount (or turbine load). Characteristic outlines for LP exhaust ports of different turbine producers with various design solutions for steam turbines with traditional "basement" condensers can be seen in Figures 2-2, 2-5, 2-6, 2-8, 2-10, 2-12, 3-1—3-6, 3-11, 3-16, 3-19, 3-20, 3-25, 3-29, and 4-12a. Rather non-conventional axial and lateral side exhaust ports are shown in Figures 3-22, 3-26, 3-28, 3-31, and 3-33.

For exhaust ports with downward and sideward steam flowing, that is, with lower (basement) and lateral (side) condensers, steam abandoning LSBs disperses in the radial and axial directions and then turns to reach the condenser. Coincidentally, the steam stream, having been annular, becomes rectangular in the cross section. If the turbine is furnished with side condensers, the space angle of the steam stream turn is less than in the case of basement condensers. For turbines with an axial exhaust and axial condenser, the steam flow after the LSBs

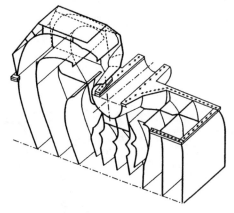

Figure 5-12. An example of structure for a characteristic LP exhaust hood. *Source: V.F. Kasilov[30]*

Last Stages and Exhaust Hoods of LP Cylinders 151

remains axially symmetrical and does not have to turn; hence their changes in the steam flow structure are much less significant.

The exhaust port efficiency is commonly characterized by the energy loss factor, ς, which, in turn, can be expressed via correlation between the steam pressure values at the port's exit and entrance (that is, in the condenser, p_c, and after the LSBs, p_2, respectively) related to the inlet kinetic energy of the steam flow, that is, the turbine's energy loss with the exit velocity, c_2. This value, called the relative decompression, or hydraulic resistance factor, can be negative or positive and is approximately estimated as follows:

$$\varsigma_{dec} = 1 - \varsigma = (p_c - p_2) / (c_2{}^2/2v_2), \tag{5.5}$$

where v_2 is the specific volume of steam after the LSBs. If the value of ς is less than 1.0 (that is, $p_c > p_2$ and $\varsigma_{dec} > 0$), this indicates that the energy loss with the exit velocity is partially compensated in the exhaust port. But if $\varsigma > 1.0$, $p_c < p_2$ and $\varsigma_{dec} < 0$, the exhaust port loss is added to the energy loss with the exit velocity.

For characteristic exhaust ports of modern LP cylinders with basement condensers, ς is commonly close to 1.2. Along with this, poor aerodynamic design of the exhaust hood can produce a very noticeable pressure drops between the LSBs and the condenser, with a remarkable growth of ς. On the other hand, aerodynamic improvement of the exhaust port, optimal configuration of a diffuser at the LSB exit, and the use of special baffles and other anti-vortex facilities in LP hoods allows decreasing their energy loss factor value to 1.0 and even lower. The diagram of Figure 5-13 presents some bench test results of MEI illustrating the influence of the dimensionless steam velocity at the exhaust port entrance on the total energy loss factor for a model exhaust hood with and without a diffuser at the hood entrance. With the increase of the steam velocity (it is supposed not varying across the entrance section), the pressure drop along the hood slightly decreases until a certain threshold value and then sharply skyrockets because of the flow separation and appearance of vortices in the steam passage. A probability of steam flow separation rises as the inlet steam velocity closes to the acoustic velocity and increases with the presence of various obstacles (piping, LP feedwater heaters, fastening elements, and so on), settled within the exhaust hood and condenser neck. The resulted vortices obstruct the passage, and the actual steam pressure at the exhaust port exit turns out to be less than it

Figure 5-13. The total loss coefficient of a model exhaust hood as a function of the dimensionless velocity at its entrance. 1: no diffuser in the exhaust hood, 2: the exhaust hood with a diffuser. *Source: A.E. Zaryankin et al.*[31]

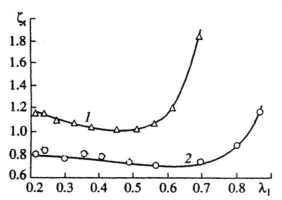

could be expected. With the diffuser, this phenomenon rather pads out, and what is more important the energy loss level turns out to be considerably lower for the entire steam velocity variation range.[31]

The aim of a well-designed diffuser is to accommodate effectively high steam velocities at the turbine exit, gradually retard the steam flow, and recover as much of its kinetic energy as possible. However, a wrongly designed diffuser itself can cause flow separation and additional energy losses. To prevent this, some turbine designers, including GEC Alsthom and Škoda, arranged steam extraction from the concave diffuser surfaces.[28,32-34] For the same purpose, it is proposed to install special annular vanes (baffles) just downstream of the LSBs with negative overlapping—internal and external diffuser contours.[31,35] In addition, exhaust hoods are provided with vortex dampers in the form of a specially arranged grid of fins in the exhaust port's bottom part, as shown in Figure 5-14.

For turbines with lateral exhaust ports and side condensers, the energy loss factor values are commonly substantially less than 1.0, making up about 0.7. The high efficiency ratings of steam turbines with lateral exhaust ports and side condensers were, in particular, confirmed by the long-term operational practice and field heat-rate performance tests of 500-MW and 1,000-MW wet-steam turbines of Turboatom with side condensers at several nuclear power plants of the former Soviet Union.[3]

Efficiency of axial exhaust ports and ways to increase it were analyzed in many details by GEC Alsthom in the process of developing a "compact" two-cylinder 315-MW turbine of a *EUREKA-Turbine* project[32,36]—see Figure 3-22. Detailed experimental and computational

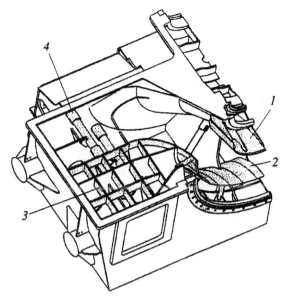

Figure 5-14. Arrangement of a antivortex grid and baffles in the bottom part of an exhaust port with downward steam flow.
1 and 2: internal and external annular contours of baffles, 3: antivortex grid, 4: stiffening rods of the exhaust hood. *Source: A.E. Zaryankin et al.*[35]

investigations of axial exhaust ports were also conducted by GE as applied to the SS CC turbosets.[37] An axial exhaust hood was also to be experimentally investigated and optimized at the experimental 105-MW steam turbine of MHI—see Figure 3-33.[38]

In the past, a common approach to the exhaust hood design was rather purely experimental and empiric. To optimize a new design, the internal flow passages of the hood were duplicated in an exact scale model, which was tested in an air flow test facilities, and the model design features were consequently varied. In doing so, any calculations bore rather secondary and qualitative character, because accessible computational methodologies and computer possibilities did not match the problem's complexity. Gradually, with the use of more advanced computational models, more powerful computers, and more refined calculation methodologies, they have become more trustworthy and turned out into a major instrument for exhaust hood improvements.

Figure 5-15 demonstrates a computational grid developed and employed by ABB for modeling mathematically steam flows in a downward-flow exhaust hood as a whole, as well as a diffuser after the 1,200-mm LSBs.[39] For most of existent computational models and programs for steam turbine's exhaust ports, the field of steam velocities at the exhaust hood inlet, that is, after the LSBs, is artificially set as the boundary conditions for each specific case of the considered operating conditions—steam flow amount and back-pressure. The next step should bring to a joint computational model of an exhaust port and the LP steam path or, at least, its last stage as a whole.

Figure 5-16 demonstrates the calculated streamlines for LP exhaust ports with the downward steam flow as applied to large steam turbines of Siemens without and with special anti-vortex baffles positioned in the upper centre of the enclosure and underneath the lower part of the diffusers downstream of the LSBs. These baffles are intended to reduce whirl of the outflowing steam and reduce the local pressure drops. This improvement was first implemented at the German nuclear power plant Emsland's 1,300-MW wet-steam half-speed turbine, but it is quite acceptable for large steam turbines of fossil-fuel power plants with great steam flow amounts to the condensers.[3,40]

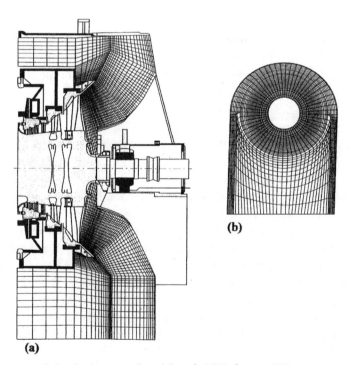

Figure 5-15. Calculation mesh grids of ABB for an LP steam path exit (a) and exhaust hood diffuser (b). *Source: E. Krämer et al.*[39]

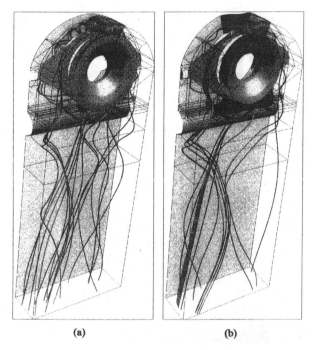

(a)   (b)

Figure 5-16. Streamlines in the LP exhaust port with the downward steam flow for large steam turbines of Siemens without (a) and with (b) special antivortex baffles. *Source: H. Oeynhausen et al.*[40]

LAST STAGE BLADE PROTECTION
AGAINST WATER DROP EROSION

The tip circumferential speed of modern LSBs reaches 700 m/s (about 2,300 ft/s) for steel ones and even 800-830 m/s (2,620-2,730 ft/s) for titanium LSBs (see Tables 1-5 and 1-6). Even for half-speed shafts of CC turbines, where the maximal speed values are relatively lower, they reach 500-520 m/s (1,640-1,700 ft/s). In addition, although the mean steam wetness in last stages of modern turbines is never more than 12-14%, the local steam wetness at the stage periphery can be significantly greater and more dangerous, because the liquid-phase contents and the size of water drops in the two-phase flow of wet steam increase toward the blade tip. The coarse-grained water, which mainly concentrates in the blade tip region and presents the main threat of blade erosion, lags behind the steam stream and, as a result, impacts the blade back at the row inlet. (More detailed consideration of these processes can be found in the author's book on wet-steam turbines,[3] as well as in special monographs devoted to turbine steam path damages.[41,42])

In order to protect steel LSBs against water drop erosion (WDE), they are shielded with Stellite laminas, or strips, brazed to the blade surface—Figure 5-17. The erosion resistance of Stellite exceeds that of stainless steel by 8-9 times and is considerably higher compared with titanium alloys. Water drop erosion mainly affects the upper third of the LSB back near the leading (inlet) edge, and the Stellite laminas are attached just at these surfaces. A disadvantage of this method is that the protection strips or shields create some discontinuity in the blade profile, resulting in increased profile losses. In addition, during the operation process, individual laminas sometimes break away from the blade surface, causing local abrasion of the profile, as well as noticeable changes in the blade's vibration characteristics. To avoid these complications, Stellite laminas are made flush with the blade surface (see Figure 5-17c) or the blade's leading edge is substituted by a bar-nose welded to the bucket body—see Figure 5-17d-f. Such a profiled nose combines an erosion resistant insert with a weld filler of Inconel 82, which is a very ductile material with high fracture toughness.[41] Welding allows complete restoration of the airfoil profile, leaving no surface discontinuity. The use of ductile weld filler also prevents the blade's parent material from crack propagation through the leading edge in the event of extensive water impact. Instead of covering with Stellite laminas, the blade surfaces are also sometimes hardened with electric-spark machining, even though this way seems to be less effective.

Typical patterns of WDE at its diverse stages as applied to the LSB's leading edge are presented below in Figure 12-22. Along with this, of importance is that, besides the area immediately adjacent to the leading edge, erosion often expands to a wider portion of the blade back commonly unprotected by Stellite strips or a bar-nose. Figure 5-18 demonstrates this pattern of erosion for the tip part of unshrouded and integrally shrouded LSBs. This erosion is mainly caused by water drops drawn by centrifugal forces to the stage periphery and then pulverized by the leading edge of the blade. Special calculations of Russian Central Boiler and Turbine Institute (TsKTI) allowed tracing the way of these drops of various sizes in the gap between the rows and within the in-

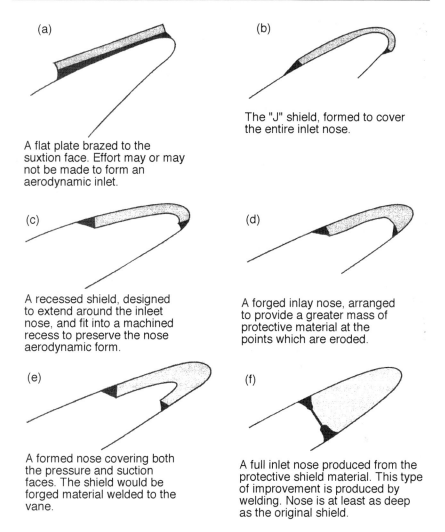

Figure 5-17. Various forms of shielding the LSB's leading edge. *Source: W.P. Sanders*[42]

terblade channel;[43] their trajectories can also be seen in Figure 5-18. The largest drops have a size of 220-230 mkm (about 9 mil) and a velocity of approximately 140 m/s (460 ft/s), which is much less then that of the main steam entering the blade, and these drops mainly hit the zone commonly protected by Stellite. Smaller drops acquire a greater velocity of about 180 m/s (590 m/s), and most of them fall onto the blade's back side downstream of the Stellite laminas. Noteworthy is that blades covered with shrouds have approximately a twofold higher rate of wear downstream of the Stellite lamina than unshrouded blades have. It is caused by larger sizes of drops and greater amount of moisture held by the shroud and, especially, the greater impact velocity. Anyway, the fulfilled analysis, as well as the operational practice, is an obvious evidence of the necessity to increase the width and thickness of the protective shields on the steel blade's back surfaces.

Along with the leading edge in the tip zone, LSBs also suffer from WDE of the trailing (outlet) edge in the root and median zones. This phenomenon is mainly stipulated by the reverse motion and vortices in the stage's root part under low-flow operating conditions (see Figure 5-8). The reverse steam streams commonly bring large-sized water drops that erode LSBs, especially their thin outlet edges. These unfavorable processes are additionally promoted by steam discharged into the condenser from the hot reheat or main steam-lines through the turbine bypasses, which come into operation exactly at low-flow transients. Discharged steam is cooled by sprays, and the resultant steam-water mixture is often injected by the reverse steam streams to the LSBs. That is why the cor-

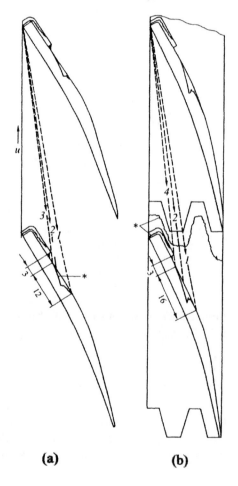

Figure 5-18. Erosion wear of the tip section of LSBs without (a) and with (b) shroud.
Calculated trajectories of water drops with diameters: 1 – 20 mkm, 2 – 50 mkm, 3 – 60 mkm, and 4 – 70 mkm. The asterisk designates the outline of erosion.
Source: N.V. Averkina et al.[43]

(a)   (b)

rect location and design of these devices in the condenser port is so important.

Titanium alloys are less prone to WDE than stainless blade steel is, even though titanium is inferior to Stellite in this regard. The first commercially employed titanium LSBs were originally protected by a nithinol coating, but long-term experience have proved the possibility of their use without anti-erosion protection.[4,5,7] Nevertheless, nowadays, for the longest shrouded LSBs with the tip circumferential speed of 700 m/s and more, some developers recommend protecting the shroud and the very tip portion of LSBs with an erosion shield made of Ti-15Mo-5Zr-3Al, which is electron-beam welded to the leading edge. An example of such a shield is shown in Figure 5-19 as applied to Hitachi's 1,116-mm long titanium LSB. Although the operational experience for these LSBs is not as long as that for conventional 12% Cr steel blades with Stellite protection, the first operational data match the predicted erosion rate for more than 120,000 operation hours—Figure 5-20.

The erosion rate at the leading edge of LSBs can be influenced by certain design

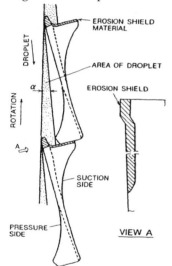

Figure 5-19. Protection against WDE of the tip section and shroud of Toshiba's 1,016-mm titanium LSB. *Source: T. Suzuki et al.*[8]

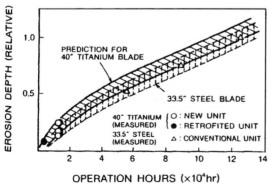

Figure 5-20. Estimation of the WDE rate for Toshiba's 1,016-mm titanium LSB. *Source: T. Suzuki et al.*[8]

measures. For example, according to Toshiba,[44] this rate noticeably reduces with an increase of the gap between the stationary and rotating blades, that gives water drops leaving the nozzles more space to be extracted from the steam flow. This effect was well confirmed by experiments at the model turbine. It is commonly combined with peripheral and intrastage water separation and removal from the steam flow—Figure 5-21. The more effective these countermeasures are, the less is the erosion rate.

Peripheral water separation and removal from the LP steam path take place thanks to the circumferential component of the wet-steam

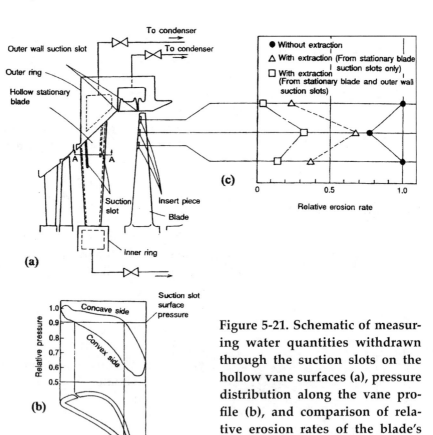

Figure 5-21. Schematic of measuring water quantities withdrawn through the suction slots on the hollow vane surfaces (a), pressure distribution along the vane profile (b), and comparison of relative erosion rates of the blade's inlet edge with and without water extraction (c). *Source: T. Sakamoto et al.*[44]

velocities—under action of centrifugal forces drawing the water drops out to the stage periphery. This separation takes place within the stages, in the gaps between the stationary and rotating blade rows, as well as after the rotating blades. The efficiency of this (peripheral) kind of water separation depends on the steam wetness, steam velocity and circumferential rotation speed at the blade tip, the share of the coarse-grained water drops in the wet steam flow, the shape, size, and location of the water-taking channels and water traps (Figure 5-22), as well as numerous other factors.

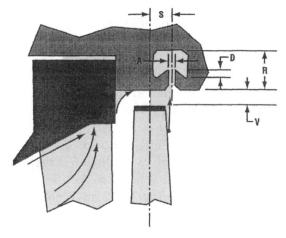

Figure 5-22. Peripheral water trap positioned after the LP stage.

The separated water is captured in the water trap belts above the rotating blades. But along with this captured water, they also withdraw some amount of steam, usually estimated as much as about 0.5% of the total steam flow amount through the stage. The energy of this steam can be partially employed if the withdrawn water-steam mixture is forwarded into the regenerative feedwater heaters. The regular steam extraction chambers for steam bleedings from the LP cylinder to the first LP feedwater heaters also play a role of water traps. Noteworthy is that open chambers of the water trap belts somewhat impair the aerodynamic properties of the steam path. For these reasons, some researchers propose not using these chambers (apart from those used for steam bleeding) and substitute them with specially fluted surfaces above the rotating blades for the deposition and subsequent removal

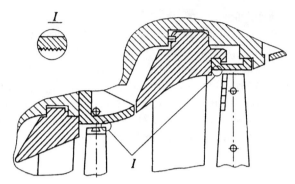

Figure 5-23. Possible design for a surface-type peripheral water trap for LP steam path. *Source: I.I. Kirillov and G.G. Spenzer*[45]

of water.[45] A possible appearance of such a surface-type water trap is shown in Figure 5-23. The captured water is to be gathered and drained from the casing ring.

Along with peripheral water separation and removal, the last and next-to-the-last LP stages of modern turbines are often provided with intrachannel water separation. For this purpose, the diaphragms of these stages are made with hollow nozzle vanes: stamped-and-welded (Figure 5-24) or solid-and-drilled (Figure 5-25). The nozzle vane surface is connected by slot, or apertures, with the internal vane space, whose ends are drained to the turbine condenser. As a result, the water film is withdrawn from the nozzle surface instead of pulverizing into drops and eroding the subsequent rotating blades. Different designers locate the water-taking suction slots at different areas of the nozzle surface: on the back surface near the inlet (as shown in Figure 5-24), on the face surface where the steam flow turns (see Figure 5-25), immediately in the exit (trailing) edge, and so on. To raise the separation efficiency, it is reasonable to have at least two slots at different areas. Yet under conditions of common internal vane space, they should be located at the places on the vane surface with the equal steam pressure, to avoid water pumping over one slot to another. Such suction slots on the back and face surfaces of the nozzle vane can be seen in Figure 5-21 as applied to Toshiba's 1,016-mm (40 inch) titanium LSB. The intrachannel water extraction decreases the relative erosion rate by as much as about twice, and additional peripheral water extraction with an increased distance between the stationary and rotating blades further decreases this rate by another 10-20%.

The separation efficiency of the intrachannel water removal also significantly depends on the amount of steam withdrawn together

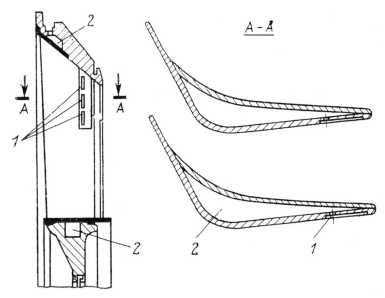

Figure 5-24. Hollow welded stationary blades with intrachannel water removal.
1 - water-taking apertures, 2 - water-taking internal channels.

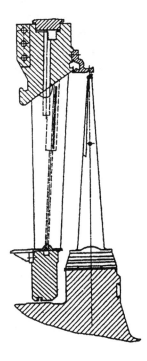

with water—this efficiency increases with the steam amount until a certain threshold and then remains almost invariable. In it, all the quantitative characteristics of these dependences substantially change with the shape of the suction slots and their position on the nozzle surface.[44,46]

Nevertheless, the water separation factor Ψ for hollow nozzle vanes with intrachannel water separation could be approximately assessed with the use of a diagram like that shown in Figure 5-26 as a function of the steam pressure at the stage inlet, p_0, and the Mach number value for the mid-section exit of the nozzle row, M_1

Figure 5-25. Characteristic LSB with intrachannel water separation and removal of LMZ. *Source: I.I. Pichugin et al.*[6]

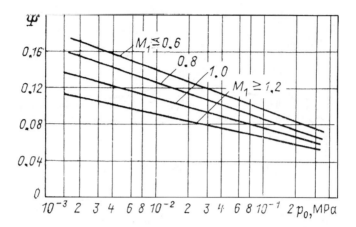

Figure 5-26. Estimation of the water separation factor for intrachannel water separation (hollow nozzle vanes with two suction slots). *Source: B.M. Troyanovskii et al.[47]*

$= (c_1/a_1)_m$. It is assumed that the vane profile have two suction holes with the slot width of about 0.7-0.9 mm (28-36 mil): on the profile back and in the trailing edge along the entire vane height for relatively short stages or in the top third or half of the stage height for relatively long stages ($l/d_m > 0.17$). Sure, there are many other factors influencing the separation efficiency (such as the geometric forms of the nozzle vanes and the slots themselves, the moisture dispersion, the velocity ratio, the Reynolds number, and so on), but they can be considered only as applied to specific stages with their actual geometry and operating conditions.[3,47]

Some disadvantage of intrachannel water separation is that a certain part of steam is withdrawn from the steam path together with the removed water, somewhat decreasing the last stage's output. Besides, a lower pressure inside the hollow nozzle vanes can cause cooling their outer surfaces with resulted additional water condensation.

In addition to, or instead of, the intrachannel water separation and removal, Siemens also proposes to warm the last LP stage's stationary nozzle vanes with steam from one of the LP steam extractions. The schematic diagram of warming the hollow nozzle vanes is presented in Figure 5-27. A similar approach was realized at one of 300-MW supercritical-pressure steam turbines of Turboatom installed at the Stavropol power plant in Russia.[48] The internal space of the hollow nozzle vanes of the last LP stages (see Figure 5-24) was provided with heating steam from the preceding LP steam extraction, and the suction slots on the vane surfaces were welded up. Special calculations showed that commonly almost all moisture with the water drop diameters of 25 mkm and more deposes on the last stage's vane surfaces and in the form of films and rivulets flows along these surfaces toward the trailing edge. Its amount for each vane accounts to approximately 2.85 g/s. Warming the vanes allows evaporating an average of about 75% of the entire moisture to be an erosion hazard for the LSBs. Distribution of the number of the LSBs by their erosion rate in the peripheral zone for the LP flow with heating the last stage's nozzles vanes compared with that with standard intrachannel water removal from the vane surfaces is presented in Figure 5-28. The average values of wear in the peripheral zone were equal to 3.5 mm (0.14 in) and 9.5 mm (0.37 in), respectively. The total amount of the heating steam was assessed as equal to approximately 520 kg/s for one LP flow, which is about 0.3% of the entire steam flow through the stage and equivalent to 30 kW of the last stage output. At the same time, a decrease of the energy loss with wetness completely compensates this value with a good excess. It seems possible to increase further the efficiency of heating by means of some improvements in the heating steam inlet and outlet design.

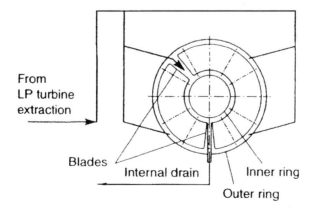

Figure 5-27. Proposed warming of hollow nozzle vanes for last LP stages of Siemens turbines. *Source: H. Oeynhausen et al.*[29]

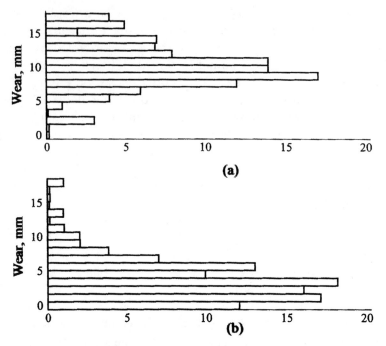

Figure 5-28. Distribution of the number of the last stage blades by their erosion rate in the peripheral zone before (a) and after (b) implementation of heating their vanes. *Source: N.V. Averkina et al.*[48]

References

1. Mujezinovic A. "Bigger blades cut costs," *Modern Power Systems* 23, no. 2 (2003): 25-27.
2. Nomoto H., Y. Kuroki, M. Fukuda, and S. Fujitsuka. "Recent Development of Steam Turbines with High Steam Temperatures," presented at the 2005 Electric Power Conference, Chicago, 2005.
3. Leyzerovich A. *Wet-Steam Turbines for Nuclear Power Plants*, Tulsa (OK): PennWell Corporation, 2005.
4. Jaffee R.I. editor. *Titanium Steam Turbine Blading, Workshop Proceedings,* New York: Pergamon Press, 1988.
5. Pichugin I.I., Y.N. Nezhentsev, and B.M. Troyanovskii. "Development of a Low-Pressure Cylinder of Increased Throughout for Large Steam Turbines," *Thermal Engineering* 37, no. 5 (1990): 225-230.
6. Pichugin I.I., A.M. Tsvetkov, and M.S. Simkin. "Features of Steam Turbine Design at the Leningrad Metallic Works," *Thermal Engineering* 40, no. 5 (1993): 355-366.
7. Pichugin I.I. and I.A. Kovalev. "Develop titanium-alloy blades for large steam turbines," *Power* 138, no. 5 (1994): 77-80.
8. Suzuki T., M. Watanabe, and M. Aoyama "Mechanical Design of a Titanium Last Stage Blade for 3600 rpm Large Steam Turbines," *The Steam Turbine Generator Today: Materials, Flow Path Design, Repair and Refurbishment*, PWR-Vol. 21, New York: ASME Press, 1993: 153-159.

9. Kishimoto M., M. Hojo, M. Mase M, et al. "Development of 3600 rpm 40 inch Titanium Blade and Actual Operating Results," *MHI—Technical Review* 31, no. 2 (1994): 61-65.

10. Kishimoto M., Y. Minami, K. Takayanagi, and M. Umaya. "Operating Experience of Large Supercritical Steam Turbine with Latest Technology," *Advances in Steam Turbine Technology for the Power Generation Industry*, PWR-Vol. 26, New York: ASME Press, 1994: 43-47.

11. Saito E., Y. Yamazaki, K. Namura, et al. "Development of a 3000 rpm 43-in Last Stage Blade with High Efficiency and Reliability," *Proc. of International Joint Power Generation Conference*, PWR-Vol. 33, New York: ASME Press, 1998, Part 2: 89-96.

12. Watanabe E., H. Ohyama, Y. Kaneko, et al. "Development of New Advanced Low-Pressure End Blades for High Efficiency Steam Turbine" [in Japan], *Mitsubishi Juko Giho* 38 (2001), no. 2: 92-95.

13. Machida M., H. Yoda, E. Saito, and K. Namura. "Development of Long Blades with Continuous Cover Blade Structure for Steam Turbines," *Hitachi Review* 51, no. 5 (2002): 143-147.

14. Troyanovskii B.M. "The New Design of the Low-Pressure Cylinders of Mitsubishi Steam Turbines," *Thermal Engineering* 50, no. 2 (2003): 170-172.

15. Hisa S., T. Matsuura, and H. Ogata. "The Improvement in Efficiency and Reliability of Last Stage Blades for Steam Turbines," *Proc. of the American Power Conference* 45 (1983): 207-213.

16. Bütikofer J., Händler M., and Wieland U. "ABB low-pressure steam turbines—the culmination of selective development," *ABB Review*, no. 8/9 (1989): 9-16

17. Bütikofer J. and Wieland U. "Modern LP Steam Turbines" [in German], *VGB Kraftwerkstechnik* 71, no. 4 (1991): 341-346.

18. Weiss A.P. "Aerodynamic design of advanced LP steam turbines," *ABB Review*, no. 5 (1998): 4-11.

19. Gloger M., K. Neumann, D. Bermann, and H. Termuehlen. "Advanced LP Turbine Blading: a Reliable and Highly Efficient Design," *Steam Turbine- Generator Developments for the Power Generation Industry*, PWR-Vol. 18, New York: ASME Press, 1992, 41-51.

20. Meier H.-J., M. Alf, M. Fischedick, et al. "Reference Power Plant North Rhine-Westphalia (RPP NRW)," *VGB PowerTech* 84, no. 5 (2004): 76-89.

21. Baumgartner R., J. Kern, and S. Whyley. "The 600 MW Advanced Ultra-Supercritical Reference Power Plant Development Program for North Rhine Westphalia—A Solid Basis for Future Coal-Fired Plants Worldwide," presented at the 2005 Electric Power Conference, Chicago, 2005.

22. Krämer E., H. Huber, and B. Scarlin. "Low-pressure steam turbine retrofits," *ABB Review*, no. 5 (1998): 4-11.

23. Shibata M., Y. Mitsuyama, and H. Fukuda. "Development of Noncontact Blade Vibration Measuring System," *Mitsubishi Heavy Industries—Technical Review* 37 (2000), no. 3: 97-100.

24. Varley J. "Modernising Mehrum: 40 green MW from a coal plant," *Modern Power Systems* 24, no. 7 (2004):11-13.

25. Cheski J.R., R. Patel, K. Rockaway, et al. "A Large Steam Turbine Retrofit Design and Operation History," presented at the POWER-GEN 2005 International Conference, Las Vegas, 2005.

26. Peltier R. "Steam turbine upgrading: Low-hanging fruit," *Power* 150, no. 3 (April 2006): 32-36.

27. Shcheglyaev A.V. *Steam Turbines* [in Russian], 2 vols., 6th edition, Revised and expanded by B.M. Troyanovskii, Moscow: Energoatomizdat, 1993.

28. Troyanovskii B.B. "Improving the Flow Path of Steam Turbines," *Thermal Engineering* 43, no. 1 (1996): 9-18.

29. Oeynhausen H., A. Drosdziok, W. Ulm, and H. Termuehlen. "Advanced 1000 MW Tandem-Compound Reheat Steam Turbine," presented at the American Power Conference, Chicago, 1996.

30. Kasilov V.F. "An Investigation of Facilities Acting on Swirl Flow in the Collection Chamber of the Exhaust Hoods for the Low-Pressure Cylinders in Steam Turbines," *Thermal Engineering*

47, no. 11 (2000): 984-990.

31. Zaryankin A.E., B.P. Simonov, A.N. Paramonov, and S.A. Chusov. "Advancement in the Aerodynamics of Exhaust Hoods of Turbines," *Thermal Engineering* 45, no. 1 (1998): 23-27.

32. De Paul M.V., M. Wallon, and A. Anis. "Twenty Years' Progress in Steam Turbine Aerodynamics," *Proc. of the American Power Conference* 51 (1989): 166-173.

33. Zaryankin A.E., V.A. Zaryankin, and B.P. Simonov. "Several Ways of Improving the Efficiency of the Flow Paths for Steam Turbines," *Thermal Engineering* 50, no. 6 (2003): 442-448.

34. Leyzerovich A. *Large Power Steam Turbines: Design & Operation*, in 2 vols., Tulsa (OK): PennWell Books, 1997.

35. Zaryankin A.E., V.G. Gribin, and A.N. Paramonov. "Applying Nonconventional Solutions to Enhance the Efficiency and Reliability of Steam Turbines," *Thermal Engineering* 52, no. 4 (2005): 267-274.

36. Tremmel D., W. Kachler, and P. Bourcier. "Development of a compact 300-MW steam turbine with single-flow LP section and axial exhaust" [in German], *VGB Kraftwerkstechnik* 72, no. 1 (1992), 33-43.

37. Coffer J.I., IV. "Advances in Steam Path Technology," *Transaction of the ASME. Journal of Engineering for Gas Turbines and Power* 118, April 1996: 337-352.

38. Watanabe E., Y. Tanaka, T. Nakano, et al. "Development of New High Efficiency Steam Turbine," *Mitsubishi Heavy Industries Technical Review* 40, no. 4 (2003): 1-6.

39. Krämer E., H. Huber, and B. Scarlin. "Low-pressure steam turbine retrofits," *ABB Review*, no. 5 (1998): 4-11.

40. Oeynhausen H., Claßen H.-P., and Riehl J. "Upgrading the Low-Pressure Turbines of the Emsland Nuclear Power Plant," *VGB PowerTech*, vol. 83 (2003), No. 1/2, pp. 85-90.

41. McCloskey T.H., Dooley R.B., and McNaughton W.P. *Turbine Steam Path Damage: Theory and Practice*, 2 vols., Palo Alto: EPRI, 1999.

42. Sanders W. P. *Turbine Steam Path. Maintenance & Repair*, in 2 vols., Tulsa: PennWell Publishing Company, 2001.

43. Averkina N.V., E.P. Dologoplosk, Y.Y. Kachuriner, and V.G. Orlik. "The Specific on Erosion Wear of the Working Blades in the Last Stages of Turbines with a Capacity of 300 MW and Higher," *Thermal Engineering* 48, no.11 (2001): 915-922.

44. Sakamoto T., S. Nagao, and T. Tanuma. "Investigation of Wet Steam Flow for Steam Turbine Repowering," *Steam Turbine-Generator Developments for the Power Generation Industry*, PWR-Vol. 18, New York: ASME Press, 1992, 33-39.

45. Kirillov I.I. and G.G. Spenzer. "Problems of Steam-Water Separation and Water Drainage When Designing Flow Path Elements of Steam Turbines," *Thermal Engineering* 40 no. 3 (1993): 191-193.

46. Tanuma T. and Sakamoto T. "The removal of water from steam turbine stationary blades by suction slots," *Proc. of the Institute of Mechanical Engineers*, 1991, C423/022, pp. 179-189.

47. Troyanovskii B.M., G.A. Filippov, and A.E. Bulkin. *Steam and Gas Turbines for Nuclear Power Plants* [in Russian], Moscow: Energoatomizdat, 1985.

48. Averkina N.V., Y.Y. Kachuriner, V.G. Orlik, et al. "Plant Experience in Introduction of Heated Nozzle Vanes to Reduce Erosion Wear of Wet-Steam Turbine Stages" [in Russian], *Elektricheskie Stantsii*, no. 2 (2004): 24-28.

Chapter 6

Thermal Expansion, Bearings, and Lubrication

ARRANGEMENT OF THERMAL EXPANSION

Traditional arrangement of thermal expansion for a multi-cylinder turbine is shown in Figure 6-1. In this case, the turbine is taken consisting of three separate cylinders: HP, IP and LP ones. Their specific internal design is of little significance as applied to the considered problem. The HP cylinder leans with the paws of its outer casing on the pedestals (chairs) of the adjacent first (front) and second bearings. The LP cylinder in this specific case has embedded bearings whose pedestals are made together with the bottom part of the cylinder's exhaust hood. The front bearing pedestal of these two is welded together with shoulders for supporting the IP cylinder's rare paws. By alternative, the bearings adjusting to the LP cylinder can be made outside, that is, not embedded into the exhaust hood, but installed directly on the foundation frame, and the inner or outer casing of the LP cylinder, depending on its design, also leans on the bearing pedestals. This design scheme seems to be preferable for modern turbines, providing more rigid support for the rotor and making assemblage and disassemblage of the turbine more convenient. The front paws of the IP cylinder's outer casing rest on the second bearing pedestal adjacent to the HP cylinder.

To be more precise, the casing paws lean on transversal keys attached to the bearing pedestals, and these keys guide lateral thermal expansion of the casings. In addition, the bearing pedestals are connected with the cylinder casings with vertical keys in the turbine axis plane. In turn, the bottoms of the bearing pedestals are furnished with longitudinal keys, which make the bearing pedestals slide on the foundation frame along the turbine axis, providing a possibility of axial thermal expansion for the turbine when its metal temperatures vary (mainly in the course of start-ups and cooling when the turbine is stopped). The starting point of axial thermal expansion is the turbine *fix-point* under the LP cylinder; it is arranged by intersection of the vertical plane through the turbine axis with the axis through additional transversal (lateral) keys. Hence the LP cylinder's casing, whose metal temperatures vary only slightly at any operating conditions and differ little from the metal temperatures at the "cold" state (after long outage), remains immoveable. If the turbine has a few LP cylinders, each of them has a separate fix-point, and pedestals of the bearings between the LP cylinders have some flexible elements allowing these cylinders to expand a bit toward one another.

With such a scheme, when in the course of start-ups the turbine is heated and the casings of its high-temperature cylinders increases their axial sizes, these casing leaning on the bearing pedestals via the transversal keys push these pedestals and make them slide along the longitudinal keys in the axial direction, as if pushing off from the fix-point(s). When the stopped turbine is cooled down, the cylinder casings shorten their length and "pull" the bearing pedestals back toward the fix-point(s).

In turn, the turbine shaft expands from the thrust bearing, which is commonly located within the intermediate bearing pedestal between the HP and IP cylinders and moves together with this pedestal. As a result of different metal temperatures of the stator's and rotor's components, different coefficients of linear expansion for the rotor's and casing's steel materials, and different directions of thermal expansion (in the considered case of Figure 6-1, this refers to the IP cylinder), in the thermal expansion process the axial clearances in the steam paths of the turbine cylinders vary. To estimate these variations, the cylinders are furnished with relative rotor expansion (RRE) sensors settled at the remote (from the thrust bearing) end of each cylinder. Along with measuring the RRE values, it is also necessary to have a possibility to monitor the "absolute" thermal expansion of the turbine upon the whole (as applied to the front bearing pedestal), as well as sometimes the IP cylinder.

The total length of modern TC five-cylinder turbines, not counting the generator, reaches about 40-48 m (130-157 ft). This primarily concerns 1,000-MW-class steam turbines, with a once-throw or loop-flow HP cylinder (or an integrated SHP-HP cylinder for USC-pressure turbines), double-flow IP one, and three double-flow LP cylinders. Along with this, there exist five-cylinder turbines of a less output with only two LP cylinders and three cylinders subjected to thermal expansion: for example, VHP, HP-IP, and IP2 as applied to the double-reheat cogeneration tur-

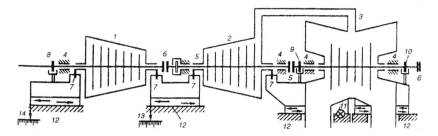

Figure 6-1. Traditional arrangement of thermal expansion for a multi-cylinder turbine.
1, 2, and 3: HP, IP, and LP cylinders, 4: journal bearings, 5: combined journal-and-thrust bearing, 6: couplings, 7: casing paws leaning on bearing chairs via transversal keys, 8, 9, and 10: RRE sensors for the HP, IP, and LP cylinders, respectively, 11: the turbine's fix point, 12: longitudinal keys, 13 and 14: indicators of absolute thermal expansion for the IP and HP cylinders, respectively.

bine shown in Figure 2-8. For such turbines, the problem of thermal expansion is especially topical. A three-cylinder turbine of a 600-MW class, with separate HP and IP cylinders and single double-flow LP one, can be about half this length—see Figure 3-6, and for a two-cylinder turbine of the same class, with an integrated HP-IP cylinder and one double-flow LP, its length can account to about 15 m (50 ft). The turbine length can be partially reduced owing to the use of common bearings for the adjacent rotors—see, for example, Figures 2-10, 3-5, and 3-6. Depending on the turbine configuration and with regard to the coefficient of linear expansion of steel equal to about $12°C^{-1}$, the total absolute thermal expansion of the turbine (as applied to its working metal temperatures compared to the "cold state") can reach 25-55 mm (1-2 in).

Modern large steam turbines of supercritical and USC main steam pressure, with their thick-wall casings and bulky rotors, feature significant weight of their high-temperature cylinders, and these entire weights lye down on the bearing pedestals, producing significant forces of friction on the sliding surfaces and hampering the axial movement of the bearing pedestals along the foundation frame. As a result, many large steam turbines in service, especially some relatively older those, can encounter serious problems with the freedom of thermal expansion. This shows itself in the broken, leap-like movement of the bearing pedestals while the turbine is being heated or cooled—Figure 6-2. Hampered thermal expansion can result in some distortion of the casings, torsion of the foundation frame's crossbars, increased vibration, damage of the turbine bearings and couplings, and so on. Problems with the freedom of thermal expansion frequently hamper the turbine's start-ups because the monitored RRE values for the HP and IP cylinders attain their limits.

Special investigations for diverse large steam turbines in service showed that the major causes why the turbine loses the freedom of thermal expansion can be: the increased friction on the sliding surfaces between the bearing pedestals and foundation frame, increased trans-

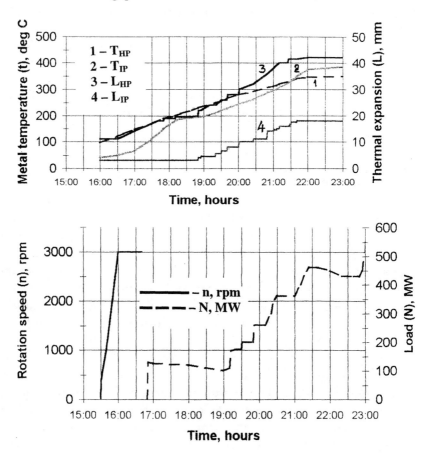

Figure 6-2. Leap-like thermal expansion of an 800-MW supercritical-pressure steam turbine in the course of a cold start-up.
1 and 2: measured characteristic metal temperatures of outer casings for HP and IP cylinders, respectively, 3 and 4: measured absolute thermal expansions of HP and IP cylinders.

Thermal Expansion, Bearings, and Lubrication 173

versal load on the turbine from steam-lines connected to the turbine cylinders, poor transfer of the axial thrust from one cylinder to another, and insufficient rigidity of the foundation crossbars. For turbines of different types with different arrangement and operational conditions, the contribution of different causes can be different.[1] The more the turbine output and the higher its main steam conditions, the more massive and rigid its adjoined steam-lines become. This especially concerns cold and hot reheat steam-lines. If their thermal expansion is ill-compensated, they vitally affect the turbine in its thermal expansion. As a result, it happens that the bearing pedestals are squeezed in their motion along the base frames. This influence is especially great if the steam-lines are settled asymmetrically relative to the turbine axis—for example, if the turbine is arranged along the turbine hall, that is, the boiler happens to be from one side of the turbine. It is difficult, if not impossible, to avoid such an arrangement for largest turbines of a great length that cannot be settled across the turbine hall in front of the boiler. The transversal forces from the steam-lines, as well as asymmetrical loads on the left-hand and right-hand paws of the high-temperature cylinders, sometimes become the main obstacle for free thermal expansion of the turbine.

As a result of conducted investigations, a set of diagnostic, design, and technological measures were developed and carried into effect at different steam turbines in service to reveal the specific causes of their problems and eliminate them.[1,2] Among the most widespread and effective countermeasures are the placing of special fluoroplastometallic bands or removable plates on the sliding surfaces under the bearing pedestals, electrochemical processing of the keys' surfaces, adjustment of the support-and-suspension system for the steam-lines connected to the turbine, and tightening of the foundation frame.

An example of using the mentioned antifriction fluoroplastometallic band is described in one of the author's previous books.[3] Such a band presents a composite antifriction structure consisting of three layers: a steel band covered on two sides with a layer of copper or brass no less than 0.01 mm thick, a porous layer of bronze granules 0.063-0.16 mm in size, which are baked to the copper or brass plated steel base, and a layer of fluoroplastic with a MoS filling agent that covers the bronze granules and fills the free spaces of the bronze layer. According to bench tests, the coefficient of friction for this material at the temperature varying in the range of 20-to-250°C (70-480°F) and under pressure of 2-5 MPa (290-725 psi) lies in the range between 0.08 and 0.13. These bands have been successfully used at many power plants with different types of large steam turbines to make their thermal expansion smoother.

Some turbine producers also furnish their turbines with special rods to pass the pushing and pulling forces directly from one cylinder to another besides the intermediate bearing pedestal, which in this case follows after the cylinders. To reduce the friction forces on the sliding surfaces, some turbine producers cover these surfaces with hardened steel plates (or strips) lubricated by solid substances.

Along with this, in increasing frequency, the problem of thermal expansion for modern large steam turbines is solved in a principally another manner. In particular, such a new scheme has been carried into effect at the 1,000-MW-class steam turbines of Siemens, including, for example, those for the power units Boxberg Q, Niederaußem K, Yuhuan, as well as turbines of a more moderate output—see Figures 2-10 and 2-22.[4-9] According to this scheme, the bearing pedestals are rigidly mounted on the foundation frame, whereas the outer casings of the HP and IP cylinders, as well as the inner casings of the LP cylinders, leaning on the bearing pedestals, are free to slide along their upper surfaces in the axial direction with the help of longitudinal keys. The common anchor point and the origin for the axial thermal expansion is the pedestal of the intermediate bearing between the HP and IP cylinders, which is designed as a combined journal-and-thrust bearing. Because in this case both the casings and rotors of the HP and IP cylinders expand together in the same directions, this decreases variations of RRE for both of these cylinders and thereby allows decreasing the axial clearances in their steam paths. As to the LP cylinders, their outer casings rest on the condensers, and in order to arrange thermal expansion of the LP rotors and inner casings in the same direction and reduce the RRE and variations of axial clearances in these cylinders, the special pushing rods in the LP cylinder bottoms connect all their inner casings together—see Figures 2-10 and 2-22. Figure 6-3 illustrates how the pushing force is applied to the LP inner casing making it expand to the generator (in the same direction as the rotor) without any effect on the assembly of the outer casing and condenser.

It might be well to note that the scheme with the fixed bearing pedestal between the HP and IP cylinder was earlier applied by LMZ for its 1,200-MW TC supercritical-pressure turbine, but then the front (first) and third bearing pedestals were remained sliding. Similar design solutions were also introduced to some other large steam turbines, including, for

Thermal Expansion, Bearings, and Lubrication 175

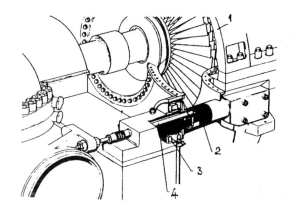

Figure 6-3. Arrangement of axial thermal expansion with pushing rods for LP cylinders of modern large steam turbines of Siemens.
1: inner casing, 2: protection shell, 3: compensator, 4: push rod. *Source: H. Oeynhausen et al.*[4]

example, 850-MW turbines of MAN operated at the U.S. power plant Martin Creek.[3]

Of significance is that the increased forces of friction, transversal load on the turbine from the adjoined steam-lines, and other obstacles hampering the thermal expansion freedom do exist as applied to the sliding surfaces between the cylinder casing paws and bearing pedestals, as they do as applied to the surfaces between the bearing pedestals and foundation frame. So the mentioned problems and ways for their solution deserve serious attention to be paid as applied to the thermal expansion arrangement with the fixed bearing pedestals as well.

IMPROVEMENT OF JOURNAL BEARINGS

With an increase of the turbine capacity, the diameters of rotor journal necks and hence journal bearings increase too, and there increases the load on them. So, for example, development of a 735-MW TC steam turbine for the Thailand power plant Ratchaburi (the maximum continuous rate equal to 841 MW) made MHI apply the rare journal bearing (at the LP cylinder's end connected to the generator) with the diameter of 535 mm (21 in), which is declared the maximum diameter for turbines with the rotation speed of 3,000 rpm.[10] Along with this, the 1,200-MW supercritical-pressure steam turbine of LMZ (as well as 1,000-MW, 3,000-rpm wet-steam turbines of LMZ for nuclear power plants) has the journal bearings with diameters of 620 and 575 mm (24.4 and 22.6 in), whereas the rare journal bearings of standard 800-MW supercritical-pressure steam turbines of the same producer have been made with a diameter of 520 mm (20.5 in).[11]

To provide vibrational reliability of large steam turbines and decrease the losses with friction in the bearings, many large steam turbines have been furnished with segment, or multi-wedge, bearings. In this case, the journal neck interacts not with an entire bush of the bearing but with a few self-adjusted segments, each of which can turn independently relative to its rib of leaning. The lubricant is brought to each segment, forming separate oil wedges. In such a way, the journal neck appears held by a system of the oil wedges. This noticeably increases rotation stability. With all these merits, segment journal bearings are much more complicated in their design and require more accuracy in their assemblage compared to more traditional bush journal bearings with an elliptical bore. Along with this, the operational practice shows that as a rule these (segment) bearings do not have an advantage in the efficiency over bush bearings. Improvement in design of bush bearings with an elliptical bore has allowed diminishing the turbine vibration to the level even lower than that with the segment bearings, simultaneously reducing the lubricant flow amount and the losses for friction in the bearings. This made it possible to replace the existent segment bearings at some large steam turbines in service with the bush bearings of the improved design—Figure 6-4. In particular, such replacement was accomplished for supercritical-pressure steam turbines of LMZ with the output of 300, 500, 800, and 1,200 MW. Results of field tests for such bearings with an elliptical bore, lubricant supply immediately to the hydrodynamic wedge, and additional removal of the spent lubricant as applied to such large turbines are presented in Table 6-1. Replacement of the existent bearings with those of an improved design decreased the specific losses for friction in the bearings from 0.28-0.4% of the turbine rating output to 0.175-0.19%.[12]

To increase the load-carrying capacity and reliability of journal bearings at low speeds of rotation, it was also proposed to cover their working surface with special micro-relief directed from the both of bearing end-faces to the middle. Such a measure was implemented at some supercritical-pressure steam turbines with the output of 300, 800, and 1,200 MW at a few power plants, as well as at a 1,000-MW wet-steam turbine for nuclear power plants, resulting in more favorable operating conditions with the turbine shaft rotated by the turning gear. As this took place, there also decreased the flow amount of the jacking oil and increased the rotor lift by the jacking oil. Effectiveness of this measure has been confirmed by

Thermal Expansion, Bearings, and Lubrication

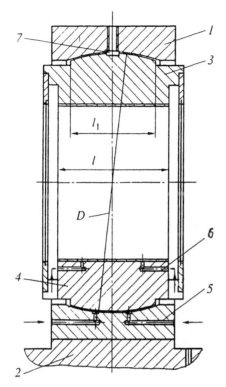

Figure 6-4. Schematic of a standard journal bush bearing of an improved design for large steam turbines of LMZ.
1 and 2: top and bottom halves of the bearing ring, 3 and 4: upper and lower semi-bushes, 5 and 6: supply passages for lubricant and jacking oil, 7: lubricant feed. *Source: A.S. Lisyanskii et al.[11]*

Table 6-1. Experimental data for bush bearings of an improved design as applied to two types of large steam turbines of LMZ.

	\multicolumn{7}{c}{Turbine}
Capacity, MW	\multicolumn{4}{c}{800}	\multicolumn{3}{c}{1,200}
Diameter of the journal necks, mm	450	475	500	520	435	450	575	620
Number of journal bearings	3	1	2	2	1	1	3	3
Total lubricant flow amount, m^3/h	\multicolumn{4}{c}{175}	\multicolumn{3}{c}{290}
Total losses for friction in bearings, kW	\multicolumn{4}{c}{1,400}	\multicolumn{3}{c}{2,350}
Temperature of bearing Babbitt, °C	\multicolumn{4}{c}{70-76}	\multicolumn{3}{c}{72-78}

Based on data of A.S. Lisyanskii et al.[11]

PREVENTION OF OIL FIRES DUE TO THE USE
OF FIRE-RESISTANT FLUIDS AS LUBRICANT

a long-term experience at several turbines with a decrease in the wear of the working bearing surfaces by about 10-15%.[12]

Turbine-generators of power plants are equipped with highly spread and ramified hydraulic and lubrication systems whose total volume in many cases reaches and even exceeds 40 m^3 (about 10,000 gallons) for a unit. The use of standard, highly combustible petroleum-based oil in these systems produces a serious hazard of oil fires in turbine halls. Of significance are a considerable length of the oil pipes, a huge number of flange joints, and their close proximity to hot surfaces of steam-lines, turbine casings, and electric facilities. If the escaping oil falls on such a surface, a severe fire is likely to occur.

According to some insurance company data, between 1971 and 1997 worldwide there occurred 83 power plant turbine-hall fires that could be traced at least in part to the turbine's oil system; other sources count 97 oil-related fires between 1972 and 1996, and these statistics likely do not cover all the incidents that actually took place.[13-15] According to EPRI's report of 1983, only six of oil fires occurring at U.S. power plants in 1985-89 required the total repair costs of $172 million, and the average duration of forced outages caused by these fires amounted to about 270 days.[14] According to the FM Global insurance company, in the period of 1988-2003 at least 19 U.S. power plants with turbines ranging from 40 to 820 MW experienced catastrophic fires involving turbine oil systems; the mean outage due to the fires was over six months, with three units having to be retired prematurely.[16] The fires primarily strike main equipment (turbines and generators), turbine-hall auxiliaries and their piping, but frequently also extend to cable trays and control room, wreck the roof and steel structures, etc.

Oil fires in turbine-generator islands of power plants can be prevented almost entirely by means of replacing petroleum oil in turbines' hydraulic and lubrication systems with synthetic fire-resistant fluids (FRFs).[15,17]

Nowadays, essentially all the world's leading turbine producers use FRFs in hydraulic lines of their turbines' electric-hydraulic governing systems, and over 1,000 steam and gas turbines have operated with

Thermal Expansion, Bearings, and Lubrication 179

diverse FRFs in their governing systems. This measure eliminates the most dangerous high-pressure oil source of oil-fires in turbine-halls. The high-pressure hydraulic and low-pressure lubrication systems are separated from each other, and petroleum oil is left as a lubricant. As a result, a probability of oil-fires remains considerably high. Moreover, this risk even increases as the average turbine capacity grows since larger turbines have more voluminous and ramified lubrication systems. The fire danger further rises for large turbines equipped with jacking oil pumps for lifting the shaft rotated by the turning gear. According to some statistical considerations, [14] if in the 1950s one oil-fire came to about 200 turbine-years of operation; in the 1960s this figure decreased to 140, in the 1970s—to 100, and now the oil-fire risk for a typical U.S. 600-MW steam turbine is assessed as equal to about one per 40 turbine-years.

The operation conditions for lubricants in the turbine-generator bearings are much more severe than those in the hydraulic systems: more heating, aeration, water contamination, and so on. Therefore, the demands on FRF properties used for lubrication are greater. At present, the FRFs are more expensive than petroleum oil, and in addition, the turbine lubrication systems are much higher-volume compared with hydraulic ones. These are the main reasons that hampered the FRF advance as lubricant. Along with this, it is well known that phosphate-ester-based FRFs feature excellent lubricating properties, and some European countries, Germany in particular, and the former Soviet Union have accumulated a wealth of experience in employing the FRFs in the turbine lubrication systems.[18-21] Notably, modern phosphate-ester-based FRFs belong to the same toxicity class as conventional turbine petroleum oils and do not require any special precautions in service.

In particular, since the mid-1980s the FRF called OMTI, developed in All-Russia Thermal Engineering Research Institute (VTI), has been used in the hydraulic circuits of governing systems for more than 200 steam turbines with the single capacity of from 140 to 1,200 MW. In doing so, some of these turbines have also employed OMTI in the lubrication systems from the very beginning of their service, while the others were converted to the use of OMTI as lubricant after years of operation with petroleum oil. At some turbines, the hydraulic and lubrication systems remain separated from each other, whereas in other cases they have common tanks. Schematic of a lubrication system with the use of OMTI for a 300-MW supercritical-pressure turbine is given in Figure 6-5.

In Germany, according to VGB, there are at least 38 power plants,

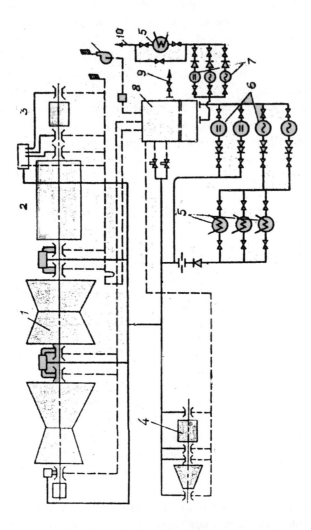

Figure 6-5. Schematic of a lubrication system with the use of FRF OMTI for a 300-MW supercritical-pressure two-cylinder turbine.
1: steam turbine with emergency lubricant reservoirs above the bearings, 2: generator, 3: exciter, 4: turbine-driven boiler-feed pump, 5 – lube coolers, 6: DC- and AC-motor-driven lube pumps, 7: DC- and AC-motor-driven lube pumps for the generator's lube seals, 8: lube tank, 9: emergency lube drain, 10: to the generator's lube seal system.

Thermal Expansion, Bearings, and Lubrication

including nine nuclear ones that use FRFs as lubricant, which are primarily triaryl-phosphate-esters.[21]

Noteworthy is a rich experience of employing the FRF as lubricant for gas turbines at gas-pumping stations, including those in the USA and Canada.[21] One of such gas turbines with the output of 5.5 MW commissioned in 1986 at Klybeck station in Bazel, Switzerland, operated its whole design lifetime of 100,000 operating hours on its initial charge of lube oil Reolube manufactured by Great Lakes Chemical Corporation as a derivative of OMTI.[22]

Fire-resistant fluids offer much better fire protection than petroleum oils because of their much higher self-ignition temperature equal to approximately 750°C (1,380°F) compared to 370°C (700°F) for petroleum oils. Even if such fluid spills on a high-temperature steam-line or other hot surfaces (the most frequent scenario of the oil fires), it does not ignite, and even once ignited by a local extra-high temperature spot or open fire, it does not keep burning.

At current production quantities, synthetic FRFs cost about $6,000 per metric ton. A characteristic 600-MW steam turbine's lubrication system holds about 35 t of lubricant, and together with turbine-driven boiler-feed-pumps—up to 50 t. With a safety margin for the case of unforeseen losses, the plant will need about 75 t over ten years of operation. This way, total expenditures will amount to about $450,000. With modern ion-exchange-based resin adsorbents, expected FRF losses should not exceed 3 t per year, resulting in additional annual expenditures below $20,000 in the subsequent years. With a fire probability of 0.025,[14] this sum does not seem significant considering potential fire-related damage costs of $20-70 million, not counting the cost of power generation interruption during the months of the ensuing outage.

Successful implementation of FRFs in the turbine lubrication systems to a great degree depends on the turbine manufacturers, which should closely collaborate with FRF developers and producers and take into account specific properties of specific FRFs, their differences from properties of petroleum oil. In particular, such thorough work was accomplished by LMZ together with VTI in the process of implementing OMTI at fossil-fuel and nuclear power plants, including standard machines of 800 and 1,000 MW.[19,23] Similarly, in Germany, it took close collaborative efforts of turbine manufacturers, power plants (VGB members), and insurance companies to implement FRFs for both lubricating and governing turbine systems.[13,20,21]

The fire resistance of FRF opens many attractive opportunities for turbine designers. It makes possible such design decisions as, for example (see Figure 6-5):

- replacing the main oil pump on the turbine shaft with motor-driven pumps incorporating a higher level of redundancy,

- arranging the emergency lube reservoirs within the bearing casing covers (see Figure 2-4) to provide the bearings with lubricant in the case of turbine coast-downs without lube-pumps (with petroleum oil, such reservoirs would present a great fire hazard),

- settling the lube tank with more freedom up to taking it away into a separate room.

These and some other measures facilitate the steam turbine designers' and power plant developers' efforts to assure the turbine's operating reliability and general safety, including removal of the fire hazard. For combined-cycle units using FRF as lubricant, both the steam turbine and gas one(s) can have a mutual lubrication system.

Noteworthy is that FRFs possess excellent lubricating properties such as stickiness, lesser coefficient of friction under incomplete lubrication, ability of standing greater specific loads, and others. These features are especially significant for large multi-cylinder steam turbines with an enhanced misalignment risk, especially at start-ups and shutdowns, as well as at coast-downs without jacking pumps.

It is also important that FRFs enable operation with higher temperatures in the bearings. With the increase of the rotor journal diameter over 500 mm (20 in), the flow of petroleum-based lubricants in the gap between the journal and the bearing shell changes its nature and becomes more turbulent. As a result, the energy losses in the bearing rise, causing additional heating of the lubricant and the need for greater lubricant flow amounts that entails a further rise of the losses. With FRFs, flow amounts can be decreased by 25-30%; higher temperatures are allowable; the lubricant can be input directly into the lube wedges, and the energy losses in the bearings remarkably decrease, enabling a 1,000-MW-class turbine to recover a few megawatts.

While passing through the bearings' lube wedges, a petroleum lubricant absorbs air. It is also contaminated with water coming through

Thermal Expansion, Bearings, and Lubrication

the end seals, especially during start-ups and shut-downs. While the lubricant is in the tank, this air and water should have time to separate and be removed. Otherwise, intensive aging of the lubricant and sludging in the pipelines begin. To prevent these problems, special additives are periodically made up to petroleum oil; however, they are to a great degree removed with the water. On-line monitoring of the additives' content in the oil is rather difficult, and if the additives are not introduced in time, separation of the air and water becomes worse. Thereby, the water (which is heavier than oil) accumulates at the tank's bottom and can reach the oil pumps and bearings. Often, the water is forcedly removed with the help of special centrifuges, but this technology entails huge oil losses. By contrast, FRFs possess technological properties, including good deaerating and demulsifying abilities and antioxidant stability in the presence of water, which avoid these problems without additives. Since FRFs are heavier than water, the latter comes off by evaporating from the surface in the tank and is withdrawn by exhausters. All these peculiarities ease maintenance of the lubrication systems, reduce the lubricant losses, and facilitate lifetime extension.

At the same time, some physical and chemical properties of FRFs to a degree differ from those of petroleum oil that should be taken into consideration while designing or modifying the turbine lubrication systems for FRF. First, FRFs are heavier than petroleum oil by approximately 30%. Correspondingly, it has a greater value of dynamic viscosity, which determines the rotor lifting and the quantity of heat given off under hydrodynamic lubrication conditions. All other things being equal, this would require greater upper and side gaps in the bearings, although in reality equal gap values are often sufficient. A greater density also reduces the speed of air removal for FRFs, making it desirable to use more efficient air separators in the lube tank. In addition, because of the greater density, the centrifugal lube pumps will produce a greater head pressure. Noteworthy is also that the specific heat capacity (related to the unit of volume) of FRFs is even higher than that for petroleum oil. It might also be well to emphasize that FRFs have much greater dissolving ability compared with petroleum oil. This refers to many sealing, insulating, and painting materials.

Finally, it is pertinent to note some electro-physical properties of FRFs that does not allow them to replace transformer oil. Coefficient of cubic resistance for FRFs is lower than that for fresh petroleum oil and close to that for water-contaminated one. Such dielectric properties could

184   *Steam Turbines for Modern Fossil-Fuel Power Plants*

present some problems for hydrogen-cooled generators with the voltage above 20 kV if FRF finds its way to the inner space. For such generators, it is preferable to have a separate small system of butt-end seals with petroleum oil. This is unnecessary if the generator voltage is less than 20 kV, and these problems do not occur at all for water- and air-cooled generators.

References

1. Avrutsky G.D., I.A. Savenkova, E.A. Don, et al. "Thermal Expansion Normalization for Large Steam Turbines in Service," *Proc. of the American Power Conference* 61, Chicago, 1999: 806-809.

2. Don E.A., G.D. Avrutsky, A.N. Mikhajlova, et al. "Improving (Restoring) the Indicators of Thermal Expansions for Steam Turbine Cylinders at Operating Conditions of Start-ups and Shut-downs" [in Russian], *Elektricheski Stantsii*, no. 2 (1999): 12-15.

3. Leyzerovich A. *Large Power Steam Turbines: Design & Operation*, in 2 vols., Tulsa (OK): PennWell Books, 1997.

4. Oeynhausen H., A. Drosdziok, W. Ulm, and H. Termuehlen. "Advanced 1000 MW Tandem-Compound Reheat Steam Turbine," presented at the American Power Conference, Chicago, 1996.

5. Oeynhausen H., A. Drosdziok, and M. Deckers. "Steam Turbines for the New Generation of Power Plants," *VGB Kraftwerkstechnik* 76, no. 12 (1996): 890-895.

6. Hoffstadt U. "Boxberg achieves world record for efficiency," *Modern Power Systems* 21, no. 10 (2001): 21-23.

7. Baumgartner R., J. Kern, and S. Whyley. "The 600 MW Advanced Ultra-Supercritical Reference Power Plant Development Program for North Rhine Westphalia—A Solid Basis for Future Coal-Fired Plants Worldwide," presented at the 2005 Electric Power Conference, Chicago, 2005.

8. "Yuhuan: a Chinese milestone," *Modern Power Systems* 25, no. 6 (2005): 27-31.

9. Wichtmann A., M. Deckers, and W. Ulm. "Ultra-Supercritical Steam Turbine Turbosets Best Efficiency Solution for Conventional Steam Power Plants," *VGB PowerTech* 85, no. 11 (2005): 44-49.

10. Ando K., E. Asada, K. Yamamoto, et al. "Design, Construction, and Commissioning of the Nos. 1 and 2 Units of the Ratchaburi Thermal Power Plant for the EGAT of Thailand as a Full Turn-key Contract," *Mitsubishi Heavy Industries Technical Review* 39, no. 3 (2002): 95-100.

11. Lisyanskii A.S., N.P. Egorov, M.I. Shklyarov, et al. "Experience in Increasing Reliability of Journal Bearings for Large Steam Turbines" [in Russian], *Elektricheski Stantsii*, no. 10 (2005): 41-45.

12. Lisyanskii A.S., N.P. Egorov, M.I. Shklyarov, et al. "Design Improvements for Journal Bearings for Large Steam Turbines of LMZ" [in Russian], *Elektricheski Stantsii*, no. 9 (2004): 15-19.

13. Schenk K., Höxtermann E., and Hartwig J. "Operation of Turbines with Fire-Resistant Fluids, Including the Lubrication System," *Proc. of the International Joint Power Generation Conference*, New York: ASME, 1998, vol. 2: 799-806.

14. Cooper T. and K. Ulmann. "Fire Loss History and Advances in Fire Protection for U.S. Based Electric Utilities," *VGB PowerTech* 80, no. 8 (2000): 19-21.

15. Fragin M., A. Leyzerovich, and M. Shapiro, "Fire Resistant Fluids Seek to Expand Application into Turbine Lubrication Systems," *Power Engineering*, #105, no. 11

(2001) 106-112.

16. Cooper T. "Avoiding lube oil fires," *Turbomachinery International* 46, no. 1 (2005): 24-26.

17. Leyzerovich A. "Don't Let the Flame Blaze up Instead of Fighting It Thereafter..." *Energy-Tech*, no. 2 (2001): 21-24.

18. Vilyanskaya G.D., V.V. Lysko, and W.D. Phillips. "Recent Operating Experience in Europe and the Soviet Union with Fire-Resistant Turbine Lubrication," *Proc. of the American Power Conference* 52, Chicago, 1990: 704-708.

19. Vilyanskaya G.D., V.V. Lysko, M.S. Fragin, and A.G. Vainshtein. "VTI Fire-Resitant Turbine Oils and the Part Played by Them in Increasing Fire Protection at Thermal and Nuclear Power Stations," *Thermal Engineering* 38, no. 7 (1991): 378-381.

20. Günther R. "Operating Experience with Synthetic Fluids in the Control and Governing Systems of Steam Turbines," *VGB PowerTech* 77, no. 12 (1997): 930-934.

12. Phillips W.D. "The Use of a Fire-Resistant Turbine Lubricant: Europe Looks to the Future," *Proc. of the ASTM Conference Turbine Lubrication in the 21st Century*, West Conshohocken, PA, 2001: 1-17.

22. "Great Lakes Fire-Resistant Lubricant for Gas Turbines in as New Condition after 100,000 Hours of Operation," *VGB PowerTech* 83, no. 1/2 (2003): 25-26.

23. Ogurtsov A.P. and V.K. Ryzhkov, editors. *Steam Turbines of Supercritical Pressure of LMZ* [in Russian], Moscow: Energoatomizdat, 1991.

Part II

Steam Turbine Transients and Cycling Operation

Chapter 7

Operating Conditions and Start-up Systems for Steam-Turbine-Based Power Units

TYPOLOGY OF OPERATING CONDITIONS

A totality of steam turbine operating conditions and possible transitions between them can be presented in a graphic form, as shown in Figure 7-1. Such a formal presentation occurred useful to develop knowledge bases for the transients of power steam turbines as applied to development of their systems for automated control and surveillance based on expert systems and elements of artificial intelligence.[1,2] In particular, such automated systems have to distinguish and identify the current operating conditions of the turbine, and this leads to a necessity of their formal definition. Distinguishing the operating conditions is also necessary for arranging on-line informative support for the operational personnel, automatic off-line (post-operative) analysis of the turbine transients, and executing some other informative functions. At the same time, some formalization seems to be very helpful for better comprehension of the turbine's operating conditions, better insight into their essence.

First of all, it is necessary to subdivide the whole totality of operating conditions into stationary (or steady-state) and transient (or unsteady-state). Strictly speaking, none actual operating conditions are absolutely stationary. In reality, they are almost always accompanied with some fluctuations caused by variations of outside conditions (for example, quality of fuel consumed by the boiler), as well as operation of automatic governing systems. This phenomenon is often neglected, but it deserves the very serious attention. In particular, stochastic fluctuations of steam temperatures during so-called stationary operating conditions

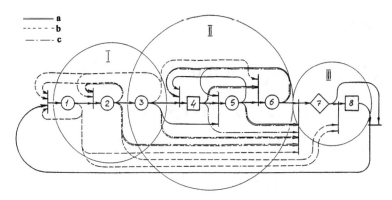

Figure 7-1. Typology graph of a steam turbine's operating conditions and their possible transitions
1 - pre-start heating up of the main steam-lines and HP stop-valve steam chests, 2 - running up to the synchronous rotation speed (including possible additional heating up of the hot reheat steam-lines at an intermediate rotation speed without supplying the IP and LP sections with steam), 3 - start-up loading with simultaneous raising the main and reheat steam temperatures and main steam pressure up to the rated values, 4 - turbine operation under stationary load, 5 and 6 - load changes within and beyond the governed range, 7 - cooling down of the stopped turbine, 8 - cold state of the turbine.
Types of transitions: a – "regular", b - "preventive", c - "accidental"

should be taken into account while assessing the turbine's long-term strength and fatigue. The values of these fluctuations and their dispersion mainly depend on the quality and current tuning of the unit's automatic governing systems and can vary over a wide range for different power units and different time periods of their operation.

It would also be wrong to consider the operating conditions stationary as soon as the turbine load and steam conditions stop varying. In reality, there exists a certain time period necessary to stabilization of the turbine's temperature state; this usually takes tens minutes to settle at the new level. This should be taken into consideration, for example, while setting the limits for load changes within the governed range, as well as arranging heat-rate performance tests.[3]

In the considered graph of Figure 7-1, three large circles embrace three major groups of operating conditions:

- start-up operations,
- subsequent turbine operation under load, including shut-down,

Operating Conditions and Start-up Systems 191

and, finally,

- outage.

Two kinds of operating conditions marked by squares can be called stationary. They are: operation under stable load and invariable (or supposed to be invariable) steam conditions and cold stable state of the stopped turbine. A diamond denotes cooling down of the stopped turbine on the stage with practically invariable external conditions for the turbine but with its unsteady temperature state. And, finally, little circles denote stages of the real transients, such as the turbine's start-up operations, load changes within and beyond the governed range, and different kinds of shut-downs.

Transitions between the different operating conditions, in turn, can be classified as:
a) "regular," corresponding to common technological sequences and marked by solid lines in Figure 7-1,
b) "preventive," executed by the operator or automated control devices in the case of potentially dangerous deviations from normal processes and drown by dotted lines, and
c) "accidental," executed by automatic defense or emergency control devices—touch-dotted lines. (It might be well to note that the employed coins for the transition types are not generally accepted, and they are used here purely conventionally.)

For some purposes, it can be also necessary to subdivide the turbine transients into the scheduled and forced ones. The first group includes start-ups from different initial temperature states, load changes according to the dispatcher's schedule of power generation in the power system, and scheduled shut-downs for stand-by or overhauls and inspections. These transients are intentional, foreseen, and fulfilled under the manual or automated control with a direct participation of the operational personnel. By contrast, the forced or occasional transients cannot be foreseen and scheduled. They are mostly carried out under control of automatic (not automated) governing systems and defense devices. The forced transients are caused by unexpected or occasional changes in the operating conditions of the power system or sudden changes in the inherent state of the power unit's equipment. This group also comprises load changes under action of automatic systems of governing the load and frequency, as well as various automatic load discharges under action of emergency control devices of the power system and unplanned automatic shut-downs or unloading under action of the unit's defense system if any of the power unit's equipment fails to operate and cannot be substituted. Hence the term of scheduled transients partially covers the regular ones, whereas all the rest transients could be considered forced.

At the same time, it is not always possible to draw a definite boundary line between the scheduled and forced transients, for example, in reference to shut-downs. So, if the unit's equipment is suddenly damaged but this damage does not demand on urgent withdrawal of this equipment from service and does not induce the action of the defense system, the unit can be unloaded and shut down by the operational personnel with some postponement, and then this shut-down is formally counted as the scheduled one, even though matter-of-factly it is forced. Similar situations arise, for example, in cases of enhanced vibration of the turbine shaft or bearings: these events can lead to the forced automatic shut down under action of the defense devices. At other units, with other principles of constructing their defense systems, the enhanced vibration will lead only to an alarm signal with a subsequent "foreseen" shut down by the operator.

As applied to the purposes of automated control and surveillance, with all the advantages of graphic forms, it can be more convenient to use matrix (table) forms. The same transitions between the turbine's operating conditions as shown in Figure 7-1 can be presented in the form of a matrix like Table 7-1. In a similar manner, the operating conditions can be distinguished and identified with the use of a sub-algorithm constructed in a matrix form of Table 7-2. In doing so, it is easier to provide and check up the completeness and consistency of the used conventions compared with a graphic form of logic block diagrams.

In the matrix of Table 7-2, each operating condition is notable for an AND-combination of the presence or absence (denoted by the signs of "+" and "–," respectively) of the marked specific conventions. In a few cases, it occurs necessary to use additionally information about the type of the preceding operating conditions to identify the current ones. In Table 7-2, there are also some symptoms pointed out by the signs "+" and "–" in brackets; they are also characteristic for the considered operating conditions but are not necessary for their identification. Checking out if these conventions are satisfied can be used for verifying the trustwor-

Operating Conditions and Start-up Systems

Table 7-1. Matrix of transitions for steam turbine's operating conditions

	Inputs							
Outputs	1	2	3	4	5	6	7	8
1	b	a	—	—	—	—	b	b
2	b	b, c	a	—	—	—	b	b
3	b	b	—	a	a	b	b, c	—
4	—	—	—	—	a	a, c	c	—
5	—	—	—	a	—	a, c	a, c	—
6	—	—	—	a, c	a	—	a, c	—
7	a	—	—	—	—	—	—	a
8	a	—	—	—	—	—	—	—

Note: designations of operating condition numbers (1-8) and transition types (a-c) are the same as in Figure 7-1.

Table 7-2. Matrix of identifying steam turbine's operating conditions

Operating conditions	$t_{HP}>150°C$ $t_{IP}>150°C$	$B_f >$ $B_{f.min}$	HP SV closed	$n>50$ rpm	FG $=1$	$N>$ $N_{gov.min}$	$t_0>500°C$ $t_{rh}>500°C$	$N'>$ N'_{min}	$\Delta t_{HR}<$ $[\Delta t_{HR}]_{gov}$	Preceding operating conditions
1	...	+	...	−	−	(−)	...	(−)	(+)	...
2	...	+	−	+	−	(−)	...	(−)	...	...
3	(+)	(+)	(−)	(+)	+	...	−	...	...	2
4	(+)	(+)	(−)	(+)	+	...	+	−	+	(3, 5, 6)
5	(+)	(+)	(−)	(+)	+	+	...	+	...	(3, 4, 6)
6	(+)	(+)	(−)	(+)	+	−	...	+	...	(3, 4, 5)
7	+	−	+	...	(−)	(−)	...	(−)	(+)	...
8	−	−	+	(−)	(−)	(−)	(−)	(−)	(+)	...

Notes: designations of operating condition numbers (1-8) are the same as in Figure 7-1; t_{HP} and t_{IP} are characteristic metal temperatures of the HP and IP sections' outer casings, respectively; B_f and $B_{f.min}$ are the fuel consumption and its minimal significant value; HP SV—the state of the HP stop valves; n is the rotation speed; FG is the sign that the generator is connected to the grid; N and $N_{gov.min}$ are the turbine load and its lower limit for the governed range; t_0 and t_{rh} are the main and reheat steam temperatures; n' and n'_{min} are the rate of load changes and its minimal; Δt_{HR} and $[\Delta t_{HR}]_{gov}$ are the effective radial temperature difference for the HP rotor and its admissible value when the turbine is operating in the governed range; "+" means the convention is satisfied, "−" means the convention is not satisfied, "..." means it does not matter if the convention is satisfied or not.

thiness of the used information.

The contents of graph Figure 7-1 and matrices Tables 7-1 and 7-2 are to a great degree tied with some specific types of power units with their start-up systems and operating technologies. As applied to other objects, these constructions can require some corrections. In addition, the list of the transient operating conditions described by the considered graph and matrices can be expanded with including some additional operations as, for example, forced cooling-down of the turbine under load and/or when the turbine is stopped, rotating the turbine by the generator as a motor during stand-by outages, and so on.

The most intricate kind of the turbine transients is start-ups. Their technological operation sequence should be subdivided into a few stages, essentially differing from one another in their contents and conditions of carrying out. The main stages of start-ups are:

- *pre-start warming up* of the turbine steam-lines, crossover pipes, valve steam-chests, and even turbine cylinders while the turbine is rotated by the turning gear (if steam is allowed to come to the turbine cylinders, the "pre-start" warming is accompanied with a rotation speed increase up to some intermediate value, but this increase remains practically uncontrollable),

- *rolling up* of the turbine with a controllable raise of the turbine's rotation speed up to the synchronous value and synchronization of the generator, and, finally,

- *loading up* the turbine to the set stable load with raising simultaneously the main and reheat steam parameters, whereas all the preceding operations are mainly run (or could/should run) with invariable start steam conditions.

This division of start-up operations into the stages can also be made more detailed, depending on the pursued aims, or even can vary, depending on the start-up technology, which, in turn, mainly depends on the unit's steam-generator type, start-up system, and the turbine's design features.

Each start-up (excepting the very first one after the turbine is assembled) is preceded by the turbine shut-down and outage, when the turbine cools down. Depending on the outage duration, start-ups are commonly subdivided into three categories named *cold*, *warm*, and *hot*

Operating Conditions and Start-up Systems 195

start-ups. As a rough guide, the start-up is classified as cold if the time period after the previous shut-down exceeds approximately three days or the maximum pre-start metal temperature of the high-temperature turbine elements (primarily, the HP and IP outer casings) does not exceed 120-160°C (250-320°F). By contrast, the start-up is considered hot if it is preceded by the outage of less than eight-to-ten-hour duration. All the intermediate start-ups are known as warm.

In a narrower meaning, the terms "hot" and "warm" often refer to start-ups after shut-downs for "nights" and "weekends," that is, after 5-8-hour and 50-54-hour outages, respectively. These time periods are most typical when power units participate in covering deep variations of power consumption and are shut down for relatively long load ebb periods.

All the three main start-up types considerably differ in the contents and duration of their technological operations. In particular, hot start-ups run without pre-start warming of steam-lines, valve steam-chests, and crossover pipes. Sometimes, it may be expedient to separate the fourth start-up type in addition to three above mentioned ones, namely *start-ups from the hot stand-by state*. They are carried out after the outage for less than two-three hours, when both the steam parameters after the steam generator and metal temperatures of the turbine have not had time to decrease noticeably.

As a matter of fact, the start-up technology is determined not by duration of the previous outage, but the pre-start temperature state of major turbine elements and steam-lines. These temperature states to a great degree depend on the technology of shutting down the turbine and quality of thermal insulation of individual turbine design elements and steam-lines. For this reason, the boundaries between the start-up types should be defined taking into account the differences in their scenario and technology, caused by a possibility to keep the main and reheat steam temperatures at the early start-up stages matching the temperature state of the most thermally stressed ("critical") turbine elements. By way of illustration, this can be considered as applied to the main steam and HP metal temperatures.

When the turbine is started up, in order to prevent excessive thermal stresses in the HP section's critical components (primarily, the HP rotor), the initial (start) main steam temperature should match their metal temperature, exceeding it by the set value of Δt_0^{in}. This value substantially depends on the turbine's design features, as well as the unit's main steam parameters, start-up system, and start-up technology. By convention, this value can be taken equal to, for example, 100°C (180°F) like it is for many supercritical-pressure steam-turbine units.[4] In addition, to be sure that the main steam temperature is well controllable and to prevent possible water induction into the turbine, the main steam temperature should exceed the saturation temperature at the initial main steam pressure by not less than 40-50°C (70-90°F).

If the value of the initial temperature difference between main steam and the HP cylinder metal is taken equal to $\Delta t_0^{in} = 100°C$, this means that, for example, in order to start up a cold turbine with the HP metal temperature of 80°C (175°F), the main steam temperature should not be higher than 180°C (355°F). However, to observe this limit and assure the required excess over the saturation temperature, the initial main steam pressure should be less than 0.35 MPa (50 psi). This pressure is evidently insufficient to raise the turbine's rotation speed to the synchronous value. In addition, as a matter of fact, modern large power boilers with their attemperators can provide a stable steam flow amount to run up the turbine to the synchronous speed with the main steam temperature not less than 280-300°C (535-570°F). So the set convention of exceeding the initial main steam temperature by $\Delta t_0^{in} = 100°C$ over the pre-start HP metal temperature can be satisfied only if the latter is more than 180-200°C (355-390°F). On the other hand, the initial main steam temperature can not and must not be higher than the rated (nominal) value. Moreover, when the boiler is operated with a respectively small initial steam flow amount, the main steam temperature does not reach the rated level, but amounts to, say, 500°C (930°F), with the rated main steam temperature of 540°C (1,004°F). In this case, the required excess of the main steam temperature over the HP metal temperature can be satisfied only if the latter is lower than 400°C (750°F). Thus, if the accessible initial main steam temperature values range between 280 and 500°C (535-930°F), the set requirement can be satisfied only if the HP cylinder's pre-start metal temperature lies in the range between 180 and 400°C (355 and 750°F). These figures are to be taken as the upper boundary of cold start-ups and lower boundary of hot start-ups, whereas all the intermediate values refer to warm start-ups, as shown in Figure 7-2.

The described approach is thought to be more substantiated in terms of start-up technology. In the case of cold start-ups, when the steam temperature is too high to match the turbine metal temperature, the turbine should be held for some time before raising its rotation speed to the synchronous value and switching on the generator to the

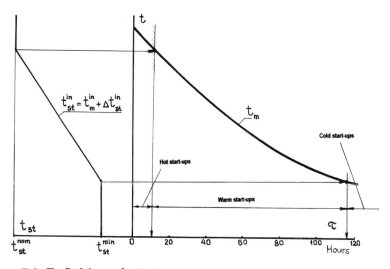

Figure 7-2. Definition of start-up types.
t_m – characteristic metal temperature of the turbine, t_m^{in} – its pre-start (initial) value after cooling for τ hours, t_{st} – steam temperature at the boiler's exit, t_{st}^{nom} and t_{st}^{min} - its nominal (rated) and minimal controllable values, t_{st}^{in} – its set initial (pre-start) value.

grid, that is, while the heat transfer conditions from steam to the turbine metal are relatively low. In addition, this prevents the turbine from an uncontrolled rise of the RREs caused by different heat transfer conditions for rotors and stator components. It is also necessary to heat the rotors (especially, the IP and LP ones) at the intermediate rotation speed over the fracture-appearance transition temperature (FATT) level to prevent a danger of their brittle fracture under combined action of tensile centrifugal and thermal stresses in the rotor depth.

By contrast, at hot start-ups, when the boiler cannot provide the steam temperature high enough to match the initial metal temperature state, the turbine should be run up as fast as possible to pass rapidly the start-up stages with significant steam temperature drops and avoid undesirable chilling of the turbine with appearance of high tensile thermal stresses on the "heated" surfaces.

As for intermediate, warm start-ups, when it is possible to maintain the steam temperature matching the turbine metal temperature, the start-up operations can run in the way most convenient to control them. In particular, the unified running-up pattern for warm start-ups commonly includes the step at the intermediate rotation speed for listening to the turbine. This step can also be used, if necessary, to heat additionally the turbine steam-lines. The presence or absence of this necessity can be used for more detailed subdivision of start-ups.

When the generator is switched on to the grid and accepts the initial load, the turbine metal temperature can be supposed matching the steam temperature provided by the boiler. This is caused by warming the turbine in the case of cold start-up or decreasing the main steam temperature drop with the steam flow amount increase. In the process of subsequent loading, the main and reheat steam temperatures are raised to the rated values with the set rates providing the admissible thermal stresses in the "critical" turbine elements, primarily the high temperature (HP and IP) rotors. The time duration necessary to raise the steam temperatures to the levels acceptable for the rated load determines the length time of loading the turbine. This time is maximal for cold start-ups and minimal for hot start-ups, when the rate of loading is limited not by unsteady-state thermal stresses in the turbine but only the operational capabilities of the personnel.

To decrease the time of loading without excessive thermal stresses in the turbine's critical elements, it is advisable to reach the turbine's rated (or set final) load at somewhat decreased final steam temperature, $t_{st}^{fin} < t_{st}^{nom}$. What is more, even at hot start-ups, when the steam temperatures after the boiler are tried to kept at the nominal (or maximally close to nominal) level from the very start-up beginning, for power units operated in a so-called "cycling," or two-shift (daily start/stop) manner with frequent hot start-ups after shut-downs for nights, it is considered reasonable to lower the steam temperatures in the process of loading to a decreased value of t_{st}^{fin}; then the steam temperatures are to be raised back to the nominal level—Figure 7-3. This makes it possible to decrease to a certain degree the maximum unsteady thermal stresses in the turbine critical elements as applied to this start-up type.

In specific cases, depending on the steam-temperature control range for the power unit's boiler with regard to its static characteristics and possibilities of its attemperators, the metal temperature range referring to warm start-ups can occur remarkably narrower. As an example, the controllable ranges for main and reheat steam temperatures, quite characteristic for some old subcritical-pressure power units, are depicted in Figure 7-4. In this case, the controllable range of the main steam temperature at the minimum steam productivity is as wide as only 140°C (250°F): between 340 and 480°C (650-900°F). Correspondingly, this nar-

Operating Conditions and Start-up Systems

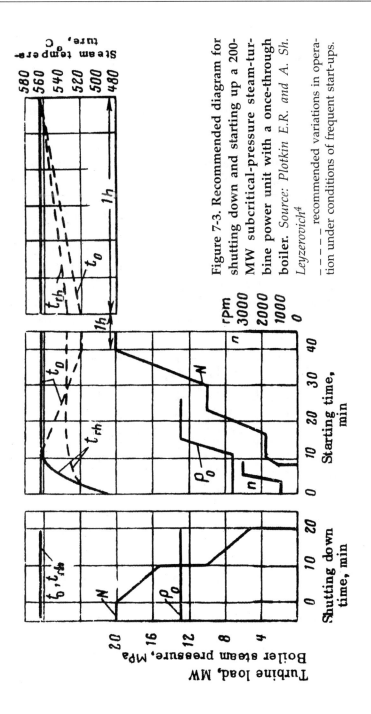

Figure 7-3. Recommended diagram for shutting down and starting up a 200-MW subcritical-pressure steam-turbine power unit with a once-through boiler. Source: Plotkin E.R. and A. Sh. Leyzerovich[4]
– – – – recommended variations in operation under conditions of frequent start-ups.

rows the actual range of warm start-ups.

It should be noted that for power units with two-bypass start-up systems the boundary between hot and warm start-ups becomes rather conventional. This is associated with the fact that turbines of such units are commonly run up to the synchronous rotation speed with the closed HP control valves (see below). Main steam is given to the HP section only when the generator is switched on to the grid and the steam flow amount through the turbine, needed to accept the initial load, does not cause substantial steam temperature drops in the HP control valves and does not result in chilling the HP rotor. In this sense, hot start-ups for these turbines are properly absent.

Once the turbine is started up, it operates under load until it is shut down. In the entire range of load variations from zero to the rated (nominal) or maximum continuous rating (MCR), it is important to distinguish a so-called *governed range*. By definition, it implies the load change strip within which the cast of the power unit's auxiliaries remains invariable and it is not necessary to change the fuel or do any switching over in technological systems. Most of scheduled turbine load changes take

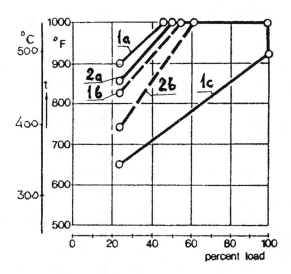

Figure 7-4. Characteristic controllable ranges of main (1) and reheat (2) steam temperatures under sliding (a) and constant (b) main steam pressure for a subcritical-pressure steam-turbine power unit with the rated main and reheat steam temperatures of 540 °C (1,000 °F)
1c: lower border of main steam temperature's controllable range at start-ups

Operating Conditions and Start-up Systems 201

place within this range.

Along with the lower boundary of the governed range, the unit's operating conditions are also characterized by the value of a so-called *minimum stable load*. Transition to this load and return into the governed range imperatively require some switching over in the unit's technological system. So, these transitions need some additional technological operations and go slower than the load changes within the governed range. Both these values, namely the lower boundary of the governed range and minimum stable load, essentially depend on the used fuel, the steam-generator type, and its features.

Together with such characteristics as the load change rate within the diverse load change ranges, start-up durations for diverse start-up types, the mentioned values determine the *flexibility characteristics* for the power unit. These characteristics mainly depend on both the turbine design and the features of the applied start-up system, but they also to a great degree depend on the qualification, skills, and experience of the operational personnel and some other non-material factors.

Start-ups, as well as scheduled and unscheduled load changes within the governed range and beyond, are the most important transient operating conditions which to a great degree influence the reliability and flexibility performances of steam-turbine power units. Modern concepts of them are to be considered in next sections of this part. As to other transients, such as deep load discharges and impulse unloading of the turbine, shut-downs for stand-by and shut-downs with forced cooling of the turbine for inspections and repairs, as well as forced cooling of the stopped turbine with air or early switching off the lube oil system and turning gear at the stopped turbine, for the last decade there have not appear any serious new notions related to them, and taking into consideration that they were sufficiently detailed in the author's previous book,[3] it seems irrelevant to return again to these subjects.

SOME FEATURES OF START-UP SYSTEMS AS APPLIED TO STEAM-TURBINE AND CC POWER UNITS

When power units are started up, their main and reheat steam temperatures have to match the HP and IP cylinder metal temperatures. In addition, the main and reheat steam-lines should be warmed before steam enters the turbine to prevent water induction into the turbine and inadmissible (or, at least, undesirable) temperature differences and thermal stresses in the turbine. For the entire time period until steam is given into the turbine and steam flow amount through the turbine becomes equal to the boiler's current steam productivity, steam after the boiler should be discharged beside the turbine through its bypasses. Controlled steam flow amounts through the boiler's superheater and, if possible, reheater surfaces make the initial steam temperatures more controllable. For units with drum-type boilers, the turbine bypasses also allow controlling the steam pressure in the drum so that the saturation temperature in the drum rises with the permissible rate. If the boiler's reheater is located in a high-temperature flue-gas zone, the bypass system should protect its tubes when the steam amount through the turbine is zero or too small to cool these tubes. In particular, arrangement of due steam flow amounts through the boiler's superheater and reheater tubes by means of the turbine bypasses at start-ups to a great degree diminishes a danger of tube scale defoliation and solid particle erosion of the turbine.

The turbine bypasses also come into operation at the turbine's emergency trips or sudden load discharges. Once the turbine valves are closed finally or temporarily, the steam pressure upstream from the valves begins to rise, and it is desirable to prevent opening the relief (safety) valves on the main and reheat steam-lines. If the turbine bypasses have a sufficiently high flow capacity and are sufficiently fast-acting, they can act like the relief valves and instead of them. Sometimes, it is said that the turbine bypasses should provide independent operation of the boiler and turbine, but such a statement seems to sound too exaggerated because thermal energy of steam discharged through the bypasses are completely lost, and it is desirable to avoid the excessive use of bypasses as far as possible.

At the early start-up stages, when steam is discharged beside the turbine, it is advisable to keep the boiler with the minimum feedwater flow amount to decrease the start-up fuel outlay. From the same reasoning, it is also desirable to utilize heat of steam brought out through the bypasses. Most commonly this is done in the turbine regenerative system.

Besides the initial matching of the steam and metal temperatures, it is not less important to control raising the main and reheat steam temperatures while the turbine is loaded. It is also substantial to hold the raising of the main steam pressure during the loading stage at the reasonable level.

Operating Conditions and Start-up Systems 203

All these functions are to be fulfilled by start-up systems of power units. In addition, these systems provide the unit's auxiliaries with steam and heat until the regular steam sources from the turbine extractions are ready.

The most characteristic part of the start-up system is the turbine bypasses. Depending on their arrangement, so-called *two-bypass* and *one-bypass* start-up systems are recognized, even though within each category these systems can considerably differ in their ideology, degree of sophistication, and numerous details.

Two-bypass start-up systems are mostly widespread in the power industry of some West European countries (primarily, Germany and France) and Japan. A skeleton chart of a characteristic two-bypass start-up system for power units with once-through boilers is presented in Figure 7-5, and the flow chart for a typical German fossil-fuel power unit with a once-through boiler is shown in Figure 7-6.

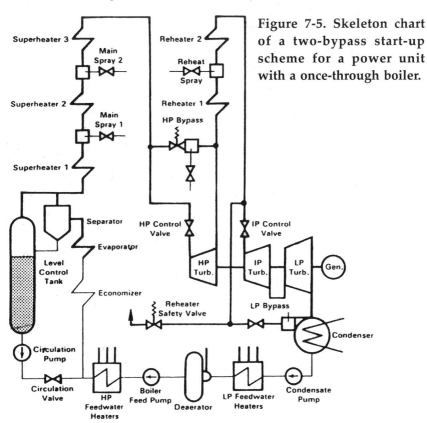

Figure 7-5. Skeleton chart of a two-bypass start-up scheme for a power unit with a once-through boiler.

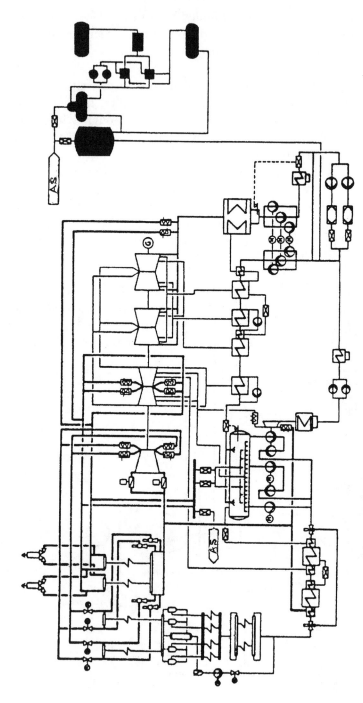

Figure 7-6. Flowchart for the 750-MW subcritical-pressure German power unit Bexbach.
Source: H. Spliethoff and G. Abröll[5]
AS – steam to the power unit's auxiliaries and for pre-start heating water added to the circuit.

Operating Conditions and Start-up Systems 205

The turbines in these units are bypassed in two steps: the first step is the HP bypass from the main steam-lines into the cold reheat steam-lines, and the second one is the LP bypass from the hot reheat steam-lines into the condenser. As a result, the steam flow amounts through both the boiler's superheater and reheater are almost the same, no matter how much the steam flow through the turbine is. This protects the reheater against overheating at the early start-up stages and load discharges. In addition, this provides favorable conditions to control the reheat steam temperature with the use of spray attemperators in the entire range of the operating conditions. This also makes it possible to warm effectively the hot reheat steam-lines before steam is given into the turbine. More often than not, the bypass flow capacities are set equal to 100% of the unit's maximum continuous rating (MCR) to provide full load discharges of the turbine without opening the relief valves. If the flow capacity of the bypasses is less than 70% MCR, the unit's start-up duration increases remarkably. The bypasses for large power units can have several strings depending on the bypass valve's accessible flow capacity. This especially concerns the LP bypasses because of a large volumetric flow amount under the reheat-steam pressure. Because the bypass valves are used not only at start-ups, but also at load discharges, they have to be quite fast-acting.

Two-bypass start-up systems can be fraught with a danger of overheating the HP exhaust of the rotating turbine when steam is discharged through the HP bypass. If the LP bypass is not sufficiently fast-operating or its flow capacity does not suffice, the steam pressure at the HP exhaust increases, and this results in heightened heat losses with fan and friction of the rotating blades in dense steam—that is, overheating because of windage. In addition, the HP bypass and its water spray sometimes become a source of water induction into the HP cylinder from its exhaust; such accidents have repeatedly taken place at different power plants. To prevent these events, the turbines are furnished with check valves at the cold reheat steam-lines (see Figure 7-6) In addition, these check valves are supplied with own bypasses to provide a possibility of controllable warming of the HP cylinder through the exhaust at cold start-ups.

One-bypass start-up systems are simpler in control compared to two-bypass systems and do not need fast-acting LP bypass valves of a large volumetric flow capacity. These are the two main reasons why such systems have received a wide acceptance at power plants of the former Soviet Union and many other countries. A simplified steam/water flowchart for a proposed Russian supercritical-pressure 525-MW power unit with a one-bypass start-up system was shown in Figure 3-17, and a steam/water flowchart for standard Soviet 300-MW supercritical-pressure units is presented in Figure 7-7. The latter is of special interest because many gas/oil-fired supercritical-pressure power units of this type, with such a start-up system, have actively participated in covering the variable part of week and day power consumption with regular shutdowns for weekend and night load ebbs with subsequent start-ups.

For units with one-bypass start-up systems, the turbine is bypassed as a whole, but there are supplementary steam-lines with fast-opening valves from the hot reheat steam-lines upstream of the IP intercept valves to the condenser. As a matter of fact, during sudden emergency turbine trips, the steam flow amount through these lines is determined by the reheater system's steam capacity, and it is much smaller than the steam flow amount through the LP bypasses for two-bypass systems.

For units with one-bypass start-up systems, until steam enters the HP cylinder the boiler's reheater is not cooled by the steam flow. That is the reason why the boilers intended for such units are designed with the reheater located in a relatively low flue-gas temperature zone.

Because the "own" steam does not reach the reheat steam-lines until it goes into the turbine, there is a problem how to warm these steam-lines before steam enters the IP cylinder. Some one-bypass start-up systems comprise special sources of outside steam for this purpose. For example, at standard Soviet/Russian 800-MW units without motor-driven boiler-feed pumps (BFPs), the hot reheat steam-lines are preheated by outside steam given to the BFP's turbine drive. (Normally, these turbine drives are fed with steam from the cold reheat steam-lines, so that there is a steam-line between them.) However, for most units with one-bypass systems, the standard start-up technology implies warming the hot reheat steam-lines with steam passed through the HP cylinder, the IP valves being closed. This is done at the intermediate rotation speed, and it is necessary to increase this steam flow amount in order to make this heating more effective and hasten it.

Along with the bypasses, start-up systems also include drainage lines which are necessary to warm the steam-lines downstream from the bypasses, as well as the crossover pipes, valve steam-chests, and cylinder casings. These lines also withdraw water condensing from steam.

Another important aim of the start-up systems concerning a turbine is to control the main and reheat steam temperatures during

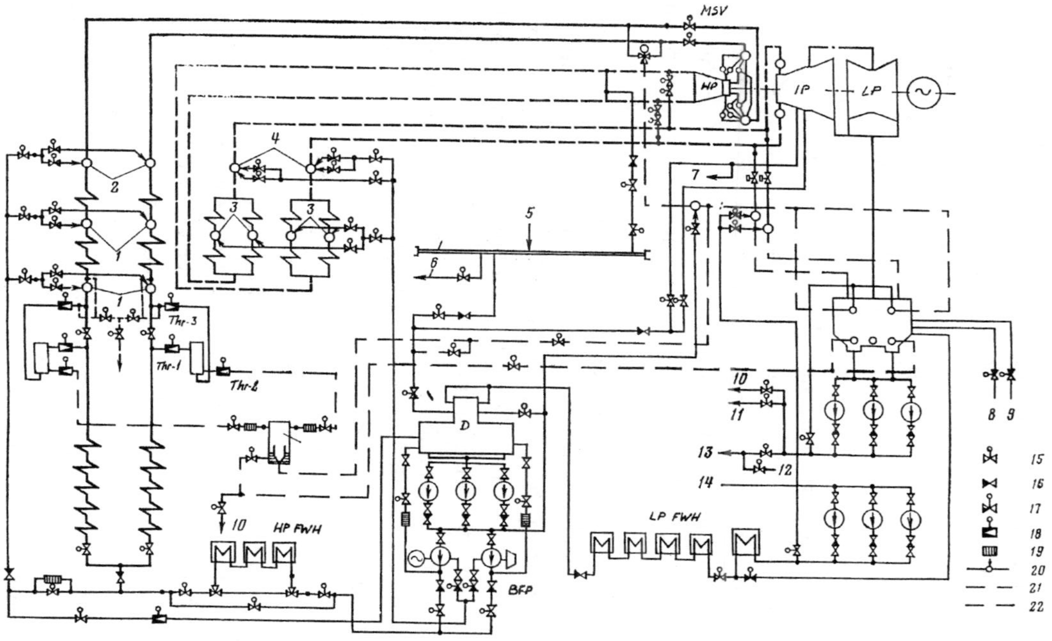

Figure 7-7. Flow-chart for a standard Soviet 300-MW supercritical-pressure power unit. *Source: After E.R. Plotkin and A.Sh. Leyzerovich[4]* 1 and 2 – regular and start-up main-steam spray attemperators, 3 and 4 – regular and emergency reheat-steam spray attemperators, 5 – steam reheater's start-up by-passes, 6 – steam for BFP's turbine-drive, 7 – steam from a plant header, 8 – steam to the turbine's end gland seals, 9 – lines of regular and emergency make-up, 10 and 11 – water discharges to the cooling-water lines and soiled-water tanks, 12 – water from the spare tanks, 13 and 14 – condensate to and from the unit's water-treatment facility, 15 – shut-off valve, 16 – back valve, 17 – governing valve, 18 – controllable throttle valve, 19 – throttling orifices, 20 – spray, 21 – charge lines, 22 – start-up lines.

start-ups. Naturally, these temperatures to a great degree depend on the boiler's current firing rate. The steam temperatures can be additionally corrected by flue-gas bypasses of the boiler's heating surfaces and some other means. However, all these measures do not exclude great steam temperature excursions during start-ups. These excursions are commonly associated with increases of the firing rate and/or steam flow amount through both the boiler's heating surfaces and the steam-lines between the boiler and turbine. The most reliable, accurate, and fast-acting measure to control the main steam temperatures is special "start-up" spray attemperators before the last section of the boiler superheater or immediately into the main steam-lines after the boiler (see Figures 3-17, 7-5, and 7-7). They supplement regular attemperators between the superheater's sections.

When the unit is started up, the spray attemperators can be used to control the reheat steam temperature as well, but they are mainly employed in two-bypass systems (see Figure 7-5). For one-bypass-system units, these attemperators can be reliably used only when the steam flow amount through the reheater is sufficiently great. Otherwise, the sprays do not provide due pulverization (atomization) of the sprayed water. This causes "cold spots" at the inner steam-line surfaces (especially near elbows) and can lead to water induction into the turbine. (Such events also take place in hot reheat steam-lines of two-bypass-system units and in main steam-lines if the sprays are misused.)

At the same time, the reheat steam temperature in one-bypass-system units is very sensitive to changes of the steam flow amount when the turbine is led to the synchronous rotation speed and accepts the initial load. This process is accompanied with a sharp increase of the reheat steam temperature after the boiler. In addition, because the steam flow amount through the hot reheat steam-lines increases, the steam-temperature drop in these steam-lines due to heat transfer to their metal reduces, and the steam temperature at the IP entrance rises rapidly. If this rise is not stopped, the radial temperature difference in the IP rotor can exceed its admissible value even by two-three times, and the spraying attemperators fail to control this process. By an example, such a process is illustrated by the start-up diagram of heating the turbine rotors in Figure 7-8. At the same time, experience shows that the rise of reheat steam temperature can be reliably controlled by the reheater's steam bypass. The steam bypasses between the cold and hot reheat steam-lines were included as an obligatory component into the standard start-up systems

Operating Conditions and Start-up Systems 209

shown in Figure 7-7, but were not used at some power plants. This just causes such processes as shown in Figure 7-8.

Many once-through boilers are furnished with full-capacity separators between the evaporating and superheating parts (see Figures 7-5 and 7-6). Such a separator functions like a drum and is not switched off at any operating conditions. In order to reduce the minimum feedwater flow amount at the early start-up stages and therefore diminish the fuel outlay, these separators are supplemented with circulating circuits to increase the flow amount through the evaporating tubes and utilize heat of water discharged from the separator.

Many other once-through boilers are also furnished with a "start-up complex" with the separator which is employed only at start-ups while the steam flow amount is relatively small (usually not more than two thirds of the MCR). In this case, as shown in Figure 7-7, the boiler's evaporating and superheating parts are separated by so-called *embedded valves* which are to be opened when the steam flow amount approaches the separator's flow capacity, and the separator is switched off. In the start-up system shown in Figure 7-7, both the water and steam flow amounts from the separator are controlled. This gives certain advantages to control the start-up process.

Subcritical-pressure power units with drum-type boilers, including those with the output of 500, 600, and 660 MW, operated at fossil-fuel power plants of the UK and actively involved in covering the variable part of week and day power consumption are equipped with a kind of two-bypass start-up systems comprising an HP bypass taken from a tee prior to the turbine's stop valves and then linking back to the main path after the HP turbine exhaust; for some power plants their start-up systems were extended by including additional LP turbine bypasses, but according to conclusion by a team of international experts this was considered rather excessive.[7]

Steam-bypass systems of combined-cycle (CC) units are used during all the operating conditions with no-load or low-load steam turbine operation, that is, start-ups and shut-downs, steam turbine trips, and temporary simple-cycle operation of the unit's gas turbine(s). In all these operating conditions, when steam production of the heat recovery steam generators (HRSGs) leaves behind the steam consumption of the steam turbine, the whole excessive steam is discharged into the steam turbine's condenser. For CC units with several gas turbines and their HRSGs and a single steam turbine, the bypass system makes it possible to start up the

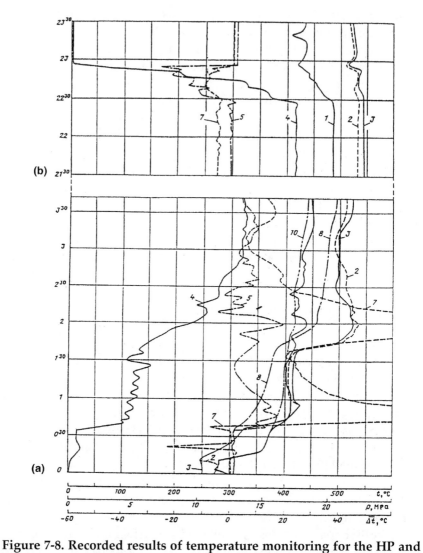

Figure 7-8. Recorded results of temperature monitoring for the HP and IP rotors of a 300-MW supercritical-pressure steam turbine of Turboatom at the Uglegorsk power plant (Ukraine) for a typical start-up after week-end (a) and shut-down (b). *Source: A.Sh. Leyzerovich et al.[6]*

Recorded variables: 1 and 4 – steam temperature and pressure at the HP control stage, 2 and 3 – steam temperature in the JP steam admission sleeves, 5 and 7 – effective radial temperature differences in the HP and IP rotors in the most thermally stressed sections, 8 and 10 – average integral metal temperature of the rotors in the same sections.

Operating Conditions and Start-up Systems 211

gas turbines subsequently. If one or more gas turbines and their HRSGs are already in operation and an additional gas turbine and HRSG should be brought on-line, the start-up by-pass system allows a gradual warming of the lagging HRSG and facilitates raising its steam temperatures up to those for the leading HRSG(s).

The place of steam bypasses in schematics of double-pressure CC units can be seen in Figures 3-24 and 3-30. The steam bypass system can also be used to operate a CC unit in a simple-cycle mode before the steam-turbine part is completed. However, this assumes that the steam turbine's condenser is available to accept steam passing through the by-pass system from the HRSG(s) if they are not plugged up.

For modern CC units, their steam bypass systems are predominantly made of two types:

- with *parallel bypasses* (also known as "direct" or "dry reheater" bypasses), which are similar to one-bypass start-up systems for "pure" steam-turbine power units, and

- with *cascade bypasses* (also known as "European" or "wet reheater" bypasses), which can be considered analogous to two-bypass start-up systems. Most of modern CC units are presently designed with cascade bypass systems.[8] Such a system for a triple-pressure reheat CC unit is shown in Figure 7-9a.

In this arrangement, the HP steam, which is generated at start-ups or other operating conditions with low steam flow through the steam turbine, is bypassed around the HP turbine section into the cold reheat (CRH) steam line. The bypass line is equipped with a pressure-reducing valve and an attemperator using spray water from the feedwater pump. The bypassed steam mixed with the steam from the IP drum(s) is sent through the reheater. The hot reheat steam line is provided with another pressure reducing/attemperating station, that is, the hot reheat (HRH) bypass that directs steam to the condenser. In addition, there can exist the LP bypass, which discharges steam from the HRSG's LP drum into the condenser. The attemperators of both the HRH and LP bypasses use spray water from the condensate pump discharge. Each bypass includes a pressure reducing device, an attemperator with a spray water control valve, and a backpressure device, also referred to as a "sparger," which is located directly within the condenser neck.[9,10]

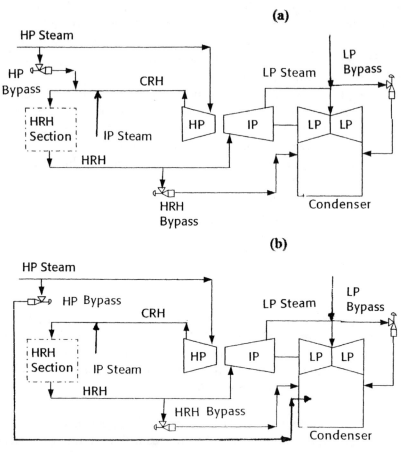

Figure 7-9. "Cascade" (a) and "parallel" (b) steam bypass systems for a triple-pressure CC unit.

As distinct from the cascade arrangement of the bypasses, in the parallel bypass system (Figure 7-9b) the steam generated in both the HP and IP drums is discharged directly to the condenser after being attemperated with spray condensate. As such, there is no steam flow through the reheater, which operates "dry" when the bypass system is in service. The steam generated in the LP drum(s) can be similarly discharged into the condenser, but instead it can be sent into the LP evaporator or deaerator. With the reheater operated in the "dry" mode, its tubes tend to overheat. This places enhanced requirements to the tube metal, adding the HRSG cost. Another drawback of the parallel bypass system is that it

Operating Conditions and Start-up Systems   213

requires the use of long steam piping from the HRSG(s) to the condenser, which also adds to the power unit's capital cost.

The main disadvantage of the cascade bypassing is that in this case the steam turbine has to be started up with a pressurized reheater, which raises concerns about overheating of the HP turbine exhaust at no-flow and low-flow operating conditions. To alleviate this problem, the turbine can be equipped with an additional blowing down line connecting the intercasing space of the HP section or cold reheat steam lines upstream of the non-return valve to the condenser. In this way, if necessary, the HP exhaust is isolated from the reheater and can be kept under a low pressure close to the backpressure in the condenser.

References

1. Leyzerovich A. "Development of Knowledge Bases for Large Steam Turbine Transients," *Proc. of the American Power Conference* 57, Chicago, 1995, Part 1: 600-605.
2. Leizerovich A. Sh. "Forming Knowledge Bases for Transients in Steam Turbines of Power Units," *Thermal Engineering* 43, no. 6 (1996): 480 – 485.
3. Leyzerovich A. *Large Power Steam Turbines: Design & Operation*, 2 vols., Tulsa (OK): PennWell Books, 1997.
4. Plotkin E.R. and A. Sh. Leyzerovich. *Start-ups of Power Unit Steam Turbines* [in Russian], Moscow: Energiya, 1980.
5. Spliethoff H. and G. Abröll. "The Coal-Fired Unit of 750-MW Output at Power Plant Bexbach" [in German], *VGB Kraftwerkstechnik* 65, no. 4 (1985): 346-362.
6. Leyzerovich A. Sh., A.D. Trukhny, V.G. Grak, et al. "Low-Cycle Thermal Fatigue of Steam Turbine Rotors at Transients" [in Russian], *Elektricheskie Stantsii*, no. 11 (1989): 61-67.
7. Starr F., J. Gostling, and A. Shibli. *Damage to Power Plant Due to Cycling*. ETD Report No. 1002-IIP-1001, London: European Technology Development, 2000.
8. Bachmann R., H. Nielser, J. Warner, and R. Kehlhofer. *Combined-Cycle Gas & Steam Turbine Power Plants*, Tulsa (OK): PennWell Books, 1999.
9. Akhtar S.Z. "Proper Steam Bypass System Design Avoids Steam Turbine Overheating," *Power Engineering* 107, no. 6 (2003): 44-52.
10. Eaton R.H., S. Clark, G. Baxter, and R. Madugula. "Design of High Energy Bypass Systems in Combined-Cycle Power Plants," *Energy-Tech*, no. 5 (2005): 17-23.

Chapter 8

Experimental and Calculation Researches of Turbine Transients

MAIN GOALS OF START-UP FIELD TESTS

According to the methodology for investigating the transients of large power steam turbines having been developed since the 1960s by E.R. Plotkin and the author, such investigations should rest on the combination of experimental researches *in situ* (so called *start-up field tests*) and calculation researches based on digital mathematical modeling of the heating-up processes for the most thermally stressed ("critical") elements of the turbine.[1,2] This methodology was first worked up and implemented in the course of investigating the transients of 200-MW subcritical-pressure power units with turbines K-200-130 of LMZ and different types of boilers. Later, it was used and further developed and improved as applied to investigating the transients of 300-MW and 800-MW supercritical-pressure turbines, wet-steam turbines with the single capacity of 220, 500, and 1,000 MW for nuclear power plants, and some other objects. It might be well to note that this methodology was being developed to a great degree keeping in mind the needs of automated control and surveillance of steam turbines at the transients.[3]

The basic distinguishing feature of this methodology lies in the fact that, instead of the search for and direct reproduction of the operating conditions which could be immediately advised for operation, the main purpose of start-up field tests becomes rather the acquisition of information needed for the mathematical modeling and calculated optimization of the turbine transients. In doing so, the field tests do not depreciate themselves. They provide researchers with necessary data used as the boundary conditions for mathematical models, and their results remain the principal criterion whether the developed and applied speculative schemas and mathematical models are valid, trustworthy, and can be applied to the considered object. In addition, the experimental measurements at the actual turbines in the course of actual transients often reveal new effects and phenomena that could not be foreseen from the previous experience. Finally, any new technological notions and approaches, intended to improve the transients, as well as their improved technological scenarios, can be put into effect only through the start-up field tests and subsequent operational practice.

The field tests should be preceded by analyzing the turbine's design, the unit's start-up system, expected operating conditions, operation and maintenance experience for turbines of similar types, and so on. Then, the critical design elements of the researched turbine are selected. The temperature and thermal-stress states of these elements should potentially limit the rate of the turbine transients. For these elements, it is necessary to find out the most representative indices of their temperature and thermal-stress states. It is just those objects that are to be modeled in calculations and investigated in experiments. In particular, such an analysis is also necessary to choose the scope of special, mainly temperature, measurements for the field tests, as well as the regular ones for continuous temperature monitoring of the turbine in operation.

If possible, field start-up tests include direct experimental measurements of stresses/strains in some turbine design elements, axial and radial clearances in the turbine's steam path, and some other variables facilitating better comprehension of the transients and possibilities of their improvement.[1,2]

Speaking of information needed for the mathematical modeling and calculated optimization of the transients, we primarily mean the boundary conditions for heating the critical design elements of the turbine, that is, changes of the heating steam temperature and heat transfer conditions for these elements depending on the current operating conditions of the turbine. Sometimes, intricate geometric shapes of the turbine elements and/or the absence of reliable data on the heat transfer conditions for them give no way for calculating temperature fields of the considered elements with all the details necessary to estimate the unsteady-state thermal stresses arising in these elements at the transients. In these cases, the experimental metal temperature measurements should provide a possibility of shaping these temperature fields, and start-up field tests should materialize this possibility. With accumulating results of start-up field tests for steam turbines of different types and generalization of these data, they can be extended to other turbine types, making it pos-

Experimental Researches of Turbine Transients 217

sible to reduce the scope of experimental researches for steam turbines of new types and, ultimately, even reduce them only to verification of new operation technologies, completely carrying over all the necessary investigations of the turbine's temperature and thermally-stressed states to the stage of calculation researches.

HEAT TRANSFER BOUNDARY CONDITIONS FOR MAJOR TURBINE DESIGN ELEMENTS

For the HP valve steam-chests, as potential critical elements, their heating steam temperatures are commonly reckoned amongst the regular turbine temperature measurements. As a rule, these temperatures are measured in the stop-valve steam-chests. It is understandable that these temperatures primarily depend on the controlled main steam temperatures at the boiler exit and are commonly close to them. However, at the transients, especially start-ups, the considered steam temperatures can significantly differ from the steam temperatures downstream of the boiler because of the heat transfer to metal of the main steam-lines. Dynamics of these processes can be mathematically modeled with confidence. (This subject is to be considered below.) If the HP stop and control valves are settled separately (not in blocks or combined units), that is, in individual steam-chests connected to one another with crossover pipes, the steam temperatures in these steam-chests can also differ at the transients because of heat losses in the crossover pipes, and this effect should be under consideration, too.

For the HP rotors, or the HP parts of the integrated HP-IP cylinder's rotors, the most dangerous unsteady thermal stresses, with regard to stress concentrators on the rotor surface, commonly arise in the vicinity of the first stage (the control stage for turbines with nozzle-group control) and at the inlet of the front gland seal (or the central, intermediate, seal for the integrated HP-IP and loop-flow HP cylinders). For turbines with throttle steam admission control, the steam temperature in this zone can be calculated with a satisfactory accuracy relying on the measured main steam parameters in the HP valve steam-chests and steam flow amount through the turbine, which can be estimated via the turbine load or steam pressure in the HP cylinder. By way of example, Figure 8-1 presents a nomogram generalizing results of such calculations as applied to the 800-MW supercritical-pressure steam turbine of LMZ (K-800-240-3) as applied to its operation mode with throttle steam control.

Such results are obtained by mean of constructing the steam expansion process in the Mollier diagram (see Figure 2-1); there exist (or can be developed) special computer programs for such calculations. However, these calculations take into account only the static steam temperature drops in the control valves and the first stage, whereas these temperature drops can be supplemented by very significant dynamic effects because of the heat transfer to metal of the valve steam-chests and crossover pipes between the HP valves and cylinder. In addition, the accuracy of

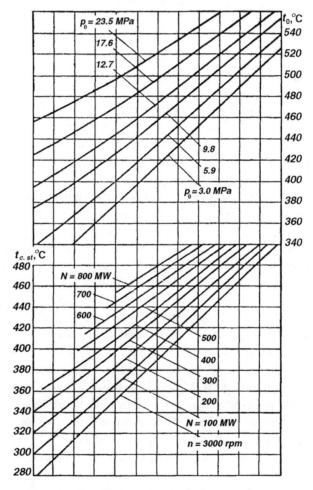

Figure 8-1. Nomogram for finding steam temperature at the first HP stage of an 800-MW supercritical-pressure steam turbine of LMZ under throttle control.

Experimental Researches of Turbine Transients 219

such steam temperature calculations sharply reduces at low-flow steam conditions of the turbine. So, for start-up tests of newly designed steam turbines, even with throttle steam control, it is very advisable to arrange little-inertial experimental measurements of the steam temperature at the first HP stage (along with the steam temperature measurements in the HP valve steam-chests), and this item is to be included in the list of necessary regular temperature measurements for all the turbines in service.

All the more, this also refers to turbines with nozzle-group steam admission control. For such turbines, the temperature of steam after the control stage practically cannot be calculated with due accuracy, but a diagram for static steam temperature drops in the control valves and control stage depending on the steam flow amount through the turbine and control valve position can be constructed on the basis of temperature measurements at start-up field tests and stationary operating conditions in the process of the turbine's operation. Such a diagram for a 300-MW supercritical-pressure turbine is shown in Figure 8-2. If the HP control valves are connected to the cylinder by crossover pipes, at the transient the static temperature drops are supplemented with dynamic reduction of the steam temperature because of heat transfer to the metal of these pipes with variable shares of the total steam flow amount passing through the individual control valves and crossover pipes.

Measuring the steam temperature at the first (control) stage can also be used for estimating the steam temperature at the front/central seal inlet with the help of a static correction. As seen from Figure 8-2, this correction can be taken practically independent of the steam flow amount through the turbine.

By contrast, similar corrections as applied to the heating steam temperatures for the IP rotors substantially vary with the steam flow amount through the turbine (see Figure 8-2), and at the transients this static temperature reduction should be supplemented with dynamic steam temperature drops mainly caused by hydraulic resistance of the channels between the mainstream steam path and rotor surfaces. Experimental measurements of the steam temperatures in the main steam path and near the rotor surfaces around the first IP stages of the 800-MW supercritical-pressure steam turbine of LMZ, as well as results of these measurements in the start-up process, are presented in Figure 8-3, and results of processing these measurements in the form of static and dynamic corrections for the temperatures of steam sweeping the rotor, relating to the steam temperature at the cylinder entrance, are shown in Figure 8-4.

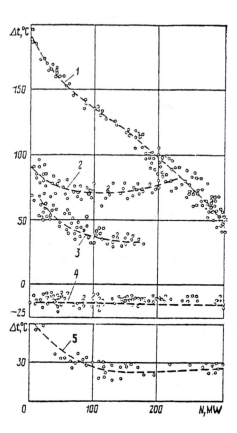

Figure 8-2. Static decrease of steam temperatures in a 300-MW supercritical-pressure turbine of LMZ on the basis of start-up field tests.
Steam temperature differences: 1 – between the stop-valve steam chests and control stage chamber under nominal throttle pressure, 2 – the same with sliding main steam pressure and four (of seven) open HP control valves, 3 – the same with completely open control valves, 4 – between the control-stage chamber and central-seal inlet of the HP cylinder, 5 – between the IP steam admission chamber and second IP stage inlet.

For the operational purposes, measuring the steam temperature in the IP steam admission chamber can be replaced with measuring the steam temperatures in the IP steam admission sleeves. Measuring continuously the steam temperatures at the IP entrance is absolutely necessary for arranging the continuous temperature monitoring of the IP rotor (or IP portion of the integrated HP-IP rotor).

Experimental steam temperature measurements similar to those used at start-up field tests of the 800-MW turbine (see Figure 8-3) were also arranged in the LP steam path of a 220-MW "nuclear" wet-steam turbine with welded disc-type LP rotors similar to those employed in some steam turbines for fossil-fuel power plants—Figure 8-5. Results of processing these measurements at stationary and transient operating conditions (Figure 8-6) to a great degree resemble those for the IP steam path of the 800-MW turbine shown in Figure 8-4.

Experimental Researches of Turbine Transients 221

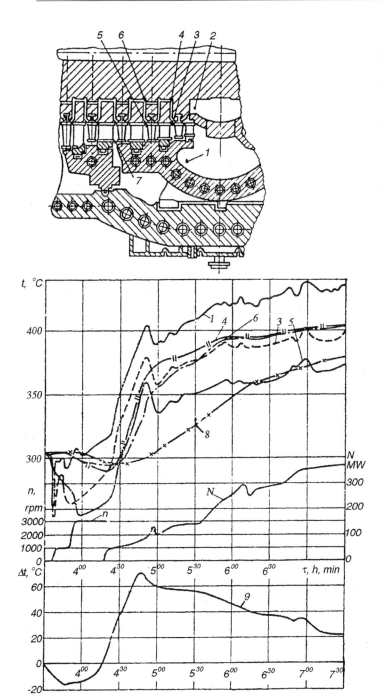

Figure 8-3. Experimental measurements of steam temperatures around the first IP stages of an 800-MW supercritical-pressure turbine of LMZ during its start-up according to field test data with results of calculating the IP rotor's temperature state. *Source*: G.D. Avrutskii et al.[4] Designations of steam temperature curves (1 - 7) correspond to the measurement point numbers at the sketch; 8 and 9 – calculated average integral metal temperature of the IP rotor in the section after the 1st stage and effective radial temperature difference in the same section

Figure 8-4. Results of processing steam temperature measurements around the first IP stages of an 800-MW supercritical-pressure turbine at field tests with shaping static (*a*) and dynamic (*b*) characteristics for the steam temperature drops. *Source:* G.D. Avrutskii et al.[4] Designations of steam temperature drops correspond to the measurement point numbers at the sketch in Figure 8-3; I and II - measurements while the turbine operated under load and idle, respectively.

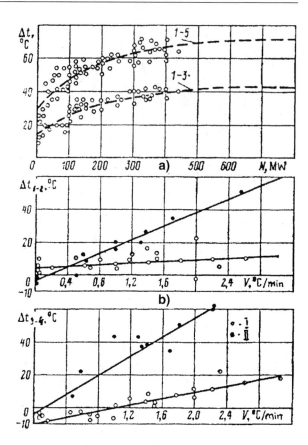

In particular, these results were later used to analyze the temperature and thermal-stress state of the welded LP rotors employed for steam turbines of fossil-fuel power plants. In this case, the entire necessary preparation of the turbine for start-up tests came to installing one additional experimental thermocouple for measuring the temperature of steam in the crossover pipe at the LP cylinder entrance.[6] In turn, the steam temperature at the LP entrance can be tied with the steam temperature at the IP entrance through a static dependence similar to those for intermediate IP stages—see Figures 8-2 and 8-4a. Characteristic temperature fields for the LP rotor calculated on the basis of the above static and dynamic characteristics for the heating steam temperatures are shown in Figure 8-7.

Unfortunately, the obtained data do not still allow generalizing and extending them to the IP and LP steam paths of an arbitrary type of

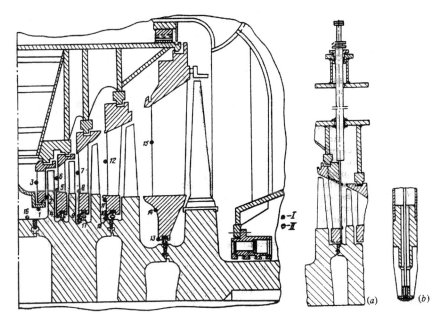

Figure 8-5. Experimental temperature measurements at the LP cylinder of a 220-MW wet-steam turbine for investigating the boundary and initial temperature conditions for its rotor; installation (*a*) and the head (*b*) of a temperature probe for measuring the rotor surface temperature of the stopped turbine. *Source: B.N. Lyudomirskii et al.*[5]

I – steam temperature measurements (numbers from 1 to 15), *II* – measurements of the rotor metal surface temperature at the stopped turbine (numbers 16, 17, and 18).

turbines. It is necessary to accumulate additional experimental data for steam paths of different steam turbines, as well as to associate these data with hydrodynamics of steam flowing through the root seals, pressure-balance holes, and diaphragm seals and sweeping the rotor.

Anyway, with results of start-up tests, the heating steam temperatures for the IP and LP rotors can be referred with due corrections to the temperature of steam entering the IP cylinder or section. At the same time, this steam temperature can essentially differ from the reheat steam temperature after the boiler to be controlled at start-ups. The case in point is the dynamic steam temperature difference lengthwise of the *steam transfer path* because of the heat transfer and accumulation in its metal. This influence can be very significant. So, for example, for power units

Figure 8-6. Static dependences of steam temperature decreases in the LP cylinder. *Source: B.N. Lyudomirskii et al.[5]*
Numbering of the temperature curves is given in accordance with designations in Figure 8-5; measurements: *I* – under stationary operating conditions, *II* at start-ups and shut-downs.

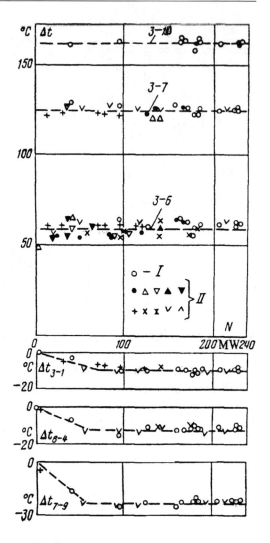

with one-bypass start-up systems (like shown in Figure 7-7), at start-up stages of rolling up and initial loading the turbine, the steam temperature at the IP cylinder entrance can differ from the reheat steam temperature after the boiler (or, more exactly, after the outlet attemperators) by 100-180°C (210-350°F). The same effects, even though in a less scale, take place for the reheat steam-lines in the case of two-bypass start-up systems, as well as the main steam-lines and HP crossover pipes.

Special methodological investigations showed that the considered steam transfer path, with several pipes and all their valves, can be treated as a single equivalent steam-line. In doing so, such factors as the heat flux lengthwise of the pipe walls because of thermal conduction of metal, heat losses outwards through the thermal insulation, transport delay, and dynamics of heating the steam-line wall across its thickness can be neglected as of secondary importance. Then, the dynamics of the temperature processes in such an equivalent steam-line can be character-

Experimental Researches of Turbine Transients 225

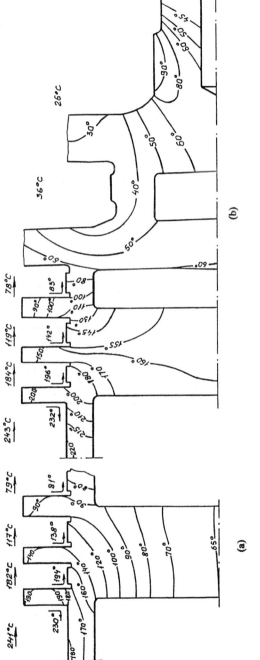

Figure 8-7. Calculated temperature fields for the first stages' disc of the welded double-flow LP rotor in the course of a turbine's cold start-up (a) and for right-hand half of the same rotor at stationary operating conditions under load (b). *Source:* A. Leyzerovich[6]

ized by only two generalized parameters.[1,3] They are:

the time constant of the steam-line as a thermal capacity

$$T_c = \frac{c_m M}{c_{st} G}$$

(8.1)

and dimensionless length of the equivalent steam-line

$$\xi = \frac{\alpha \psi F}{c_{st} G}.$$

(8.2)

Here, M and F are the metal mass of the equivalent steam-line and area of its inner heat transfer surface; G is the variable steam flow amount; c_{st} and c_m are the specific heat capacity of steam and metal, respectively, and α is the heat transfer coefficient from steam to metal averaged along the surface F.

Factor ψ in (8.2) approximately takes into account the temperature unevenness across the steam-line wall thickness. It represents the ratio of the temperature difference between the heating steam and heated surface to the difference between the heating steam temperature and average wall metal temperature. Generally, it depends on the Biot and Fourier number values: $\psi = \psi \ (Bi, Fo)$. However, for practical aims, it can be taken as $\psi \approx 1/(1+0.35Bi)$, where the Biot number is determined using the steam-line wall thickness: $Bi = \alpha \delta / \lambda_m$.

For each specific steam transfer path, the values of T_c and ξ are almost completely determined by the steam flow amount (Figure 8-8). Sometimes, especially for HP crossover pipes, it can be advisable to correct these characteristics for the steam pressure, influencing the specific heat capacity of steam, c_{st}. For reheat steam-lines, steam pressure is already taken into account automatically, because it is in proportion to the steam flow amount.

With the taken assumptions, dynamics of the steam temperature changes at the steam-line exit, t_{st}'' referring to the inlet steam temperature, t_{st}', providing with $G = \text{const}$, is characterized by a transcendent transfer function:

$$W(s) = \frac{L\left(t_{st}''\right)}{L\left(t_{st}'\right)} = \exp\left(-\xi \frac{T_c s}{\xi + T_c s}\right).$$

(8.3)

Experimental Researches of Turbine Transients 227

As calculations prompt, for characteristic ranges of varying the T_c and ξ values, this function can be substituted for a more convenient and simple rational transfer function

$$W(s) \approx \frac{1 + T_d s}{1 + T_i s},$$ (8.4)

where

$$T_i = T_c / (1 - e^{-\xi}) \text{ and } T_d = T_c e^{-\xi} / (1 - e^{-\xi}) = T_i e^{-\xi}.$$ (8.5)

The use of the transfer function (8.4) together with the dependences similar to those shown in Figure 8-8 opens the opportunity to obtain the helpful approximate characteristics of the steam temperature changes at the steam transfer path exit, t_{st}'', as applied to some important laws of the steam temperature changes at the steam path inlet, t_{st}':

- at jump-like changes of t_{st}' referring to the initial metal temperature t_0

$$\frac{t''_{st} \pm t_0}{t'_{st} \pm t_0} = 1 \pm k(\xi) \times \exp\left[\pm k(\xi)\eta\right],$$ (8.6)

where $k(\xi) = 1 - \exp(-\xi)$ and $\eta = \tau / T_c$;
- at harmonic oscillations of t_{st}' with the frequency ω and amplitude $\Delta t_{st}'$

$$\Delta t''_{st} = \Delta t'_{st} \sqrt{\frac{k^2(\xi) + e^{\pm 2\xi}\omega^2 T_c^2}{k^2(\xi) + \omega^2 T_c^2}}$$ (8.7)

- and at linear changes of t_{st}' in time with the rate W_t (at $\tau > 3\ T_c$):

$$t_{st}' - t_{st}'' = W_t T_c.$$ (8.8)

Heat transfer conditions for steam-lines and characteristic surfaces of turbine rotors can be obtained with due confidence using empirical dimensionless, criterion equations based on bench tests with static and rotating models. Such equations for the most characteristic cases are cited in some monographs on steam turbines.[1,2,8,9] So the heat transfer coefficients, α, for steam-lines can be found from the criterion equation for turbulent flow of liquids and gases in tubes, with the Reynolds num-

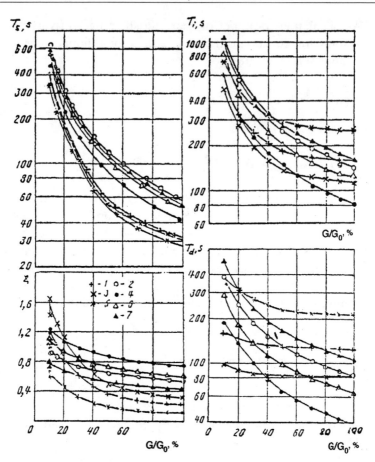

Figure 8-8. Dependences on the steam flow amount for heating-up characteristics of the main and reheat steam-lines of Soviet supercritical-pressure steam-turbine units. *Source: A.Sh. Leizerovich and A.D. Melamed[7]*

1 - main steam-lines of 300 MW single-furnace steam-turbine units, 2 - the same for 300-MW twin-furnace units, 3 - the same for 800-MW units, 4 - hot reheat steam-lines of 300-MW single-furnace units, 5 - the same for 300-MW twin-furnace units, 6 - the same for 800-MW units with two reheat steam-lines, 7 - the same for 800-MW units with four reheat steam-lines.

ber, Re, over 5×10^3:

$$Nu = 0.023 \times Re^{0.8} \times Pr^{0.43} \times C_l, \tag{8.9}$$

Experimental Researches of Turbine Transients 229

where the Reynolds and Nusselt numbers are counted relying on the inner diameter of the tube and the average fluid flow velocity, and the length coefficient, C_l, for steam-lines can be taken equal to 1.0. Results of such calculations for steam-lines can be presented as functions of the steam flow amount like those in Figure 8-8.

As to the rotors, at the rolling-up stage, the heat transfer conditions for different surfaces can be presented in the form of dependencies upon the rotation speed, which can be corrected for the steam pressure if it varies independently of the rotation speed. Characteristic surfaces for turbine rotors are the labyrinth seals of diverse types, smooth cylindrical surfaces in free space or across another (fixed) cylindrical surface, side disc surfaces in free space or across the fixed wall, end-walls of the rotating blade rows, and so on. When the turbine is operating under load (with the invariable rotation speed), of significance is that for each surface and specific case (geometric dimensions, rotation speed, and so on) the results of heat transfer calculations can be approximated in the form of dependencies on the current pressure of the heated steam. In doing so, this value characterizes not only the steam density and viscosity, but also, if necessary (for example, as applied to the gland seals), the steam flow amount and hence the forced steam velocity relative to the rotor surface. The influence of steam temperature variations is commonly of the second importance.

The example of such approximations for "critical" sections of the HP and IP rotors, where the largest unsteady-state thermal stresses are expected at turbine start-ups, is presented in Figure 8-9. Usage of such static dependencies on the turbine operating conditions, as well as the above mentioned static and dynamic dependencies for the heating steam temperatures, makes it possible to calculate the metal temperature fields of the rotors like those in Figure 8-7.

For welded turbine rotors with great internal cavities (as in Figure 8-7, as well as for rotors shown in Figures 2-5b, 3-5a, and 3-16), of importance is a problem of setting the proper boundary conditions at their surfaces. The medium (air or residual gas left after welding) that is locked within these cavities, being heated at the cavity periphery, is forced out by heavier cold gas and moves toward the rotor axes, where it transfers its heat to the metal of the rotor's colder central part, thus reducing in this way the radial temperature differences in the rotor. This thermal convection process takes place in the field of artificial gravity under action of centrifugal forces. For high-speed turbine rotors with the rotation speed of 3,000 rpm or 3,600 rpm and the outer diameter of the

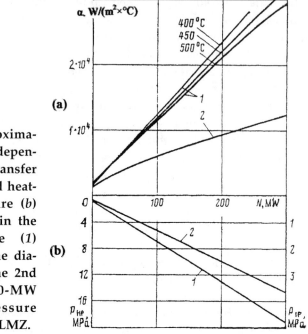

Figure 8-9. Approximation of calculated dependences for heat transfer coefficients (*a*) and heating steam pressure (*b*) for the HP rotor in the central seal zone (*1*) and IP rotor in the diaphragm seal of the 2nd stage (*2*) for 300-MW supercritical-pressure steam turbines of LMZ.

internal cavity of approximately 1 m (40 in), which is quite typical for such turbines, the centrifugal acceleration at the periphery is more than 6×10^3 g. Some preliminary assessments showed that the heat transfer with thermal convection in the field of centrifugal forces can significantly influence the dynamics of heating the rotor, and with regard to this effect the temperature difference along the rotor radius could noticeably decrease, but this phenomenon needed further studies.

Solution of the problem for free convection in internal rotor cavities is seriously hindered by several circumstances: 1) the temperature distribution at the cavity boundaries varies and is itself determined by the heat transfer conditions in the cavity, which makes necessary to solve a conjugate non-stationary two-dimensional problem of free convection in the cavity coupled with thermal conduction in the adjacent rotor body; 2) the high acceleration of centrifugal forces in the rotor cavities corresponds to the Rayleigh number (*Ra*) values beyond the range for which experimental and theoretical investigations have been ever conducted, and 3) the acceleration of centrifugal forces changes significantly over the cavity height. The considered problem was subjected to thorough analysis with the use of specially developed mathematical models.[10] The calculations showed that, if the Rayleigh number values characterizing free convection within the rotor's cavity do not exceed 10^{10}, the influence of the considered effect can be neglected, and the heat fluxes on the cavity surfaces can be accepted equal to zero. At the same time, if the free convection within the rotor cavity is artificially intensified (for example, by means of filling the cavity by special gas or compressed air), it can significantly influence the rotor's temperature fields.

Contrary to the steam-lines and turbine rotors, with their pretty definite patterns of the steam flow relating to the heated surfaces, for the valve steam-chests and cylinder casings, their steam flow patterns are much more variable and intricate, less studied and need special experimental researches at start-up field tests. Our knowledge of the heat transfer conditions for the stator turbine elements mainly comes from

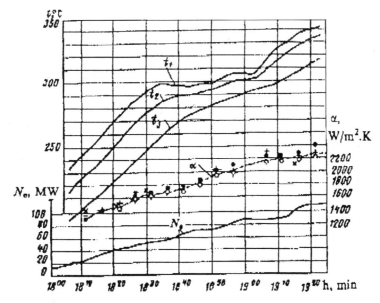

Figure 8-10. Assessing the heat transfer conditions for the loop-flow HP cylinder's intercasing space of a 300-MW supercritical-pressure steam turbine of LMZ based on the steam and metal temperature records during start-up. *Source: A.Sh. Leizerovich and E.R. Plotkin*[13]

t_1 – heating steam temperature, t_2 – metal temperature of the outer casing's wall near the heated surface (at the depth of 7 mm), t_3 – metal temperature on the wall's external insulated surface. Different symbols for the heat transfer coefficients, α, designate results of calculations by different methods.

232 *Steam Turbines for Modern Fossil-Fuel Power Plants*

assessments based on temperature measurements of the heating steam and heated metal.[1,2,11-13] Figure 8-10 presents some results of calculating the heat transfer conditions for the intercasing space of the loop-flow HP cylinder (like that of a 800-MW turbine shown in Figure 2-4a) of a 300-MW supercritical-pressure turbine based on the measured steam and metal temperature changes in the process of a cold start-up.

Accumulation of such experimental data opens a possibility of constructing empirical criterion equations for most characteristic design solutions. So, for example, on the basis of processing results of special experimental temperature measurements at start-up field tests of different-type steam turbines, E.R. Plotkin described the heat transfer conditions on the internal surfaces of the HP valve steam-chests with the side steam admission by an empirical expression:

$$Nu = 0.046 x Re^{0.8} x Pr^{0.43}, \tag{8.10}$$

with the dispersion of processing not exceeding 20-25%. Here, the inlet sleeve diameter and steam velocity in the sleeve were taken as a characteristic size and velocity to calculate the Nusselt and Reynolds numbers.

In the same way, processing special experimental temperature measurements at start-up field tests of three different-type steam turbines (200-MW subcritical-pressure turbine K-200-130, 300-MW supercritical-pressure turbine K-300-240, and 500-MW half-speed wet-steam turbine K-500-60/1500) allows us to obtain a general criterion equation for heat transfer conditions in the IP steam admission chambers with tangential steam inlet in the same traditional form of

$$Nu = 0.021 x Re^{0.8} x Pr^{0.43}, \tag{8.11}$$

where the Reynolds and Nusselt numbers are calculated on the basis of the conventional circular steam velocity in the chamber and its cross-section's equivalent diameter.[3,13]

Along with this, in many cases there do not exist any criterion equations that can be used for estimating the heat transfer conditions for steam turbine stator elements, especially for turbine casings. Then, there is no other way to get such knowledge but start-up field tests. Figure 8-11 shows results of calculating the heat transfer conditions on the basis of experimental temperature measurements for the loop-flow HP

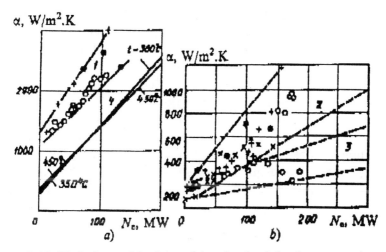

Figure 8-11. Variation with the turbine load of the heat transfer conditions in the intercasing space of the loop-flow HP cylinder (a) and the steam extraction chamber of the IP cylinder (b) of the 300-MW supercritical-pressure steam turbine of LMZ. *Source: A.Sh. Leizerovich and E.R. Plotkin*[13]
1-3 – experimental data based on temperature measurements of start-up field tests at different operating conditions (2 – with the HP feedwater heaters connected for steam and 3 – with the HP feedwater heaters shut off), 4 – calculated data using the criterion equation based on bench test results[8].

cylinder's outer casing and the third steam extraction chamber of the IP cylinder of a 300-MW supercritical-pressure turbine of LMZ (see Figure 12-15). Similar data were obtained in the process of start-up field tests for 200-MW subcritical-pressure turbines[1,11,12] and, then, for 800-MW supercritical-pressure turbine. It is very instructive that the experimental heat transfer condition data for the intercasing space of the loop-flow HP cylinder (Figure 3-4a) are 1.5-3 times higher than those calculated with the help of the empiric criterion equation based on the data of bench tests.[8] This can indicate the presence of factors not taken into account in the used physical model—for instance, twisting the steam flow at the intercasing space inlet, after the inner casing's last stage. Corroboration of those data by results of some start-up field tests[14] does not seem very convincing because of shortcomings in arranging those researches.

Unfortunately, all the existent data on heat transfer for turbine casings are not still sufficient for their generalizing, and new experimental

234 *Steam Turbines for Modern Fossil-Fuel Power Plants*

researches are needed at start-up field test of diverse turbines. Therewith, in doing so, it is vital to observe accurately all the requirements guaranteeing the reliability of the gotten results. Of special interest would be the heat transfer conditions for the intercasing space of the integrated HP-IP cylinders of different types (see Figures 3-13—3-15). Presently, such data are practically absent that does not make it possible to calculate accurately their temperature fields. It should be supplemented that for some intercasing chambers of such cylinders without a forced steam flow it is also difficult to set definitely the dependency between the steam temperatures in the mainstream and within the chamber. All these circumstances need thorough investigations.

METAL TEMPERATURE FIELDS FOR TURBINE DESIGN ELEMENTS

Along with temperature measurements intended for exploring the boundary conditions of heating the turbine's major design elements, of importance are direct measurements of metal temperatures for these elements—outer and inner casings, valve steam-chests, and, if possible, rotors. By way of illustration, a stationary temperature field in the steam admission sleeve of the HP cylinder's outer casing for the 300-MW supercritical-pressure turbine of LMZ, built on the basis of experimental metal temperature measurements, is shown in Figure 8-12; the main purpose of these measurements was to confirm the effectiveness of shielding the outer surface against the high-temperature steam entering the cylinder. Many design solutions intended to improve the turbine reliability, decrease the level of the maximum metal temperatures and thermal stresses can be approved only relying on results of such experimental metal temperature measurements.

If the metal temperature measurements are aimed at estimating the thermal-stress state of the turbine elements, the measurements should correspond with the assumed scheme of the acting factors and provide sufficiently complete temperature fields for calculating thermal stresses with regard to these factors. So, for example, Figure 8-13 suggests a principle scheme of acting forces and bending moments for estimating the thermal-stress state of an HP stop-valve steam-chest beneath the flange joint, where the thermal fatigue cracks were revealed, and results of such estimations on the basis of the measured metal temperatures in the steam-chest, its

Experimental Researches of Turbine Transients 235

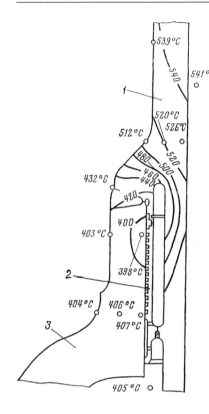

Figure 8-12. Stationary temperature field for the steam admission zone of an HP cylinder of a 300-MW supercritical-pressure steam turbine of LMZ according to experimental data of start-up field test. *Source: A.P. Ogurtsov and V.K. Ryzhkov[15]*
1 – steam admission sleeve, 2 – thermal shield, 3 - outer casing.

flange cover, and stud bolts.

For estimating thermal stresses in the walls and flanges of cylinder casings, it is not sufficient to know the maximum temperature differences across the wall thickness and the flange width and between the flange and bolts—it is also necessary to take into consideration the circumferential temperature differences between the wall and flanges, as well as the axial temperature unevenness, especially in the cases of cylinders with a small distance between the chambers swept by steam with substantially different temperature and heat transfer conditions (as, for example, the IP steam admission and exhaust chambers and the intercasing space between the outer and inner casings of the HP-IP cylinders).

The measured metal temperature fields can be also employed to verify the used mathematical models and calculation schemes. An example of such a comparison as applied to the HP rotor of a 300-MW supercritical-pressure turbine of LMZ is shown in Figure 8-14a. Results of such direct thermometry of this rotor at start-ups were also employed for verification of simplified calculation schemes and assumptions used at construction first computing devices for continuous temperature monitoring of high-temperature turbine rotors.[1-3,16]

Likewise, metal temperature measurements in the casing's flange joints of the HP cylinders of wet-steam turbines for nuclear power plants were used for verification of calculation schemes developed for mathematical modeling of their heating up at start-ups,[2,17] and so on.

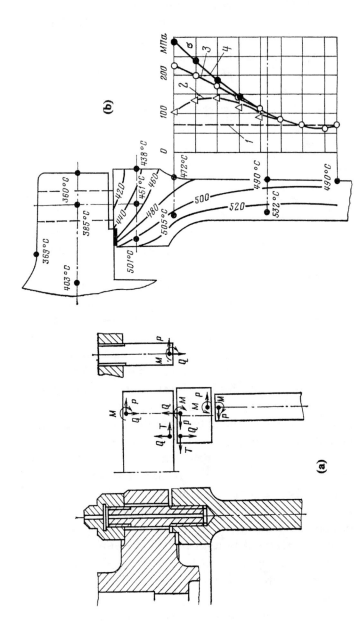

Figure 8-13. Scheme of acting forces and bending moments for estimating the thermal-stress state of an HP stop-valve steam-chest based on an engineering approach (a) and results of such estimations for an actual metal temperature field (b).

1 – stresses on the external surface of the steam-chest wall taken as a hollow cylinder with one-dimensional temperature field, 2 and 3 – tangential and axial stresses with regard to the actual geometric form and two-dimensional temperature field, 4 – axial stresses with accounting the influence of stud-bolts.

Experimental Researches of Turbine Transients

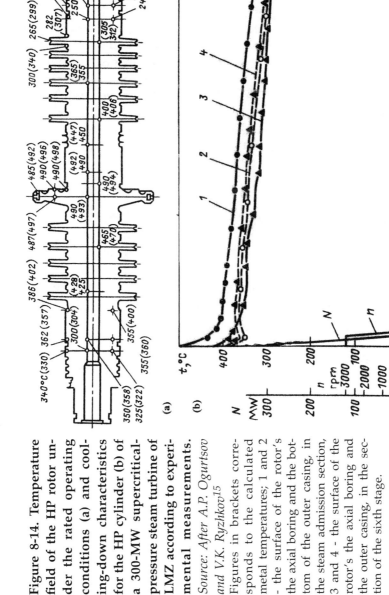

Figure 8-14. Temperature field of the HP rotor under the rated operating conditions (a) and cooling-down characteristics for the HP cylinder (b) of a 300-MW supercritical-pressure steam turbine of LMZ according to experimental measurements.
Source: After A.P. Ogurtsov and V.K. Ryzhkov[15]
Figures in brackets correspond to the calculated metal temperatures; 1 and 2 - the surface of the rotor's the axial boring and the bottom of the outer casing, in the steam admission section, 3 and 4 - the surface of the rotor's the axial boring and the outer casing, in the section of the sixth stage.

COOLING DOWN CHARACTERISTICS OF THE TURBINE

One of the simplest but, at the same time, very important aims of the start-up field tests is comprehension of the turbine's cooling down characteristics. As far as the start-up duration depends on the initial (pre-start) temperature state of the turbine, the rate with which the stopped turbine cools down during the outage occurs among the most influential factors determining the turbine's flexibility. Not only the metal temperatures of the HP and IP cylinders are significant in themselves, but of importance are also the disagreements between the cooling down rates of different turbine components and steam-lines. To the greatest extent, this refers, first, to the HP cylinder, valve steam-chest, and crossover pipes (if exist) and, second, the IP cylinder and hot reheat steam-lines.

In theory, for a solid body of a simple form and uniform initial temperature state, its cooling down process can be described by an exponential equation:

$$t_m = t_a + (t_{m.in} - t_a) \times \exp(-\tau/K), \tag{8.12}$$

where t_a is the ambient temperature, assumed to be invariable, and t_m and $t_{m.in}$ are the current (for the time instant τ after beginning of the cooling process) and initial (for $\tau = 0$) values of the considered metal temperature. Here, K and $m = 1/K$ are conventionally called the cooling time constant and cooling-down constant. Their values depend on the body's mass heat capacity, ratio of its mass and heat transfer area, and thermal insulation quality.

The cooling-down constant m can be found by processing results of measuring the considered metal temperature for two time instants of the cooling process, τ_1 and τ_2:

$$m = \left[\ln \frac{t_m(\tau_1) - t_a}{t_m(\tau_2) - t_a} \right] / (\tau_2 - \tau_1) \tag{8.13}$$

Practically, this can be done even easier by presenting graphically the curve of the decrease with time for the value of

$$\theta = (t_m - t_a)/(t_{m.in} - t_a) \tag{8.14}$$

in semi-logarithmic coordinates. In this case, the cooling-down curve

Experimental Researches of Turbine Transients 239

turns into the straight line, and the value of m can be counted as the tangent of its incline. Such a cooling-down process is called regular.

In reality, because the shapes of turbine components are complicated, their metal temperature fields in the operation process are not uniform, and in the cooling-down process there takes place heat transfer by thermal conduction between different parts of the turbine elements and adjoined steam-lines, the actual cooling-down processes are not regular. This is especially noticeable for thick-wall outer casings with steam admission in the central part. Nevertheless, after a certain initial period of regularization, whose duration can last up to 20-30 hours, the process becomes nearly regular, which makes it possible to find the cooling-down constants for major turbine components and steam-lines. This can (and must) be done even with the use of regular metal temperature measurements, without resorting to special experimental measurements, Figure 8-15. Such cooling-down characteristics should be recorded and analyzed periodically (not rarer than yearly) to reveal possible undesirable changes in the quality of thermal insulation for individual turbine components.

According to data of operation practice and field tests at numerous power plants with diverse steam turbines, it became possible to estimate the average figures of the cooling-down constants for major turbine design components of different types, as well as steam-lines. Such data are combined in Table 8-1. If the specific cooling-down characteristics of certain components or steam-lines for the considered turbine in service significantly differ in their constants from the average values to the worse (that is, the cooling-down constants are too large or the cooling time constants are too small), this should be a serious reason to improve the quality of thermal insulation for such turbine components or steam-lines, cooling down too fast. Otherwise, the start-up duration can occur to be unduly great and, as a rule, the start-up operation technology can be needlessly complicated because of too uneven initial temperature states for individual components and steam-lines. What is more, as it is noted in the European Technology Development's Report, "Poor [thermal] insulation can have unforeseen effects."[18]

Another sign of insufficient quality of thermal insulation for high-temperature (HP and IP or HP-IP) cylinders can be increased temperature differences between the upper and lower halves of the outer casings—temperature differences "top-bottom," $\Delta t_{t\text{-}b}$. As a rule, the casing cover occurs to have a higher metal temperature, entailing thermal bend

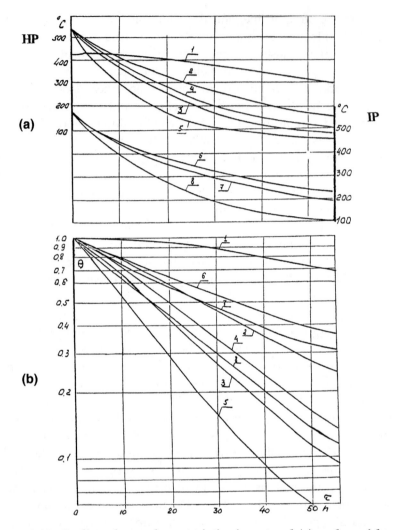

Figure 8-15. Cooling-down characteristics in natural (a) and semi-logarithmic (b) coordinates for major turbine components and steam-lines of a 300-MW supercritical-pressure steam turbine of LMZ. *Source: E.R. Plotkin and A.Sh. Leyzerovich[1]*

Measured HP metal temperatures: 1 - outer casing of the cylinder in the central (steam admission) zone, 2 and 3 - main steam-lines near the tee-joint to the turbine bypass and at the mid-length, respectively, 4 - stop-valve steam-chest, 5 - crossover pipe downstream of the control valve; measured IP metal temperatures: 6 - outer casing of the cylinder in the front (steam admission) zone, 7 and 8 - hot reheat steam-line at the mid-length and after the boiler (near the start-up attemperator), respectively.

Experimental Researches of Turbine Transients 241

Table 8-1. Average cooling-down constants for major turbine components of different types.

Turbine components and steam-lines	Cooling-down Time Constants (K, in hours)/Cooling–down Constants (m, in h^{-1})
Main steam-lines	
of subcritical steam pressure	30 – 50/0.029 – 0.02
of supercritical steam pressure	60 – 80/0.017 – 0.0125
Hot reheat steam-lines	40 – 80/0.025 – 0.0125
HP valve steam-chests	
for individually settled valves	35 – 60/0.029 – 0.017
for combined units	60 – 90/0.017 – 0.011
HP crossover pipes	20 – 35/0.05 – 0.029
Single-flow HP cylinder's outer casing	90 – 110/0.011 – 0.009
Double-flow HP cylinder's outer casing	100 – 140/0.01 – 0.007
Integrated HP-IP cylinder's outer casing	120 – 160/0.008 – 0.006
Double-flow IP cylinder's outer casing	110 – 150/0.009 – 0.0065

of the casing upward, which is often named "cat's back." It results in a decrease of the lower radial clearances and increase of the upper ones in the steam path between the rotor and stator elements. Naturally, these variations are largest near the cylinder middle, where the casing sag is maximal. Its value can be approximately estimated as

$$\delta \approx \frac{\beta L^2}{8D} \times \overline{\Delta t}_{t-b'} \tag{8.15}$$

where L is the distance between the casing supports (its paws); D is the average diameter of the casing, and $\overline{\Delta t}_{t-b}$ is the top-bottom temperature difference averaged lengthwise of the casing. It can be said that each 10°C of the top-bottom temperature difference cause the casing sag estimated as approximately 0.08-0.13 mm depending upon the casing design, which corresponds to 1.8-2.8 mil per each 10°F.[1]

More often than not, insufficient quality of thermal insulation is caused not by insufficient thickness of insulation, but rather its low integrity and poor adhesion to the metal surface, especially in the bottom part. This results in air cavities between the insulation and the metal surface and inside the insulation body, between its layers. Heat fluxes with thermal convection in these cavities make the thermal insulation quality remarkably worse. For turbine casings and valve steam-chests, their cooling-down characteristics can be also significantly affected by insufficient thermal insulation of the adjoined steam-lines and pipes (for example, crossover pipes, steam extraction steam-lines, pipes for sealing steam, and some others), as well as heat fluxes toward non-insulated parts, as, for example, casing paws, valve servomotors, and so on.

Existent methodologies and special computer programs allow calculating the cooling-down characteristics for high-temperature (HP and IP) cylinders, steam-chests, and steam-lines with quite satisfactory accuracy provided their thermal insulation and hence heat fluxes outwards through it are modeled properly. Results of such calculations for steam turbines of different types occur in quite satisfactory correlation with data of metal temperature measurements at actual steam turbines in the process of their cooling, as shown, for example, in Figure 8-16.

According to some widespread recommendations,[9,19] while calculating cooling-down processes for high-temperature cylinders, steam-chests, and steam-lines, any heat transfer within the internal chambers of these elements can be neglected. As to high-temperature (HP and IP) rotors, their cooling is considered to be mostly conditioned by the axial heat fluxes with thermal conduction to the end seals and bearings, whereas heat transfer between the rotors and cylinder casings, casing

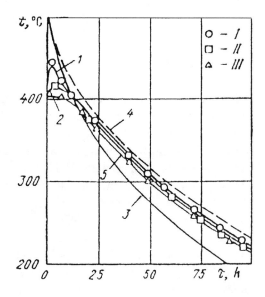

Figure 8-16. Cooling-down characteristic of the HP cylinder for a 500-MW supercritical-pressure steam turbine of Turboatom, according to calculated and experimental data. *Source: V.M. Kapinos et al.[19]*
Calculated data: 1 - inner casing's wall, 2 - outer casing's wall, 3 - rotor, 4 - maximum temperature along the rotor's axial bore, 5 - outer casing's flange; experimental data: I - inner casing's wall, II - outer casing's wall, III - outer casing's flange.

Experimental Researches of Turbine Transients 243

rings, diaphragms, and so on can be ignored.

In practice, metal temperatures measurements for the outer casings of the HP and IP cylinders in the cooling-down process can be taken as an approximate notion of the temperature state of their rotors. This can be concluded relying on results of calculations, as well as special field tests with experimental thermometry of the rotors. The number of such field tests for large power steam turbines at power plants in service executed worldwide by turbine producers and research institutions can be counted on fingers, so these results are especially important. By example, Figure 8-14b presents the data of such measurements for the HP rotor of a 300-MW supercritical-pressure steam turbine (K-300-240) of LMZ.

For some modern large steam turbines with solid or welded LP rotors, especially for turbines with the elevated reheat steam temperature and integrated HP-IP cylinder with a rather short IP section, and, as a result, a pretty high temperature of steam entering the LP cylinders, the LP rotors can happen to be among the critical (most thermally stressed) turbine components, and it is extremely advisable to monitor their temperature state during start-ups to prevent these rotors from excessive combined thermal and centrifugal tensile stresses.

As well as for the HP and IP rotors, continuous temperature monitoring for the LP rotors can be arranged based on mathematical modeling of their heating-up. However, this requires knowledge of the initial, pre-start temperature state of the rotors. As distinct from the high-temperature (HP and IP) rotors, in this case it is impossible to count on measuring metal temperatures of the cylinder's outer casing—the casing and rotor too differ in both their masses and thermal conduction and heat transfer conditions. It is sufficient to recall that the outer casings of the LP cylinders welded from thin metal sheets are not thermally insulated at all. So the only possibility to estimate the pre-start thermal state of the LP rotors is the mathematical modeling of not only their heating-up, but their cooling-down, too, with the use of some simplified models. Their trustworthiness should be confirmed by experimental data of temperature measurements at actual turbines under actual operational conditions, but until relatively recent years such data had been absent at all.

Unique temperature measurements of the LP rotor metal temperatures during the turbine outages were fulfilled at a 220-MW wet-steam turbine of Turboatom at the Kola nuclear power plant—see Figure 8-5. The LP rotors of these turbines are similar to those employed for many large steam turbines of fossil-fuel power plants. Comparison of the got-

244 *Steam Turbines for Modern Fossil-Fuel Power Plants*

ten experimental data for the metal temperatures of such rotors in the cooling-down process with calculation data showed that for the LP rotors, as distinct from the HP and IP ones, it is absolutely necessary to take into account the heat removal from the rotor surfaces caused by natural convection within the casing, above the condenser space. Otherwise, the calculated cooling curves will lie perceptibly higher than the true, experimental, ones. This especially significant for LP rotors welded of relatively thin discs, for which the heat fluxes with thermal conduction through thin cylindrical crosspieces between discs do not play such an important role. With due regard to the heat convection processes in the cylinder's intercasing space, the calculated data well correlate to the experimental data—Figure 8-17, and that makes it possible to set the cooling-down time constants for the LP rotors based on calculations. Such calculation data for the welded LP rotors of two different types are presented in Figure 8-18. It is understandable that the same approach should be applied to forged LP rotors as well, even though for them the heat transfer by thermal conduction to the end seals and bearings is more influential than it does for welded LP rotors.

References

1. Plotkin E.R. and A. Sh. Leyzerovich. *Start-ups of Power Unit Steam Turbines* [in Russian], Moscow: Energiya, 1980.
2. Leyzerovich A. *Large Power Steam Turbines: Design & Operation*, 2 vols., Tulsa (OK): PennWell Books, 1997.
3. Leyzerovich A. Sh. *Technological Fundamentals for Steam Turbine Start-up Automation* [in Russian], Moscow: Energoatomizdat, 1983.
4. Avrutskii G.D., A. Sh. Leizerovich, V.I. Nakhimov, and R.M. Ostrovetskii. "Variation in Steam Temperatures in the IP Steam Path of K-800-240-3 Steam Turbine during Start-ups," *Thermal Engineering* 28, no. 1 (1981): 21-24.
5. Lyudomirskii B.N., A. Sh. Leizerovich, and Yu.N. Kolomtsev. "An Experimental Investigation of the Condition of Heating up the Welded Rotor of the LP Cylinder of a Nuclear Steam Turbine when Starting," *Thermal Engineering* 26, no. 11 (1979): 664-668.
6. Leizerovich A. Sh. "Evaluating Fatigue Accumulation in the LP Rotor Metal of Steam Turbine in the Course of Its Operation," *Thermal Engineering* 30, no. 10 (1983): 587-591.
7. Leizerovich A. Sh. and A.D. Melamed. "Calculation of the Influence of the Boiler-Turbine Steam-Lines on the Increase in Steam Temperature with Automated Starts of Generating Units," *Thermal Engineering* 29, no. 10 (1982): 542-546.
8. Zysina-Molozhen L.M., L.V. Zysin, and M.P. Polyak. *Heat Transfer in Turbomachinery* [in Russian], Leningrad: Mashinostroenie, 1974.
9. Safonov L.P., K.P. Seleznev, and A.N. Kovalenko. *Thermal State of Highly Flexible Steam Turbines* [in Russian], Leningrad: Mashinostroenie, 1983.
10. Brailovskaya V.A., V.R. Kogan, A. Sh. Leyzerovich, and V.I. Polezhaev. "Influence of Free Convection within the Internal Cavity on Heating of Welded Rotors of

Experimental Researches of Turbine Transients 245

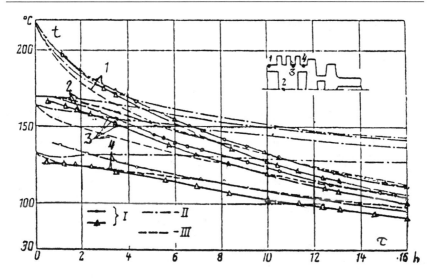

Figure 8-17. Measured and calculated cooling-down characteristics for a welded LP rotor. *Source: B.N. Lyudomirskii et al.*[5]
Designation of temperature curves correspond to the temperature measurement point numbers in the cut in sketch: I - experimental (measured) data, II and III - calculation results without and with regard to natural heat convection within the intercasing space of the cylinder, respectively.

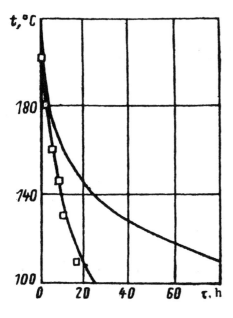

Figure 8-18. Calculated cooling-down characteristics for welded LP rotors of two different types. *Source: V.M. Kapinos et al.*[19]
Points denote measured data from experiments of VTI at the Kola nuclear power plant.

Large Steam Turbines at Start-ups" [in Russian], *Izvestiya Akademii Nauk SSSR. Energetika i Transport (Proc. of the Academy of Sciences of the USSR. Power Engineering and Transport)*, no. 5 (1980): 109-116.

11. Plotkin E.R., A. Sh. Leizerovich, and I.V. Muratova. "Investigation of Heat-Transfer Conditions in the K-200-130 Steam Turbine," *Thermal Engineering* 18, no. 5 (1971): 41-45.

12. Aleshin A.I., A. Sh. Leizerovich, and E.R. Plotkin. "Conditions of Heat Transfer and Gasdynamics in Steam Turbine Extraction Chambers," *Thermal Engineering* 23, no. 8 (1976): 37-42.

13. Leizerovich A. Sh. and Plotkin E.R. "Correlating the Results of Experimental Investigations into Heat Transfer Coefficients on Steam Turbine Stator Elements," *Thermal Engineering* 38, no. 10 (1991): 550-553.

14. Zabezhinskii L.D., V.M. Pashnin, and V.L. Pokhoriler. "Investigation of Heat-Transfer Conditions in Casings of Turbines with a Loop System of Steam Flows," *Thermal Engineering* 28, no. 8 (1981): 461-463.

15. Ogurtsov A.P. and V.K. Ryzhkov, editors. *Supercritical-Pressure Steam Turbines of LMZ* [in Russian], Moscow: Energoatomizdat, 1991.

16. Leizerovich A. Sh., B.D. Ivanov, Y.N. Vezenitsyn, et al. "Commercial Tests of a Prototype for Operational Monitoring of the Warming of Rotors of High-Capacity Steam Turbines," *Thermal Engineering* 25, no. 11 (1978): 29-33.

17. Leyzerovich A. *Wet-Steam Turbines for Nuclear Power Plants*, Tulsa (OK): PennWell Corporation, 2005.

18. Starr F., J. Gostling, and A. Shibli. *Damage to Power Plant Due to Cycling*. ETD Report No. 1002-IIP-1001, London: European Technology Development, 2000.

19. Kapinos V.M., Y.Y. Matveev, V.N. Pustovalov, and V.A. Paley. "Modeling the Cooling Processes for Large Steam Turbines," *Thermal Engineering* 35, no. 4 (1988): 196-200.

Chapter 9

Start-up Technologies as Applied to Different Start-up Systems

PRE-START WARMING AND ROLLING UP OF THE TURBINE

Pre-start warming of the main steam-lines commences as soon as the boiler is ignited, begins producing superheated steam, and the temperature of steam at the boiler's exit reaches the steam-line metal temperature. The generated steam is discharged through the HP bypass into either the cold reheat steam-lines (for units equipped with two-bypass start-up systems) or the condenser (with one-bypass start-up systems). It is a gold rule of thumb that the bypass should be connected to the main steam-lines as close as possible to the turbine's HP valves to involve the entire length of the main steam-lines into the pre-start warming by the entire steam flow produced by the boiler. However this simple rule is not always fulfilled in a reality, which makes the preheating of the steam-lines butt-ends less effective and requiring more time.

Traditionally, the stage of pre-start warming of the main and reheat steam-lines, HP valve steam-chests, and HP crossover pipes is definitely separated from the subsequent start-up stage of rolling up the turbine. The former is conducted with the closed control valves, and if in the course of this stage the turbine rotation speed spontaneously increases (because the control valves occur somewhat leaky), this increase is considered undesirable and is strictly limited. According to more recent approaches, such separation of these two stages has become less definite, and the pre-start warming stage often merges with the beginning of the rolling up, even though in this case the turbine speed can be controlled only indirectly and roughly—mainly by means of varying the turbine backpressure in the condenser.

According to traditional approaches, if the power units is equipped with the main steam valves (MSVs) upstream of the HP stop valves (see, for example, Figures 7-6 and 7-7), the MSVs are opened as soon as the main steam-lines are heated up to the metal temperature of the HP stop-valve steam-chests, and then the HP valve steam-chests, as well as the crossover pipes between the stop and control valves (if exist), are heated together with the main steam-lines. In doing so, the HP stop valves are opened, whereas the HP control valves remain closed, and their steam-chests are heated only up to their saddles.

Often, the operational personnel begin heating the HP valves with some delay, when the steam pressure after the boiler is quite high. In this case, if the MSVs are opened at once and the initial metal temperature of the heated surfaces is lower than the saturation temperature corresponding to the steam pressure after the MSVs, the pre-start warming goes with condensation of the heating steam, that is, with very intense heat transfer conditions. In this case, the valve steam-chests and crossover pipes undergo a temperature shock, and the arising temperature differences across the thickness of the heated elements reach practically the differences between the saturation temperature and initial metal temperatures. This is especially typical for power units with once-through boilers featuring relatively high initial steam pressure even at cold start-ups. The resulted thermal stresses in the valve steam-chests and crossover pipes attain remarkable values. By contrast, the subsequent warming (over the saturation temperature) goes very slowly with the heat transfer conditions determined by steam flow amounts through the drainage lines of the main steam-lines downstream of the tee-joint to the HP bypass and crossover pipes between the HP stop and control valves.

In addition, because the control valves are not absolutely tight, the pre-start warming of their steam-chests is often accompanied with chilling their sleeves after the saddle. In particular, this process is obviously seen in the diagram of Figure 9-1. This phenomenon is caused by accumulation of the heating-steam condensate over the saddle and pressing it through the valve by steam pressure. Because the space after the valve is under vacuum, the leaking condensate rapidly evaporates and cools the metal down to the low saturation temperature. Earlier similar phenomena were repeatedly recorded at steam turbines of different types.[1,2] Such a deep cooling downstream of the HP control-valve saddles causes high thermal stresses in the valve steam-chests and their sleeves, however these events usually remain unnoticeable for the operation personnel because of the absence of regular metal temperature measurements at the control-valve sleeves downstream of the saddle. In many cases, these temperature differences and resulted thermal stresses should be accepted responsible for cracks revealed at the control-valve steam-chests just in the region of their saddles and exit sleeves. Luckily, such phenomena are not characteristic for turbines with the combined stop-and-control HP valve units placed in the horizontal plane (see, for example, Figures 2-8, 2-10, 3-5b, and 3-8).

Some undesirable events associated with pre-start warming of the HP valve steam-chests can be eliminated if these steam-chests are heated together with the main steam-lines and the MSVs are opened from the very beginning. However, this is acceptable only if the initial, pre-start metal temperatures of the main steam-lines and valve steam-chests are close. In particular, this condition was provided for standard Soviet supercritical-pressure power units with improved thermal insulation of the

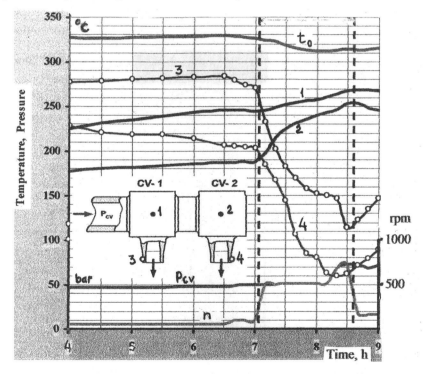

Figure 9-1. Chilling the HP control-valve steam-chest sleeves downstream of the saddle during pre-start warming up with the closed control valves. *Source: E. Plotkin et al.*[3]
Metal temperatures 1-4 are marked according to the cut in sketch

main steam-lines (see Figure 8-15). As a result, for these units the MSVs have been used permanently open and closed only for repairs of the boiler or turbine components. In this case, the stop-valves are opened with the closed control valves immediately, as soon as the boiler is ignited and superheated steam enters the main steam-lines. Nevertheless, in this case there retains a danger of the above mentioned chilling downstream of the control-valves saddles. In addition, if the HP valve steam-chests are located separately from the HP cylinder and connected with it by crossover pipes, the latter ones should be also preheated, which demands for some additional technological operations.

Many problems can be resolved if the pre-start warming of the main steam-lines, HP valve steam-chests, and crossover pipes is conducted with open HP both stop and control valves and closed IP valves and the permission to increase partially the turbine's rotation speed. Therewith, the heating steam from the HP cylinder can be discharged into the IP cylinder through the special blowing-down line, which makes it possible to preheat additionally the HP and IP rotors at cold start-ups to the level above the fracture appearance transition temperature (FATT) to prevent possible brittle fracture of the rotors under the combined action of centrifugal and thermal stresses. In particular, this technology was worked up and implemented at the Israeli power plant Eshkol with four 228-MW reaction-type subcritical-pressure steam turbines.[3]

Each turbine consists of separate single-flow HP and IP cylinders and one double-flow LP cylinder. Both the HP and IP cylinders are of a double-casing design with additional nozzle boxes in the HP cylinder. All three rotors (HP, IP, and LP) are made forged with central bores. The turbine has two HP valve units, each one comprising one stop valve and two control valves. Each pair of control valves is disposed in series within a common steam-chest, that is, a steam chamber after the stop valve,—see the sketch in Figure 9-1. Leaving the control valves, the steam passes to the HP cylinder through the crossover pipes. Because of cracks revealed in the control-valve steam-chests, the turbine producer required preheating them at cold and warm start-ups for an hour with the opened stop valves and closed control valves. Then, the HP cylinder is to be preheated if its pre-start metal temperature is lower than 200°C (390°F). This process runs with open HP valves and closed IP valves. The IP cylinder is preheated in parallel by steam taken from the HP cylinder's intercasing space. This procedure is ordered to last at least six hours, but in reality it usually takes about nine hours. Such great time

Start-up Technologies as Applied to Different Start-up Systems 251

expenses are tied with limits set by the turbine producer and presented in Table 9-1.

Table 9-1. Pre-start warming limits as applied to the Eshkol power plant's 228-MW turbines

Parameters	Original Limitations of the Turbine's OEM	New Limitations
Steam pressure at the HP control stage, bar	< 5 – 6	< 18 - 20
Rotation speed, rpm	≤ 200	≤ 800
Condenser backpressure, mm Hg	≤ 150	≤ 300
Resultant pre-start warming duration, hours	≥ 6	≤ 2.5

Source: E. Plotkin et al.[3]

Analysis of these processes allowed improving substantially this technology and making the pre-start procedure much shorter and more effective. At the beginning stage, the warming process of the HP cylinder goes quite intensely with condensation of the heating steam on the heated surfaces, and the measured metal temperatures closely follow the saturation temperature corresponding to the steam pressure in the cylinder. When the heating steam pressure reaches its final value or if the initial metal temperature exceeds the saturation temperature, the subsequent heating process sharply slows down because of a rather stagnant heat transfer by convection. This is especially characteristic for the HP outer casing because of low steam velocities in the intercasing space. With the limited rotation speed, it is practically impossible to raise the metal temperature of the HP outer casing over 220°C (428°F). To warm up the HP cylinder to a higher temperature level, it was advised to increase the steam pressure in the HP cylinder, even if this causes some increase of the rotation speed. However, this process can be accompanied with a rapid growth of the HP relative rotor expansion (RRE). Keeping this in mind, the steam pressure rise in the cylinder was limited by the value of approximately 2 MPa (290 psi). In this case, the metal temperature of the HP outer casing grows to about 200°C (390°F) in about half an hour after attaining the final pressure and then gradually rises due to the convection heat transfer process. The rate of raising the steam pressure in the HP cylinder is set depending on its initial metal temperature. A higher steam pressure level in the HP cylinder increases the steam flow amount and heat transfer intensity for the both the HP and IP cylinders. As a result, the metal temperatures of the IP casings reach the aimed final value about 140°C (285°F). Limitation of the rotation speed increase is provided by an increase of the condenser backpressure.

As applied to the considered turbines, it is sufficient to keep the condenser backpressure at the level of approximately 200 mm Hg (27 kPa, or 8 psia). In this case, with the steam pressure at the control stage of about 2 MPa, the turbine's rotation speed is kept within the range of 600-to-800 rpm, which is well apart from the first critical speed equal to 1,100 rpm. A rich experience with 300-MW supercritical-pressure turbines with 960-mm (38 in) LSBs proves that even under the condenser backpressure of 250-300 mm Hg (33-40 kPa, or 5-6 psia) the rotation speed of up to 900 rpm does not make them be overheated. In order to have the IP cylinder preheat by condensation to a higher level, after long outages it is also advisable to begin the pre-start heating process with even a higher condenser backpressure of about 300-350 mm Hg (40-47 kPa, or 6-7 psi) with subsequent lowering it to the above mentioned value. In this way, according to new recommendations, the turbine cylinders are preheated with the increased steam pressure at the HP control stage, increased rotation speed, and worsened vacuum in the condenser—see Table 9-1. Eventually, it allows decreasing duration of preheating the turbine cylinders to about 2-2.5 hours—Figure 9-2.

When the heated steam pressure in the HP cylinder reaches its final value, the HP and IP cylinders are heated only by the convective heat transfer. Under these circumstances, it was found reasonable to begin raising the main steam temperatures at this stage gradually for two hours from 300-320°C to 350-360°C (from about 600°F to 670°F). In doing so, the RREs of the cylinders remain within the allowable range. A relatively low main steam pressure combined with an enlarged steam pressure inside the HP cylinder decreases throttling in the HP control valves and, correspondingly, increases the heating steam temperatures in the turbine. This results in higher final metal temperatures of the HP cylinder casings—by approximately 30-50°C, or 55-90°F.

At cold start-ups, the HP stop-valve casings are gradually warmed up while the boiler is fired and the main steam pressure rises. The HP

Start-up Technologies as Applied to Different Start-up Systems   253

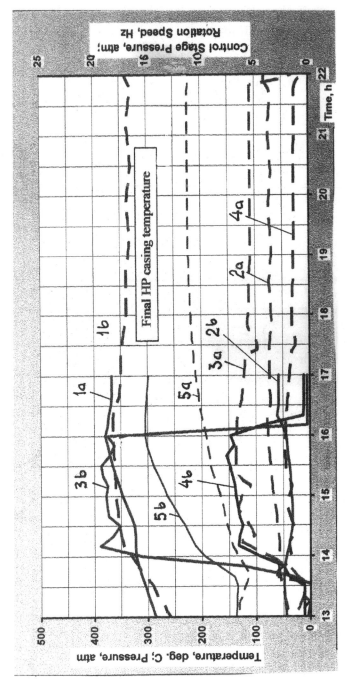

Figure 9-2. Comparison of preheating procedures for 228-MW subcritical-pressure steam turbines according to old (a) and new (b) technologies. *Source: E. Plotkin et al.*[3]
1 – main steam temperature, 2 – main steam pressure, 3 – steam pressure at the control stage, 4 – rotation speed, 5 – metal temperature of the HP cylinder's outer casing;

control valve steam-chests are warmed in parallel with preheating the HP cylinder. The pattern of raising the steam pressure in the HP cylinder ensures acceptable thermal stresses not only in the HP and IP cylinders but in the HP valve steam-chests, too. Due to an increase of the main steam temperature by the end of the pre-start warming, the HP valve steam-chests are heated up to approximately 345°C (650°F). As a result, the highest thermal stresses in the heated HP components (valve steam-chests, rotor, inner and outer casings) are approximately half as much as they were during the pre-start warming according to the former instruction. For cold start-ups, the pre-start warming operations can begin before the main steam conditions after the boiler reach their pre-start values. The HP valves can be opened and the steam can be given into the turbine when the boiler is ignited and the main steam parameters reach the values of about 2 MPa (290 psi) and 260°C (500°F). In the process of raising the heated steam pressure, the main steam conditions have time to reach their pre-start values.

Such an improved technology allowed turbine operators to decrease the duration of pre-start warming to 2-2.5 hours even after long outages instead of average nine hours, as it used to be with the former manual. The total start-up duration is additionally reduced (by at least 0.5 h), if the pre-start warming begins when the main steam parameters are only rising to its pre-start values. What is more, the higher metal temperatures of the HP and IP cylinders by the end of the preheating process additionally make it possible to shorten the subsequent duration of loading the turbine by about 2-3 hours. These considerations make the described technology advisable not only for cold start-ups but also for warm start-ups after weekend outages to improve the power unit's flexibility at the stage of loading the turbine.

It should be noted that, if necessary—for turbines with bulk welded or forged LP rotors at cold start-ups, the time of pre-heating the main steam-lines, valve steam-chests, crossover pipes, and cylinders can be also used for some pre-start warming of the LP rotors by steam given to the end seals. At this stage, the LP rotors can be effectively heated only up to the saturation temperature at the turbine backpressure, and from this standpoint lowering the vacuum in the condenser occurs very useful. This process requires the time amount in proportion to the squared rotor radius and usually takes about 3-3.5 hours—Figure 9-3.

For units with two-bypass start-up systems, pre-start warming of the main steam-lines runs in parallel with warming of the hot reheat

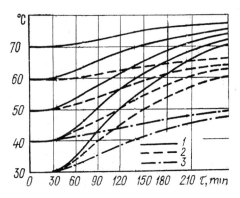

Figure 9-3. Calculated characteristics of pre-start warming for a typical welded LP rotor (of the external body diameter equal to 1,200 mm) with various initial temperature states and vacuum in the condenser.
Source: A. Leyzerovich[4]
1, 2, and 3 – for the vacuum values of 400, 540, and 650 mm Hg (the backpressure values – 48, 30, and 15 kPa)

steam-lines. In doing so, steam goes through the HP bypass into the cold reheat steam-lines, passes through the reheater, and goes away into the condenser through the LP bypass just before the IP valves. It is commonly assumed that before steam enters the IP cylinder the hot reheat steam-lines have to be pre-heated up to the mean temperature level differing from the initial metal temperature of the IP cylinder by not more than 50-100°C (90-180°F). In addition, the minimum metal temperature must be higher than 100°C (212°F).[1]

Steam directed to the reheat system can also be used to heat the HP cylinder and crossover pipes downward of the HP control valves if these valves are installed for a distance away from the HP cylinder. The HP cylinder and its crossover pipes are heated through the exhaust, and the temperature of steam given into the cold reheat steam-lines is kept matching the metal temperature of the HP cylinder. If the HP cylinder is furnished with check valves at the cold reheat steam-lines, steam is given into the cylinder through the governing valve at their bypass.

The standard rolling-up technology for steam-turbine units with two-bypass start-up systems implies giving steam first into the IP cylinder (section) with the closed HP control valves. For cold and warm start-ups, in order to shorten their total duration, steam can be given into the IP cylinder with simultaneous continuation of warming the HP cylinder and its crossover pipes from the HP exhaust side. However, as soon as the turbine reaches a certain intermediate rotation speed, the bypass valve of the check valves at the cold reheat steam-lines is closed, and the HP cylinder is set under vacuum through a special blowing-down line connecting the cylinder to the condenser. This is necessary to prevent the overheating of the HP blading because of the windage effect (friction and fanning).

Usually, the turbine is led to the synchronous rotation speed, and the generator is switched on to the grid without opening the HP control valves. They are opened automatically with accepting the initial load. For this reason, before the generator is synchronized, it is necessary to check the readiness of the HP cylinder to accept steam. In the first place, this refers the temperature consistence between the main steam and HP cylinder metal.

For units with one-bypass systems, the rotation speed increase during the entire rolling-up stage is controlled by the HP control valves. At hot and warm start-ups, great steam temperature drops with throttling in the control valves at low steam flows can cause a significant decrease in the temperature of steam entering the HP cylinder with resultant negative temperature differences in the HP rotor. This process is often aggravated by steam chilling in the insufficiently heated HP crossover pipes. With the steam flow amount increase, this effect diminishes and disappears—Figure 9-4. That is why it is so important not to stay at the intermediate rotation speed when the turbine is started up after short outages.

As to cold and warm start-ups, if the metal temperatures of the hot reheat steam-lines are inconsistent with the metal temperature of the IP cylinder (that is, if their difference is more than the set value of about 50-100°C, or the lower steam-line metal temperature is less than 100°C), the rolling-up step at the intermediate rotation speed can be used to heat the hot reheat steam-lines without giving steam into the IP cylinder. For this aim, the IP control valves are forcedly held closed while the HP control valves are opening. (This can be done, for example, with the use of a special line which relieves the oil pressure in the hydraulic control circuit of the IP valves.) In this case, steam passing through the HP cylinder and reheat system is discharged into the condenser before the intercept valves. If the turbine's first critical rotation speed is over 1,050-1,100 rpm, the hot reheat steam-lines are heated at the rotation speed of up to 900-1,000 rpm. The steam flow through the reheat steam-lines at these conditions is not so great, and the whole process takes more than 40-50 minutes. In order to somewhat intensify this process it is advisable to conduct it with a lower vacuum in the condenser. Sometimes, especially at start-ups after the week-end outages, the operator does not have enough time and patience and opens the IP valves, giving steam into the IP cylinder, before the heating is finished. This results in undesirable chilling of the IP rotor (see Figure 9-4).

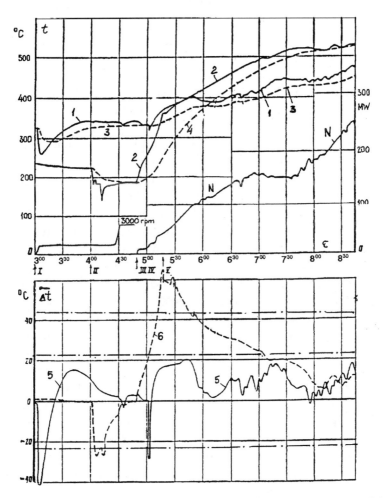

Figure 9-4. Heating of the HP and IP rotors of a 300-MW supercritical-pressure steam turbine with a one-bypass start-up system at a warm start-up. *Source: A. Leyzerovich[5]*
1 and 2 – measured steam temperatures at the HP control stage and in the IP steam admission chamber, 3 and 4 – average integral metal temperatures of the HP and IP rotors in their most thermally stressed sections, 5 and 6 – effective radial temperature differences in the HP and IP rotors in the same sections.
I – giving steam into the HP cylinder with closed IP intercept valves, II – opening the IP intercept valves, III – switching on the generator to the grid, IV – switching on the main spray attemperator, V – opening the bypass valve of the boiler's reheater

START-UP LOADING AND RAISING THE STEAM CONDITIONS

When the generator is synchronized and switched on to the grid and the turbine accepts the initial load, the amount of steam flow through the turbine sharply increases, and the processes of heating the turbine components become more intense compared with the previous stage. This concerns all the turbine's critical elements.

For power units with one-bypass start-up systems, reaching the synchronous speed by the turbine and acceptance of initial load makes the reheat steam temperature after the boiler rise rapidly. In addition, steam temperature drop lengthwise of the hot reheat steam-lines quickly reduces, too, since these steam-lines begin to be heated more intensively. This results in a rapid increase of the temperature difference and hence thermal stress in the IP rotor. To stop this increase, the steam bypass of the reheater has to be open in two-five minutes after the generator is switched on. This makes it possible to keep reliably this temperature difference within the admissible range (see Figure 9-4); otherwise, the temperature difference and thermal stress in the IP rotor can exceed their upper admissible values several times—see Figure 7-8.

When the turbine accepts the initial load, the turbine bypasses can be closed, and further loading of the turbine can be conducted by increasing the boiler's firing rate. Along with this, at many power units, the turbine bypasses remain open during almost the entire turbine start-up. This really makes it possible to load the boiler and turbine independently with discharging the excessive steam amount generated by the boiler into the condenser. In this case, the rate of heating the turbine can be partially corrected by fast-response changes of the load, and requirements to the quality of the steam temperature raise control can be somewhat lowered. At the same time, these steam flows besides the turbine essentially increase the start-up fuel outlay, and this technology does not seem reasonable.

Heating the turbine during its start-up loading mostly depends on how smoothly the main and reheat steam temperatures are raised. Any excursions of these temperatures immediately make the temperature differences in the critical elements change sharply because of high heat transfer conditions from the heating steam to the heated surfaces.

Ultimately, the temperature of steam entering the turbine should vary linearly in time or according to an exponential law to provide keeping the temperature differences in the turbine's critical elements

Start-up Technologies as Applied to Different Start-up Systems 259

near their upper admissible levels. At the same time, variations in the steam temperature should also take into account the changes in the HP control valve position and resulted changes in the steam temperature drop at the turbine entrance. To hasten accepting the full turbine load, it can be reached at the decreased main and reheat steam temperatures, so that these temperatures can be raised to the nominal level when the turbine already works under the full load. Because both the initial (pre-start) levels and admissible rates of raising the main and reheat steam temperatures differ, they reach their nominal levels for different time periods. So, if for the specific start-up its rate is predominantly limited by, for example, the temperature difference in the IP rotor, the main steam temperature can reach its nominal value before loading is finished.

A growth of the steam temperatures goes with raising the boiler's firing rate and is corrected by the start-up attemperators. While the boiler's firing rate is rising, the main (regular) attemperators come into operation, too, necessitating corrections for the start-up attemperators. These intricate processes often cause excursions of the steam temperatures and appearance of increased temperature differences and thermal stresses in the rotors (see Figures 7-8 and 9-4), affect the turbine reliability because of low-cycle thermal fatigue, and are extremely advisable to be automated.

Many start-up systems contain the embedded valves between the evaporating and superheating surfaces, as in Figure 7-7, or shut-off valves with bypassing throttle valves between the primary and secondary superheaters, as for many U.S. units. The pressure upstream of these valves is kept nominal, whereas the throttle steam pressure before the turbine depends on position of the control or throttle valves. The embedded (or shut-off) valves are to be opened when the flow capacity of the throttle valves or the valves between the separator and superheater is exhausted. In this case, the pressure difference across these valves should be sufficiently small, that is, the throttle steam pressure before the turbine's control valves has to be close to the nominal value. According many manuals, it is desirable to maintain all the control valves during the start-up operations fully opened to provide the uniform heating of the HP casings and valve steam-chests. In this case, before the embedded valve is opened, it is necessary to close partially the HP control valves to increase the throttle steam pressure. It is understandable that such an operation make the steam temperature at the HP cylinder entrance decrease perceptibly due to throttling in the control valves. As a result, there appear significant negative temperature differences in the HP rotor and cylinder casing wall, what is quite undesirable from the standpoint of thermal fatigue strength of these elements. This process is most intensive for supercritical-pressure units. To compensate for this temperature drop, the main steam temperature has to be additionally raised. Yet this brings about the increased temperature differences and thermal stresses in the stop-valve steam-chests. For all these reasons, such units are desirable to be loaded with the partially closed control valves of the turbine, so that the main pressure nears the nominal value by the instance when the flow capacity of the start-up complex exhausts. In this case, the process of heating the turbine is not disturbed and remains smooth and continuous.

For power units with two-bypass start-up system and full-capacity separators, operated with sliding steam pressure in the entire steam/water path of the boiler, the main steam pressure achieves its rated value by the end of loading. An example of such a start-up, as applied to a 1,000-MW turbine of Hitachi with the steam conditions of 24.1 MPa, 593/593°C (3,495 psi, 1,100/1,100°F) at the Japanese Matsuura Unit 2 is shown in Figure 9-5.

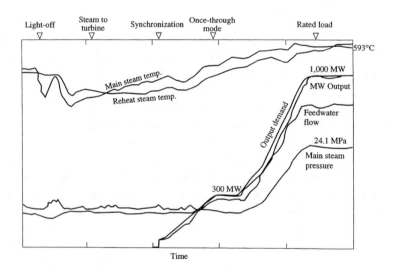

Figure 9-5. Hot start-up of the 1,000-MW Matsuura Unit 2 with Hitachi's steam turbine. *Source: K. Sakai et al.*[6]

Start-up Technologies as Applied to Different Start-up Systems   261

PECULIARITIES OF STEAM-TURBINE
START-UP TECHNOLOGIES FOR CC UNITS

The start-up procedure for CC units can be subdivided into three major phases:[7]

- purging of the HRSG(s),
- rolling-up, synchronization, and loading of the gas turbine(s), and
- rolling-up, synchronization, and loading of the steam turbine.

In order to prevent the explosion of unburned hydrocarbons left in the HRSG path from earlier operation, it is advisable, especially for CC units with oil-fired gas turbines, to purge the steam generator(s) before igniting the gas turbine(s). This is done by running the gas turbine at the ignition speed (approximately 30% of the synchronous value) with the generator as a motor, or with any other starting devices, in order to blow air through the HRSG. The purge time depends on the HRSG volume, which has to be exchanged by up to a factor of five with "clean air" before the ignition takes place. After purging, the gas turbine is ignited, run up to the nominal (synchronous) rotation speed, and loaded to the desired load. Gas turbines can be started up sufficiently fast and their start-up duration and technology can be considered independent on the outage duration. At the phase of rolling up the gas turbine, at its beginning stage, until the turbine ignites, its rotation speed is increased by the generator driving the turbine as a motor and switched on to the grid through the static frequency converter (SFC). This rotation speed is further increased by increasing the fuel fed to the turbine's combustion chambers. From about 2,000 rpm, the speed is increased to the synchronous value without the support of the SFC. The rate of loading the gas turbine is limited by thermal stresses in the thick-wall components of its HRSG (mainly, its HP drum) and the steam turbine.

It is supposed that the start-up duration, without a phase of purging the HRSG(s), for modern CC units with the output of up to 450 MW lies in the following ranges:

- for hot start-ups (after 8-hour outages)—from 40 to 50 min,
- for warm start-ups (after 60-hour outage)—from 75 to 110 min, and
- for cold start-ups (after 120-hour outage)—from 75 to 150 min.

Model diagrams of hot and warm start-ups for a 250-MW-class CC unit (with a single gas turbine) are presented in Figure 9-6.

Appropriate steam conditions matching the pre-start temperature state of the steam turbine are reached at approximately 50-to-60% gas

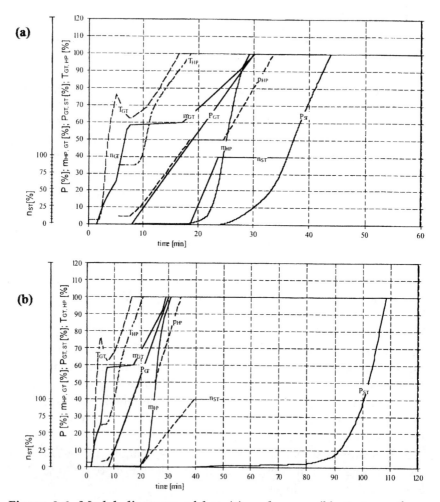

Figure 9-6. Model diagrams of hot (a) and warm (b) start-ups for a 250-MW-class CC unit (after 6-hour and 60-hour outages, respectively). *Source: R. Kehlhofer et al.[7]*

n_{GT} – gas turbine rotation speed, n_{ST} – steam turbine rotation speed, m_{GT} – gas turbine exhaust mass flow, T_{GT} – gas turbine exhaust gas temperature, T_{HP} – HP main steam temperature, m_{HP} – HP main steam flow, p_{HP} – HP main steam pressure, P_{GT} – gas turbine load, P_{ST} – steam turbine load.

Start-up Technologies as Applied to Different Start-up Systems   263

turbine load. This means 40-to-60% of the nominal steam pressure and a sufficient degree of superheating (that is, around 50°C, or 90°F). Before the steam turbine is begun to start-up, the gland seal steam system must be in operation and the condenser must be ready for steam consumption. Until the steam turbine takes over the steam flow, it is discharged across the turbine steam bypasses into the condenser. With the "parallel" bypass system (see Figure 7-9b), the steam generated in the HP and IP drums of the HRSG after being attemperated is sent directly to the condenser. With the "cascade" system (see Figure 7-9a), the HP steam is bypassed around the HP section of the turbine to the cold reheat line and, mixed with the steam from the IP drum, is sent through the reheater and then discharged to the condenser. The last scheme, with a "pressurized" or "wet" reheater, is presently considered more preferable and has gained wider acceptance for modern CC units. In particular, it provides a less danger of solid particle erosion for the steam turbine and somewhat shortens the start-up duration compared to the "parallel" scheme with a "dry" reheater. Along with this, it makes the steam turbine's start-up procedure somewhat more complicated, as it is in the case of steam-turbine power units with two-bypass start-up systems compared to those with one-bypass scheme.

According to Siemens, after the temperature of steam generated by the HRSG reaches the level matching the average metal temperature of the HP valve steam-chests, the start-up procedure for steam turbines of CC units with a "cascade" or "pressurized reheater type" bypass system includes the following operations: [8]

- warming up the turbine's HP valve steam-chests, with the open stop valves and closed control valves by the main steam discharged through the blowing down line between the stop and control valves,

- increasing the rotation speed to the intermediate level and then to the synchronous speed by opening the IP control valves, with the HP control valves keeping the work split between the HP and IP-LP sections to ensure maximum flow to the latter and HP exhaust check valves closed until the HP casing is pressurized to the cold reheat pressure,

- accepting the minimum load by opening the HP and IP control valves up to the set position and subsequent ramp loading the turbine up to the set level by opening the HP and IP valves in the mode providing the chosen loading rate, with the IP control valves completely open at 70% admission demand,

- transition of the HP control valves to the operation mode of controlling the main steam pressure (the steam turbine following the gas turbine/HRSG load) or sliding pressure with fully opened valves and closing the HP and IP bypasses.

Until the turbine accepts the initial load with the increase of the steam flow through the HP section, the steam pressure in the HP casing is controlled to prevent the temperature from rising above 500°C (930°F) because of the windage effect. In addition, the turbine is equipped with a "Turbine Stress Evaluator" (TSE) estimating the unsteady-state stress states of the most stressed ("critical") thick-wall turbine elements: the HP valve steam-chests, the HP casing and rotor, and the IP rotor. It is supposed that the thermal stresses in these components are in proportion to the differences between the surface and the average integral metal temperatures. These metal temperatures are either measured directly or calculated based on the measured steam temperatures, as it is done for the rotors. The various temperature differences are subtracted from the current allowable values to obtain a "free margin," which should not be negative. When the free margin becomes equal zero, it means that the component is stressed to its maximum allowable level for the given conditions. The actual operation diagram of a cold start-up for a CC steam turbine with results of TSE control is presented in Figure 9-7.

For multi-shaft CC units, with several gas turbines, they can be started up in parallel or successively with some delays between them and the steam turbine load following the steam production of the HRSGs. Steam bypass systems allow the gas and steam turbines to be started up independently. An example of a hot start-up (after 11 hours of outage) for a multi-shaft CC unit with three gas turbines and one steam turbine of Hitachi with the total capacity of 670 MW (the Himeji Daiichi Unit 6: 3x140+250 MW) is shown in Figure 9-8. Three gas turbines start first one after another, and then the steam turbine starts up. The unit's load increases in cooperation with the gas and steam turbines after the steam cycle changeover completes.

The described technology was primarily developed for CC units with separate gas and steam turbines, but it remains as well acceptable

Start-up Technologies as Applied to Different Start-up Systems 265

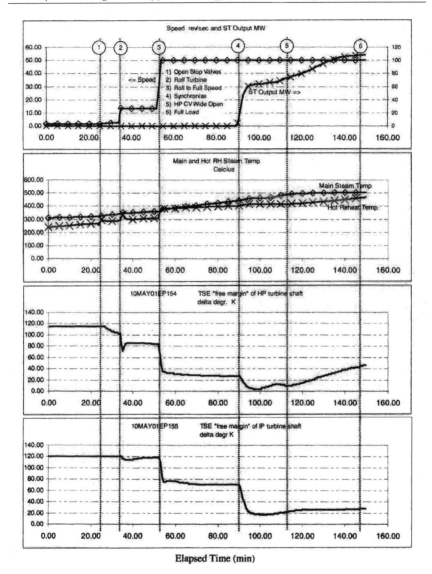

Figure 9-7. Actual cold start-up of a CC steam turbine of Siemens with Turbine-Stress-Evaluator (TSE) control. *Source: L. Bize et al.*[8]

for single-shaft (SS) CC units with a self-shifting and synchronizing (SSS) clutch connecting the steam turbine with the generator common for both the gas and steam turbines and rigidly connected to the former—see Figure 3-30. While the unit is shut down, the entire turbine shaft-line,

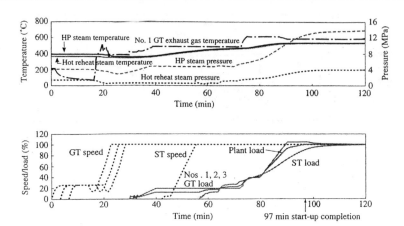

Figure 9-8. Hot start-up of a multi-shaft 670-MW CC unit of Hitachi (Himeji Daiichi Unit 6). *Source: A. Kawauchi et al.*[9]

with the clutch engaged, is rotated by a steam turbine's turning gear. The gas turbine is started up by disengaging the clutch and motorizing the generator with a variable-frequency static converter. As the gas turbine is rolled up, synchronized and takes electrical load, its exhaust gas provides steam from the HRSG. The steam is routed through a steam-bypass system, while the steam turbine's HP valves are warmed up. When sufficient steam pressure is available, the steam turbine is rolled up and accelerated to the synchronous speed. The clutch automatically engages when the steam turbine's shaft reaches the synchronous speed, and steam turbine accepts the load and then is loaded by increasing the steam flow amount through the turbine. When the unit is to be stopped, the steam turbine's valves are closed; the clutch is disengaged, and the steam turbine's shaft speed drops to the turning-gear speed, where it is maintained by the gas turbine running at the rated speed owing to lubricant film friction in the clutch. When the gas turbine is shut down and its rotation speed drops, the turning gear starts, and the clutch is engaged to keep the entire shaft-line rotating. Such arrangement of SS CC units, with the SSS clutch between the steam turbine and the generator, is primarily intended to provide a possibility of operating the gas turbine in a simple-cycle mode, without the steam turbine. It is also considered preferable for SS CC units intended for operation in a cycling mode with frequent (daily) shut-downs and restarts.[10]

Along with this, there exist SS CC units with the steam turbine settled at the shaft-line between the gas turbine and generator, with permanent engagement of all their shafts (see, for example, Figures 3-26 and 3-29a). For such units, the steam turbine is rolled up together with the gas turbine, even though at the initial start-up stages it is rotated without steam in its steam path. An example of a hot start-up after eight-hour outage for a 245-MW SS CC unit (Kawagoe Unit 3: 151+ 84 MW) is shown in Figure 9-9. In this case, the steam generated from the HRSG is introduced into the steam turbine after completion of rolling up and synchronization. In doing so, the steam turbine is not chilled because the steam temperature drops in the HP control valves are not great. On the other hand, while the gas turbine is rolling up, with the steam turbine rotating in no-flow conditions, for relatively large steam turbines with long LSBs there arises a danger of their overheating because of the windage effect. Under these conditions, it may occur necessary to hold the rolling up process at an intermediate rotation speed until the HRSG begins generating steam acceptable for feeding the steam turbine.

Such a cold start-up profile as applied to a SS CC unit STAG (see Figure 3-29a) is shown in Figure 9-10. These units are equipped with steam-cooled gas turbines S107H/S109H, depending on the grid frequency, and steam turbines of GE or Toshiba with the LSBs of 1,016 mm or 1,067 mm long for the rotation speed of 3,600 rpm or 3,000 rpm, respectively. The H-technology gas turbines are cooled by steam taken from the outlet of the HRSG's IP superheater, supported as necessary with steam from the turbine's HP exhaust; cooling steam is returned to the cold reheat steam lines. At start-ups, while the HRSG's productiv-

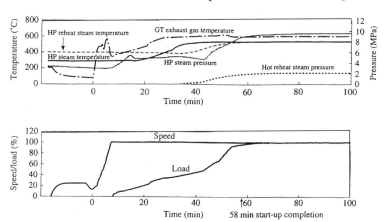

Figure 9-9. Hot start-up of a single-shaft 245-MW CC unit of Hitachi (Kawagoe Unit 3). *Source: A. Kawauchi et al.*[9]

ity is not sufficient, the gas turbine is cooled by air extracted from the compressor discharge. The turbine is hold at the rotation speed of about 60% of the synchronous value until steam from the HRSG can be given to the LP steam path, preventing the LSBs from overheating. This function could be provided by steam generated by an auxiliary boiler, but in this case its productivity is not sufficient, and it is used only for providing the turbine's end seals with steam. After switching on the generator to the grid, there is no longer need in additional steam for cooling the LP steam path, and the gas turbine's cooling system is transferred to steam.

All the above mentioned CC units are equipped with drum-type boilers. Along with this, there appear first CC units equipped with boiler with once-through (Benson) HP and IP sections, without drums. This not only makes it possible to increase the main steam pressure after the HRSG (ultimately, up to a supercritical level), heightening in this way the unit's efficiency, but also improves the unit's operating flexibility and cost effectiveness thanks to eliminating thick-walled steam drums and replacing them with small separator vessels. As a result, while starting up, the rate of loading gas turbine gives up being limited by thermal stresses in the HRSG, and the start-up duration can be reduced, especially for cold start-ups—Figure 9-11.

The first CC unit with a Benson HRSG was developed by Siemens and since 1999 has been successfully operated in the Cottam Develop-

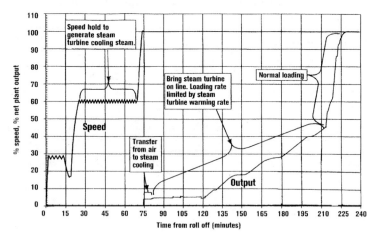

Figure 9-10. Typical cold start-up profile for a single-shaft CC unit **STAG.** *Source: "Structured steam turbines..."*[11]

Start-up Technologies as Applied to Different Start-up Systems 269

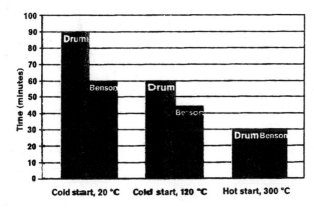

Figure 9-11. Comparison of start-up times for CC units with drum-type and Benson HRSGs. *Source: J. Franke et al.*[12]

ment Center, the UK.[12-15] The 390-MW SS CC unit with V94.3A gas turbine of Siemens has a design lifetime of 25 years with about 250 start-ups per year. The unit's automatic control system provides a possibility of starting up the units with a capability to bring it from the standstill to full load in about 50 minutes in daily start-stop operation, without risk of reducing the lifetime of the HRSG.[14]

It is also supposed to use once-through (Benson) boilers at multishaft CC units of heightened flexibility, specifically intended for operation in a cycling mode. In particular, such a 2.W501F Fast Start unit, developed by Siemens Westinghouse with two W501F gas turbines, is assumed to have about 200 hot and 50 warm start-ups per year. The diagram of Figure 9-12 presents a start-up diagram for such a unit after a 16-hour outage as compared to a common CC unit. The unit's start-up begins with concurrent ignition of both gas turbines at the time instant designated as *1*. Their generators are synchronized at the time instant *2*, and both the gas turbines are loaded in parallel up with the rate not limited by their HRSGs. When the gas turbines are loaded to their rated output (*3*), steam is given to the steam turbine, and it begins rolling up. The steam turbine's generator is synchronized at the time point *4*; the turbine bypasses are closed at *5* and the steam turbine accepts the entire generated steam flow. At this point the unit reaches approximately 97% of its rated power output.

References
1. Plotkin E.R. and A. Sh. Leyzerovich. *Start-ups of Power Unit Steam Turbines* [in Russian], Moscow: Energiya, 1980.
2. Leyzerovich A. *Large Power Steam Turbines: Design & Operation*, 2 vols., Tulsa (OK): PennWell Books, 1997.

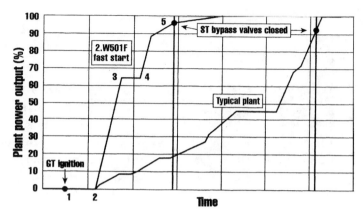

Figure 9-12. Power output during start-up after 16-hour outage for a "fast start-up" CC unit. *Source: "Fast work for the Benson..."*[16]

3. Plotkin E., Y. Berkovich, A. Leyzerovich, et al. "Improvement of Pre-Start Heating Technology for Steam-Turbine Units," presented at the Electric Power Conference 2005, Chicago, 2005.
4. Leyzerovich A. Sh. "Heating up Welded LP Rotors of Steam Turbines at Start-ups" [in Russian], *Energetik*, no. 10 (1982): 34-35.
5. Leyzerovich A.Sh. *Technological Fundamentals for Steam Turbine Start-up Automation* [in Russian], Moscow: Energoatomizdat, 1983.
6. Sakai K., S. Morita, T. Yamamoto, and T. Tsumura. "Design and Operating Experience of the Latest 1,000-MW Coal-Fired Boiler," *Hitachi Review* 47, no. 5 (1998): 183-187.
7. Kehlhofer R., R. Bachmann, H. Nielsen, and J. Warner. *Combined-Cycle Gas & Steam Power Plants*. 2nd ed. Tulsa (OK): PennWell, 1999.
8. Bize L., H. Martin, N. Henkel, and E. Gobrecht. "Combined Cycle Steam Turbine Operation," presented at the 2001 International Joint Power Generation Conference, New Orleans, 2001, PWR-19127.
9. Kawauchi A., F. Hirose, and M. Musashi. "High-Efficiency Advanced Combined-Cycle Power Plants," *Hitachi Review* 46, no. 3 (1997): 121-128.
10. "Single-train power unit stands out at King's Lynn," *Electric Power International*, Summer 1998: 48-50.
11. "Structured steam turbines for advanced combined cylces," *Modern Power Systems* 20, no. 10 (2000): 41-46.
12. Franke J., U. Lenk, R. Tauchi, and F. Klauke. "Advanced Benson HRSG makes a successful debut," *Modern Power Systems* 20, no. 7 (2000): 33-35.
13. Emberger H.-M., P. Mürau, and L. Beckman. "Siemens Reference Power Plants—Translating Customer Needs into Plants," *VGB PowerTech* 85, no. 9 (2005): 78-83.
14. Jeffs E. "New combined cycle tackles cycling," *Turbomachinery International* 41, no. 2 (2000): 31-33.
15. Smith D. "HRSG advances keep GTCC competitive," *Modern Power Systems* 24, no. 8 (2004): 31-35.
16. "Fast work for the Benson once-through HRSG," *Modern Power Systems* 24, no. 2 (2004): 23-25.

Chapter 10

Start-up Instructions for Steam-Turbine Power Units And Their Improvement

DIFFERENT APPROACHES TO CONSTRUCTING START-UP INSTRUCTIONS

Until the first reheat steam turbines appeared, there had not existed such a notion as a "steam-turbine power unit," with a permanently joined turbine and boiler. As a rule, steam turbines at power plants were fed with steam from a common header and were started up with invariable, nominal steam parameters. Under these circumstances, the existence of separate, independent start-up instructions for turbines and boilers developed by their manufacturers was absolutely logical, especially as turbines and boilers were delivered by different producers. This situation still persists up to now at power and cogeneration plants with non-reheat steam turbines of a moderate capacity.

Boilers and turbines of early steam-turbine power units were also started up successively and independently, so the boiler was fired with discharging the generated steam through turbine bypasses into either the atmosphere or turbine condenser, and the turbine's start-ups were run with invariable, nominal steam parameters. Only since the 1940s, it has become widely practiced to start up the power unit's boiler and turbine in parallel, with sliding steam parameters. In this case the turbine's start-up profiles occur to be dependent to a great degree on the boiler's capability of controlling the steam conditions and their raise. Nevertheless, even presently, more often than not, the boiler's and turbine's producers develop their start-up instructions independently as applied to some general characteristics of the used start-up systems and other equipment of the supposed power unit.

As a result, power plant operators have to deal with individual start-up instructions for the boiler and the turbine, which are not always properly agreed to one other. Sometimes, start-up instructions for boilers and turbines are copied from one power unit to another (even if their start-up systems are somewhat different), are not renewed for years, and as a result become rather obsolete.

Figure 10-1 shows the instruction diagram for cold start-ups as applied to a 500-MW supercritical-pressure steam turbine like those operated at Korean power plants as, for example, the Taean, Samchonpo, Hadong, Tangjin, and Young Heung. Even the very superficial analysis of this diagram shows it is rather far from optimal. Below are only a few notes concerning this diagram. First, it is not reasonable to raise vacuum in the condenser up to the nominal level at once, before steam is given to the turbine. It would be much more effective to warm the turbine rotors with an increased backpressure in the condenser, and this procedure would bring much better results and take much less than eight hours according to the diagram (see the previous chapter). Second, any turbine operator would prefer to listen to the turbine at an intermediate rotation speed much lower than 3,000 rpm ordered in the diagram, and this time period could be used to warm additionally the rotors. Third, if it is possible to keep the main steam temperature after the boiler at the level of about 300°C before steam is given into the turbine, it is advisable to keep the steam temperature at this level until the generator switches on to the grid. The rise of the main steam temperature in parallel with rolling up the turbine and accepting the initial load inevitably causes an uncontrollable growth of the temperature differences and thermal stresses in the HP steam admission zone. Fourth, a quick uncontrolled increase of the main and reheat steam temperatures in the process of loading unavoidably makes this process slower. If the steam temperatures were raised with smaller rates, this would allow the operator to increase the rate of loading the turbine. In addition, the final (nominal) turbine load could be reached with the steam temperatures less than their nominal values. Similar vices are inherent in the instruction diagrams for other start-up types.

As to the instruction itself, it does not tie the increase of the turbine load and main steam pressure with controlling the turbine bypasses and HP control valves. It also does not describe how to control the main and reheat steam temperatures during start-ups depending on the turbine metal temperatures, but rather forms the rolling up and loading patterns depending on the steam-metal temperature mismatch. Such an approach is not optimal. In addition, all the recommendations for shaping the

Start-up Instructions for Steam-turbine Power Units

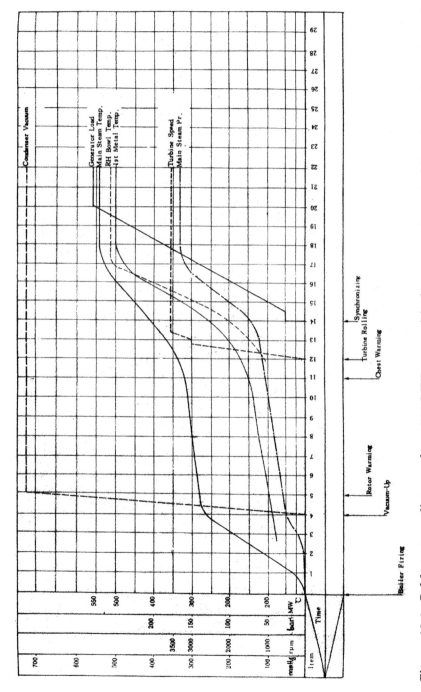

Figure 10-1. Cold start-up diagram for a 500-MW supercritical-pressure steam turbine. *Source: By courtesy of Doosan Heavy Industries and Construction*

274 *Steam Turbines for Modern Fossil-Fuel Power Plants*

start-up diagrams are based only on the temperature mismatches for the HP section of the turbine, completely ignoring the temperature and thermal-stress state of the IP section, its temperature mismatches. This is typical for many start-up instructions. It is not correct, especially for the considered turbine with its single-casing design of the IP section (see Figure 3-15a). For such a turbine, the rates of raising the reheat steam temperature and loading the turbine are to a great degree limited by the thermal stresses in the HP-IP cylinder's rotor and outer casing in the IP steam admission zone. In addition, the considered instruction does not call the main indices of the temperature and thermal-stress states of the turbine to be monitored by the operator, as well main countermeasures to be taken if these indices go beyond the admissible borders. Radial temperature differences in the HP-IP rotor at the first stage zones of the HP and IP sections have to be among these indices to be monitored. In addition, a quick rise of the reheat steam temperature during rolling up and initial loading can cause significant total (centrifugal plus thermal) tensile stresses in the LP rotors. So such a turbine would need the temperature monitoring for the LP rotors, too.

Contrary to the existent situation with individual start-up instructions for the boiler and turbine, the power plants should be provided with complex instructions embracing both the boiler and turbine together and based on thorough analysis of their operating capabilities and optimization of their transients. In particular, precisely this approach was lied in the basis of developing start-up instructions for all the standard steam-turbine units in the former Soviet Union, including particularly 300-MW, 500-MW, and 800-MW supercritical-pressure ones, as well as 200-MW subcritical-pressure units with drum-type and once-through boilers of diverse types.[1-5] These instructions were mainly developed by researchers and engineers of the Adjusting Institution for the Power Industry (ORGRES), including their regional branches, and All-Union Thermal Engineering Research Institute (VTI) with participation of projecting institutions and turbine and boiler producers. These instructions, describing and regulating start-up and shut-down operations for the power unit as a whole, were constructed based on results of field tests for the specific types of power units and analysis of their operational practice. With appearance of new field test results, as well as development and implementation of new technologies, the existent instructions were renewed and republished.

By an example, Figure 10-2 shows an instruction diagram for a warm

Start-up Instructions for Steam-turbine Power Units 275

start-up after a one-day (32-42-hour) outage as applied to oil/gas-fired 800-MW supercritical-pressure units with LMZ turbines K-800-240-3 according to the standard instruction[5] developed mainly on the basis of field start-up tests at the Uglegorsk power plant in Ukraine.[6] The instruction also comprises similar diagrams for other characteristic initial pre-start thermal conditions of the turbine. As a general approach, each one has single, unified rolling-up and loading diagrams for the certain start-up type together with multiple diagrams of raising the main and reheat steam temperatures for diverse initial metal temperatures for the HP and IP turbine cylinders which can take place for the taken outage duration.

Besides diagrams for different start-up types, the start-up instruction should comprise description in details of all the technological operations fulfilled during start-ups and their sequence according to the start-up technology developed and chosen for the taken start-ups. In the first place, it concerns the consistency of so-called *logical*, or *discrete*, *operations* with discrete-position objects, like motors, shut-off valves, and governors, which are switched on/off or closed/opened. Most of these operations and their sequence remain the same at all start-ups, and their beginning is clearly stipulated by the current conditions (turbine rotation speed, main steam pressure, and so on).

For hot and warm start-ups, some of these operations are skipped without violation of the applied technology and its logics. So, for example, for one-bypass-system units, their hot reheat steam-lines are to be pre-heated if their metal temperature state is not consistent to the IP cylinder's temperature state. So, before steam is given into the HP cylinder, the IP valves are forcedly closed, and they are to be opened at the intermediate rotation speed when the reheat steam-lines temperature reaches the assigned level. Along with this, all these discrete operations are omitted if the pre-start temperature state of the hot reheat steam-lines matches the IP cylinder temperature (at hot start-ups or warm start-ups for units with well-insulated reheat steam-lines).

Governing the throttle steam pressure, at the turbine entrance, also refers to some discrete operations: the pre-start level of the pressure is conditioned by the initial firing rate of the boiler and opening of the bypass valves. The bypass valve position can either remain invariable during the turbine rolling-up (in this case, the steam pressure decreases with opening the turbine's control valves) or, more often, the bypass valve is gradually closed keeping the constant throttle steam pressure. The last version is more favorable for turbines with nozzle-group con-

276 Steam Turbines for Modern Fossil-Fuel Power Plants

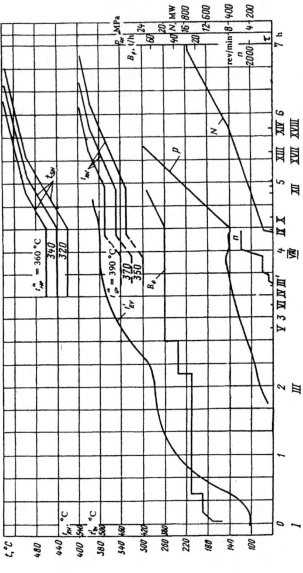

Figure 10-2. Instruction diagram of a warm start-up after 32-42-hour outages for a standard oil/gas-fired 800-MW supercritical-pressure power unit.

t_{HP}^{in} and t_{IP}^{in} – initial (pre-start) measured metal temperatures of the HP and IP cylinders (their outer casings near steam admission), B_f – fuel flow amount, t'_{EV} – steam temperature after the boiler's evaporator (upstream of the embedded valves)

Some technological operations: I – boiler ignition, III-III′ – warming up the hot reheat steam-lines, IV – introducing start-up attemperators of main steam with spraying feedwater into the main steam-lines, V – opening the HP stop valves, VI – opening the IP intercept valves, VIII – giving steam to the heating systems of the HP and IP flanges, IX – introducing the start-up spray attemperators of reheat steam, XII – giving steam for warming the second BFP's turbine drive, XIII – closing the embedded valves between the separators and flash tanks, XIV – connection of the second BFP, XVII – closing the start-up sprays to the main steam lines, XVIII – closing the start-up sprays to the reheat steam-lines trol because it allows passing the steam flow amount into the turbine through a minimum number of the control valves and their crossover pipes and decreasing in this way the steam temperature drop caused by heat transfer to the heated metal. When the generator is switched on to the grid, the bypass valves are to be completely closed, and the turbine's control valves are to be opened to the pre-assigned position, the same for all start-up types. When the main steam pressure reaches the nominal level, the turbine's valves automatically turn to maintaining this level upstream of themselves.

Such unification of the start-up technology seems to be very important for the operational personnel, facilitating their labor and decreasing a probability of erroneous actions. To a great degree, this also simplifies automation of start-up control and makes interaction between the operator and automation means easier.[7,8] On the other hand, sometimes, it is expedient to break this unification and alternate technology for individual start-up types. This mostly concerns cold start-ups, whose technology can differ from all other start-up types. In addition, cold start-ups after overhauls, repairs, or merely long outages are usually combined with some special technological operations like steam-valve tests, checking the overspeed trip system, drying the thermal insulation, and so on.

The same approach, with the quest for unifying technology as applied to different kinds of start-ups, is extended to start-up *continuous operations*, too. This primarily refers to raising the rotation speed. If the rolling-up technology is unified, it is brought to only a few operations: 1) rolling the turbine from the turning gear up to the intermediate rotation speed, 2) step at this level whose duration is conditioned by the initial and current temperature state of the turbine and its steam-lines (for hot start-ups, this stage is completely omitted without violation of the general logics), 3) raising the rotation speed to the synchronous level with the invariable rate independently of the start-up type, and 4) holding at the synchronous speed until readiness to synchronization and acceptance of the initial load, but without special holding for heating the turbine. The resultant rolling-up diagram favorably differs from much more intricate former diagrams, where, in addition, all the pieces differ for different kinds of start-ups—Figure 10-3.

The same quest for unification is applicable for constructing start-up diagrams of loading and raising the main and reheat steam temperatures. Until the generator is switched on to the grid and the turbine accept an initial load, the steam temperature is kept at the level with a set excess over

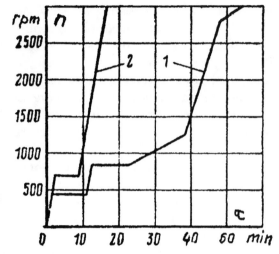

Figure 10-3. Former (1) and advanced (2) instruction diagrams for turbine rolling-up (with the rated rotation speed of 3,000 rpm) at warm start-ups. Source: A. Leyzerovich[7]

the initial (pre-start) metal temperature of the corresponding cylinder:

$$t_{st}^{in} = t_m^{in} + \Delta t_{st}^{in}, \tag{10-1}$$

where Δt_{st}^{in} is set the same for all the start-up types in such a way as to prevent an appearance of inadmissible thermal stresses in the critical turbine design elements when the turbine accepts the initial load. Then, in the process of loading the turbine, the steam temperature is raised linearly or piecewise-linearly in time to keep the thermal stresses in the critical elements near the admissible level. The loading process is also set linear or piecewise-linear in time in such a way that both the main and reheat steam temperatures reach their nominal levels by the end of loading.

All the start-up diagrams like those considered above are constructed on the assumption that the operator chooses the proper diagram corresponding to the pre-start temperature state of the turbine. If the actual pre-start temperature differs from all those foreseen by the instruction diagrams, the start-up is run according to the diagram for the nearest "colder" state. It is assumed that thereafter the operator will run the start-up process strictly following the diagram chosen beforehand.

Along with this, some turbine producers prefer to construct start-up instructions for their turbines in the form of nomograms, implying that the operator will be looking for the recommendations on how he/she has to run the start-up process on-line, that is, directly in the course of control. By example, such diagrams, developed by Westinghouse for

Start-up Instructions for Steam-turbine Power Units 279

turbines with the steam conditions of 16.5 MPa, 538/538°C (2,400 psi, 1,000/1,000°F), are presented in Figures 10-4 and 10-5.

The first of these nomograms helps the operator to find the time of rolling up and holding the turbine at the initial load as applied to the actual current steam conditions and metal temperature of the integrated

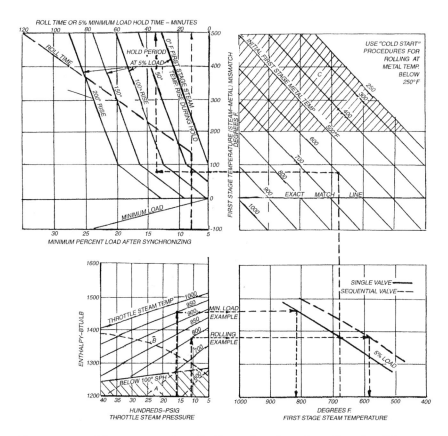

Figure 10-4. Nomogram for finding the time length of rolling up and holding at minimum load for warm start-ups of a Westinghouse subcritical-pressure steam turbine. *Source: By courtesy of Siemens Westinghouse*

Area A - operation in this area is inadmissible because of main steam superheat less than 100°F (56°C); curve B - operation above this curve is desirable to insure the proper warming of the steam-chests for transfer from throttle control to nozzle-group control, but not necessary if steam-chest metal temperature is sufficient; area C - operation in this area is not recommended because the first-stage steam temperature is more than 200°F (112°C) above the metal temperature.

280   Steam Turbines for Modern Fossil-Fuel Power Plants

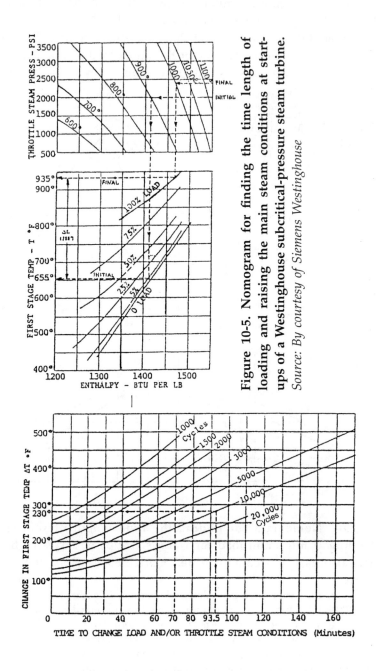

Figure 10-5. Nomogram for finding the time length of loading and raising the main steam conditions at start-ups of a Westinghouse subcritical-pressure steam turbine.
Source: By courtesy of Siemens Westinghouse

Start-up Instructions for Steam-turbine Power Units 281

HP-IP cylinder (see Figure 3-13) at warm start-ups (with the initial metal temperature over 250°F, or 120°C). So, for example, if the initial main (throttle) steam conditions are 1,000 psi, 800°F (6.9 MPa, 427°C), this corresponds to the steam temperature at the first HP stage under single-valve 5% load about 680°F (360°C). If the initial metal temperature in the first-stage zone is 600°F (316°C), the determined steam-metal mismatch of 80°F (44°C) gives the roll time value equal to 10 min. To determine how long it is necessary to remain at the 5% load level after synchronization with the main steam conditions of 1,500 psi, 950°F (10.34 MPa, 510°C), the first-stage steam temperature is found as 810°F (432°C); the rise of this temperature makes up to 130°F (72°C), and the same mismatch line extended to the line of 130°F in the mesh of the first-stage steam-temperature rise indicates a hold time of 34 min.

Similarly, the operator can estimate the change of the steam temperature at the first HP stage at the loading stage and the time length necessary to raise the turbine load and main steam conditions up to the nominal values, depending on the expected number of such start-ups for the turbine lifetime, that is, the admissible thermal stress level for the critical turbine elements—see Figure 10-5. So, for example, for a warm start-up with the initial steam conditions of 2,000 psi, 900°F (13.8 MPa, 482°C), the change in the steam temperature at the HP first stage is equal to 280°F (156°C), and, with the expected number of warm start-ups for the turbine lifetime equal to 5,000, the time of loading and raising the steam conditions accounts to 70 min. However, if the expected number of such start-ups is estimated as 10,000, the time of loading and raising the steam conditions increases to 93.5 min.

There also exist some more complicated nomograms (as, for example, those developed for 500-MW turbines of LMZ intended for cycling operation[9]), which, according to their authors, should facilitate the manual control for the turbine operator and, with the accuracy sufficient for the engineering practice, enable finding all the possible required start-up parameters and evaluating the resulted thermal fatigue of the turbine's critical elements, that is, the HP and IP rotors. Indisputably, such nomograms look very smart and sophisticated, but in their present form they are practically useless for power plant operation. In the process of control, the operator merely does not have time and conditions to trace all the interrelated variable quantities at the diagrams. The usage of such nomograms could be done easier with the help of computer-based systems of informative support for the operational personnel at

282   *Steam Turbines for Modern Fossil-Fuel Power Plants*

start-ups, but in this case, too, it seems reasonable to present required information to the operator in graphic forms which would be more obvious and convenient for perception. At the same time, the mere idea of finding the reasonable or admissible changes of the operating conditions based on the current temperature and thermal-stress state of the turbine is very valuable and must not be skipped.

It seems so important, because the actual operating conditions, especially start-ups under manual control, always somewhat differ from those ordered by instructions. In particular, the actual process of start-up loading is often interrupted because of some delays with the boiler or auxiliaries. In addition, the power system not always is ready to consume the power the turbine is ready to generate. The start-up instructions cannot take into account all the numerous kinds of deviations from the assigned schedules and give recommendations for all the possible instances, and once these deviations appear, the personnel have to continue the start-up process relying on their own view of the equipment state. As a result, the final characteristics of actual start-ups (their duration, deviations of the monitored indices beyond the admissible values, resultant thermal fatigue of critical elements, and possible change to the worse of the turbine's efficiency) to a great degree depends on the operational personnel's comprehension, as well as informative support for them. (The last subject is considered in more details in the next part.) With regard to these circumstances, there arise absolutely new requirements to the start-up instruction—for each specific object, that is, the power unit with its peculiarities, the start-up instruction should be additionally related to the specific computerized system of informative support for the operational personnel, as well as to the specific start-up system and design features of power equipment, and should comprise very specific directions: which diagrams or other graphic forms should be requested by the operator at each start-up stage, and all these graphic forms should be constructed realistically taking into account the perception abilities of the personnel and his/her other duties and needs in the start-up process.

ESTIMATION OF START-UP DURATION

While constructing start-up diagrams, it is necessary to decrease the start-up duration as much as possible without sacrifice of the turbine reliability and with regard to actual operational capability of the opera-

Start-up Instructions for Steam-turbine Power Units 283

tional personnel. First of all, this refers to the start-up stage of loading, because just the rate of loading determines the power unit's ability to cover the needs in the power system. The average rate of loading at start-ups after outages of the most characteristic time length is one of the most obvious and important characteristics of flexibility for steam turbines and power units upon the whole. In addition, reduction of the start-up duration noticeable improves the power plant efficiency, decreasing the energy losses with parasitic fuel expenditures during start-ups. Sure, these energy losses substantially vary with the employed start-up technology, features of the start-up system, and so on. Nevertheless, roughly speaking, the specific start-up fuel expenditures for modern steam-turbine power units at the stage of loading the turbine can be estimated as equal to about 0.0008-0.001 t/min of equivalent fuel per MW. So, for a 600-MW unit, reduction of the loading stage duration by about 20 minutes saves approximately 100-120 tons of equivalent fuel. (More detailed data on energy and fuel start-up expenditures for different power units at the different start-up stages can be found, in particular, in the author's previous book.[8]) According to Siemens, reduction of only ten minutes in the start-up time for a 700-MW power units after a weekly shut-down yields increased earnings of the order of DM 0.5 mln.[10]

As said before, the start-up rate is mainly limited by so called "leading" indices of thermal stresses in the most thermally stressed ("critical") turbine elements. If such critical elements are the high-temperature (HP and IP) rotors, as it is in reality for modern large steam turbines, the leading indices can be taken in the form of so called "effective" radial temperature differences, and the upper admissible values of these temperature differences are set from the reasons of low-cycle thermal fatigue of the rotor metal.[*]

As shown analytically and with computations,[7,8] in order to keep the temperature difference $\overline{\Delta t}$ at the taken admissible level $[\overline{\Delta t}]$, set invariable or decreasing linearly with the temperature, with regard to an

---

*The effective temperature difference for the rotor is determined as the radial difference between the metal temperature on the external (heated) surface and the average integral temperature in the supposed most thermally stressed section of the rotor. So this temperature difference is reckoned as being in proportion to the thermal stress on the rotor surface with the given radial temperature distribution for the rotor as a long cylinder. In the operational practice, these temperature differences for high-temperature (HP and IP) rotors of large power steam turbines should be continuously monitored by means of mathematical modeling of the heating up process for the rotors based on the measured heating-steam temperatures. In more details, this subject is considered in the next chapter of this book.

approximately linear change of the metal's temperature conduction with the metal temperature, the heating-steam temperature should vary in time by an exponential law. If the turbine is loaded under sliding pressure with the invariable control-valve position, the steam-temperature drop in the control valves remains approximately the same, and the steam temperature before the turbine should also be raised by the exponential law:

$$t_{st} = t_{st}^{in} \times \exp\left(-k_2\tau\right) + \frac{k_1}{k_2}\left[1 - \exp\left(-k_2\tau\right)\right] \le t_{st}^{nom}.$$

(10.2)

Here, τ is the time value counted from the instance when the generator is switched on to the grid and the process of loading and raising the steam conditions commences; k_1 and k_2 are the invariable quantities, and t_{st}^{in} and t_{st}^{nom} are the initial and nominal steam temperatures before the turbine. The initial steam temperature is supposed to be set according to (10-1) at such a level that the effective temperature difference reaches about its admissible level $[\overline{\Delta t}]$ with synchronizing the generator and accepting the initial load. Constants k_1 and k_2 form the linear dependence between the rate of raising the steam temperature and its current value:

$$W_t = \frac{dt_{st}}{d\tau} = k_1 \pm k_2 t_{st}$$

(10-3)

and the loading duration can be found as

$$\tau_N = \frac{1}{k_2} \times \ln\left(\frac{k_1 - k_2 t_{st}^{in}}{k_1 - k_2 t_{st}^{fin}}\right).$$

(10-4)

In the last expression, $t_{st}^{fin} \le t_{st}^{nom}$ is the final steam temperature level at which the turbine can accept the nominal load.

All these expressions are gotten based on the linear dependence of the metal's temperature conduction on the metal temperature and, if necessary, linear changes of the admissible radial temperature difference in the rotor with its temperature, because the rate of raising the heating-steam temperature W_t should be in proportion to these variables:

$$W_t = \frac{2a}{k_{2D}k_0 R^2}[\overline{\Delta t}]$$

(10.5)

Here, the current metal temperature is assumed close to the heating steam temperature; a is the metal's thermal conduction; R is the rotor thickness; k_0 is the coefficient for the rotor's shape that is as a first approximation defined by the ratio, ξ, of radii for the bore and outer surface:

$$k_0 = 0.25 \pm \xi^2 \left(\frac{\xi^2}{1 \pm \xi^2} \ln \xi + 0.75 \right), \tag{10.6}$$

and k_{2D} is the correction for the axial heat flux with thermal conduction from the considered section of the rotor, that is, for its two-dimensional temperature field.

Since the pre-start temperature states of the HP and IP cylinders usually considerably differ from one another, and the admissible values of the effective temperature differences for their rotors, $[\overline{\Delta t}]$, also substantially differ, the loading duration should be estimated separately for both the HP and IP rotors (or, for turbines with the integrated HP-IP cylinders, for most thermally stressed sections of its rotor in the HP and IP zones). By way of illustration, Figure 10-6 presents the loading time variation with the pre-start metal temperatures of the HP and IP cylinders of a 300-MW supercritical-pressure steam turbine of LMZ. These dependences were calculated from conditions of both the main and reheat steam temperatures changing in the range between 280-300°C and 520°C (535-570 and 970°F) and the maximum loading rate of 7 MW/min. The last value is determined not by the turbine, but the boiler's flexibility and the real operating capability of the operational personnel. In practice, more often than not, the loading duration is determined by the IP cylinder, which when the turbine is stopped, cools down faster than the HP cylinder does.

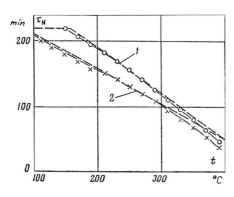

Figure 10-6. Time length of start-up loading for a 300-MW supercritical-pressure steam turbine of LMZ depending on the pre-start metal temperatures of the HP (1) and IP (2) cylinders.

CALCULATED OPTIMIZATION OF START-UP DIAGRAMS

While constructing instruction start-up diagrams, their developers mainly rest on some general, simplified notions—as, for example, an assumption of the approximately constant temperature difference between the steam upstream of the turbine HP valves and the heating steam within the HP section when the turbine is operated under the sliding main pressure. Similarly, the steam temperature drops at the IP steam entrance is also supposed invariable. In particular, these concepts are laid at the heart of equation (10.2) for raising the main and reheat steam temperatures. The actual correlations between the steam temperatures after the boiler to be controlled and the heating steam temperatures within the turbine, determining its heating, are more complicated—see Figures 8-1, 8-2, and 8-4a. It is understandable that start-up diagrams constructed with the use of the aforesaid notions can present only some approximation to the optimum.

At least, a closer approximation can be found by means of calculated optimization of start-up diagrams as a special case of a digital calculated optimization of the turbine transients. The optimization problem, set by definition as the problem of fast-acting, is solved as the search for the diagrams of changing the turbine's operating parameters at which the "leading" indices of the turbine temperature and thermal-stress states are kept at their upper admissible levels, not exceeding them. Further, the start-up diagrams are verified experimentally in the process of field tests at power units in service.

It might be well to recall that the leading indices of the turbine state are those that limit (actually or potentially) the rate of the turbine transients and, in addition, whose variations depend only on the main outside controlling actions or, in other words, changes of the main controlled operating parameters: steam flow amount through the turbine, main and reheat steam parameters. It means that the operator cannot influence on the leading indices with the use of any additional, "secondary" control actions, like, for example, switching on the steam heating systems for the flange joints of the high-temperature cylinders. For modern large steam turbines, the leading indices are generally the radial "effective" temperature differences in the HP and IP rotors. Depending on the turbine design, output, and steam conditions, and the unit's start-up system features, some other indices can play the role of leading ones as, for example, temperature differences in the HP stop-valve steam-chests, temperature differences across the flange width, and RREs of the high-

Start-up Instructions for Steam-turbine Power Units 287

temperature cylinders, or temperature differences in the LP rotors. It is expected that, if for the whole transient process the leading indices are kept at their top admissible levels, the process duration will be minimum. However, with several leading indices and several controlling actions influencing them, such a definition of the problem is not absolutely correct, and it is necessary to use some supplementary conditions to obtain the single-valued solution.

It seems most reasonable to reach the set purpose by means of a combined digital solution of the direct and reverse problems of unsteady heat conduction for potentially critical turbine elements. In doing so, the temperature differences across the thickness of each considered element is taken as a potentially leading index, and the reverse problem is solved in reference to the heating-steam temperature under the boundary conditions of keeping the taken temperature difference at its upper admissible level. Solution of the direct problem gives the change of the temperature field for the considered element. In turn, this makes it possible to solve the reverse problem referring to the heating steam temperature. Both the direct and reverse problems are solved in parallel for all the potentially critical elements.

The current upper admissible value of the heating-steam temperature can be looked for as the following:

$$[t_{st}] = t_m + [\Delta \tilde{t}] \times \left(1 + \frac{\lambda}{\alpha} \times \frac{\partial t}{\partial x}\Big|_{x=0}\right) = t_m + [\Delta \tilde{t}] \times \left(1 + \frac{\Phi}{Bi}\right). \tag{10.7}$$

Here, λ is thermal conduction of the metal; α is the heat transfer coefficient for the considered heated surface; $Bi = (\alpha \tilde{H})/\lambda$ is the dimensionless heat transfer factor (Biot number) for the considered element, related to its characteristic size $\tilde{H}$; $\Delta \tilde{t} = t_s - t_m$, and t_m are the metal temperatures of the considered element on the heated surface and at the distance $\tilde{H}$ from this surface. This point can be fixed (for example, on the external insulated surface) or sliding (if it refers to the average integral temperature). In (10.7),

$$\Phi = \frac{\tilde{H}}{[\Delta \tilde{t}]} \times \frac{\partial t}{\partial x}\Big|_{x=0} \tag{10-8}$$

is the dimensionless temperature gradient on the heated surface. In general, all the values in (10.7) are the variable quantities. For unsteady, nonstationary processes, $\Phi = var$, too. However, as the calculation experience prompts, if the process is close to optimal and the heat removal from the insulated surface is much less than the heat flux from steam to metal, it is possible and advisable to take $\Phi \approx$ const. In particular, this assures a stable solution of the optimization problem in the case of potential disturbances of the process. For forged rotors with a central bore considered as hollow cylinders, the value of Φ in reference to the admissible effective radial temperature difference $[\Delta t\,]$ is approximately equal to 1.17.

Finding the optimal change of the heating-steam temperature in time for every potentially critical design element does not still mean that the problem of optimizing the turbine transients is resolved. First, it is necessary to turn from the heating-steam temperature changes to the changes of the turbine's operating parameters, that is, the controlling actions influencing the heating steam temperatures for different elements. Second, different leading indices of the turbine state are influenced by different controlling actions in different extents. As this takes place, the change of some indices which are influenced by the same outside actions features in their dynamic, and the change of one index can depend on two different actions. Hence, it is necessary to apply some additional criterion of optimization. Finally, it is necessary to take into account an existence of some technological limitations for changing the turbine operating parameters. These limitations can relate to their absolute values, rates of variations, and functional interrelations.

Start-ups of fossil-fuel power units are characterized with, at least, four independent controlling actions. They are: the changes of the steam flow amount through the turbine (that is, its rotation speed and load), main and reheat steam temperatures, and position of the turbine's control valves, even not counting control of bypass valves.

Since the temperature differences in the turbine's critical elements, as leading indices, are much more sensitive to changes of the main and reheat steam temperatures than to other controlling actions, it is reasonable to control their changes varying the steam temperatures. Under these conditions, the rest changes (including those for the steam flow amount through the turbine) turn out to be indefinite. In order to decrease the start-up fuel outlay and increase the power generation during the loading stage, it might be well to keep the load at each start-up instance at the maximum level acceptable with the current steam temperature values. Along with this, to simplify operating the unit, especially under manual control, it is probably more expedient to set the loading diagram as a linear or piecewise-linear program in time.

Start-up Instructions for Steam-turbine Power Units 289

In any case, the loading duration is determined by the time taken to reach the main and reheat steam temperature levels that allows adopting the nominal value. For most large turbines of both subcritical and supercritical steam pressure, this level is set by the turbine producers as being approximately 20-25°C (35-45°F) below the rated value. They substantiate this limitation by steam wetness after the last LP stages and some other factors, but in reality it seems possible to increase this figure up to 40-45°C to attain the rated load at approximately 500°C (930°F) if the rated steam temperatures are equal to 540-545°C (1,004-1,013°F), and so on.

For the HP cylinders, the static temperature drop between the heating steam for the HP rotor and main steam before the turbine essentially depends on the main steam parameters and position of the control valves. Depending on the mode which the HP control valves are controlled with, this temperature drop can be calculated based on the main steam conditions and steam flow amount through the turbine (see Figures 8-1 and 10-4, 10-5) or taken from the experimental static dependences like that shown in Figure 8-2. As to the IP cylinders, for them it is also expedient to use the experimental static dependences of the steam temperature drop on the steam flow amount through the turbine like those presented in Figure 8-4a, but, in addition, for the heating steam near the rotor surface, it could occur necessary to take into consideration some dynamic delay (see Figure 8-4b). The same refers to the LP steam path—see Figure 8-6.

Transition from the optimized steam temperature before the turbine (in the stop-valve steam-chests) to the steam temperature changes after the boiler (downstream of the start-up attemperators) should take into account heat transfer and accumulation in the main and reheat steamlines. This is especially important for the reheat steam temperature at start-ups of units with a one-bypass start-up system.

The start-up schedules of raising the steam parameters are to be optimized at the loading stage. During the previous stage of rolling up, at cold and hot start-ups the steam temperature is merely kept at the minimum or maximum possible level. As for warm start-ups, any optimization at the early start-up stage is not effective because of comparatively small duration of this stage. In addition, the process can hardly be optimized because the steam flow amount through the turbine at this stage materially depends on various outside circumstances, such as vacuum in the condenser, lube oil temperature, etc. These steam flow variations powerfully influence the heat transfer conditions for the critical design elements. This being so, the initial steam temperature is set constant from reasoning that the critical elements are not cooled excessively when steam is given into the turbine and will not have caught the inadmissible temperature differences by the idling and initial load.

All these considerations were employed as applied to an 800-MW supercritical-pressure steam turbine of LMZ (see Figures 2-4 and 2-5a).

As previously stated, optimization embraces only the loading stage. At running-up, the steam temperature after the boiler is set invariable according to the pre-start thermal conditions of the turbine. The reverse task in reference to the heating-steam temperature is solved with expression (10-7). The effective radial temperature differences for the HP and IP rotors are taken as the leading indices, and the average integral metal

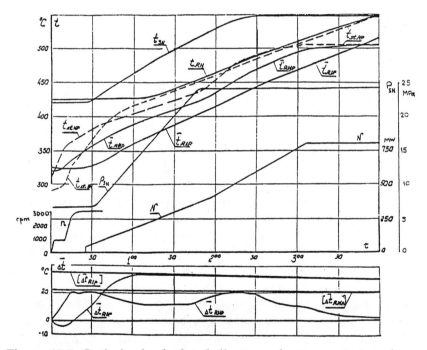

Figure 10-7. Optimized calculated diagram of a warm start-up for an 800-MW supercritical-pressure steam turbine of LMZ.

t_{stHP} and t_{stIP}—steam temperatures in the HP control stage and IP inlet chambers, $\bar{t}_{RHN}$ and $\bar{t}_{RIP}$—average integral metal temperatures in the most stressed sections of the HP and IP rotors, $\Delta \bar{t}_{RHP}$ and $\Delta \bar{t}_{RIP}$—effective radial temperature differences in the same sections, $[\Delta \bar{t}_{RHP}]$ and $[\Delta \bar{t}_{RIP}]$—upper admissible values of the aforesaid temperature differences, respectively.

temperature values, $\bar{t}$, in the most stressed section of each rotor are used as the characteristic metal temperature, t_m, in this expression. These temperature values are defined at each calculation step in time by means of solving two-dimensional unsteady heat conduction problem for the rotors, taking into account the change of metal temperature conduction with temperature. Changes of the heat transfer coefficients for the rotor surfaces and decrease of the heating-steam temperature in reference to the steam temperature at the cylinder inlets during start-ups are assigned by the static dependences on the steam flow amount through the turbine with correction for the steam temperature dynamics in the crossover pipes. Changing the main steam pressure is set in line with the current turbine load according to the program of controlling the turbine valves. In turn, the loading schedules are formed either with the permanent rate depending on the turbine pre-start temperature conditions (as before) or upon the condition of the maximum load value at each moment according to the current values of the main and reheat steam temperatures with some additional restrictions caused by technological requirements.

Simultaneously with the turbine rotors, the metal temperature field of the hot reheat steam-lines is also calculated, as well as the steam temperature at the steam-line inlet, whereas the exit steam temperature is assigned. The dynamic correction $\Delta\theta_{set}$ for the required inlet steam temperature in reference to the desired steam temperature before the turbine at each "subsequent" time moment is evaluated from the equation:

$$\Delta\theta^{set} = \Delta\theta \times \frac{\theta_{ex}^{set} - t_m}{\theta_{ex} - t_m},$$

$$(10\text{-}9)$$

where $\Delta\theta$ is the decrease of the steam temperature along the steam-line in the "current" time instance; θ_{ex} is the current steam temperature before the IP cylinder (at the steam-line exit); θ_{ex}^{set} is the assigned (set) value of this temperature by the subsequent time instance, and t_m is the average metal temperature or metal temperature at the characteristic point of the steam-line length.

The optimized start-up diagram for a warm start-up of the considered turbine is exposed in Figure 10-7. For the taken initial temperature conditions and admissible values of the leading indices, the start-up rate is mainly limited by the temperature difference in the IP rotor that is close to its upper admissible level during almost the entire duration of loading, whereas the temperature difference in the HP rotor reaches its admissible level only at certain spaces of time. This situation can varies for other initial conditions and other limits for the leading indices.

The optimized start-up diagrams usually cannot be directly recommended for manual control—they should be somewhat simplified. In particular, this refers to the diagram of raising the reheat steam temperature at the IP cylinder entrance. It is set from conditions to heat in the optimal way the IP rotor; however, to get this diagram, raising the steam temperature after the boiler (at the hot reheat steam-line inlet) should be arranged with a substantial initial over-shooting and its sharp decrease with the increase of the steam flow amount. It can be hardly materialized with manual control, but occurs very useful at arranging automated start-up control.[11] By way of illustration, Figure 10-8 demonstrate variation of the reheat steam temperature for an 800-MW supercritical-pressure unit at a warm start-up with limiting the effective radial temperature difference in the IP rotor; the loading process is finished as soon as the reheat steam temperature reaches its rated value. Automated control of this temperature after the boiler with allowance for the influence of the reheat steam-lines makes it possible to reduce the loading duration by about 20 minutes.

By the way, this example demonstrates us another possible approach to getting start-up diagrams close to theoretically optimal—through mathematical modeling of the automated (or automatic) start-ups with the governors built and tuned to keep the leading indices of the turbine at their admissible level.[12]

Using advanced start-up technologies, optimizing start-up diagrams, and automation of start-up control make it possible to decrease the start-up duration by up to 10-20%. In so doing, such a hastening is not achieved at a sacrifice in more intense heating of the critical elements and increasing temperature differences and thermal stresses in them, but only due to their more even heating within the settled limitations.

References

1. *Standard Instruction for Starting up from Diverse Initial Thermal Conditions and Shutting down 200-MW Power Units with Drum-Type Boilers TP-100* [in Russian]. *Moscow: SPO ORGRES, 1977.*
2. *Standard Instruction for Starting up from Diverse Initial Thermal Conditions and Shutting down 200-MW Power Units with Once-Through Boilers PK-33* [in Russian]. *Moscow: SPO ORGRES, 1977.*
3. *Standard Instruction for Starting up from Diverse Initial Thermal Conditions and Shutting down 300-MW Power Units with Steam Turbines K-300-240 of LMZ* [in Russian]. *Moscow: SPO ORGRES, 1980.*

Start-up Instructions for Steam-turbine Power Units 293

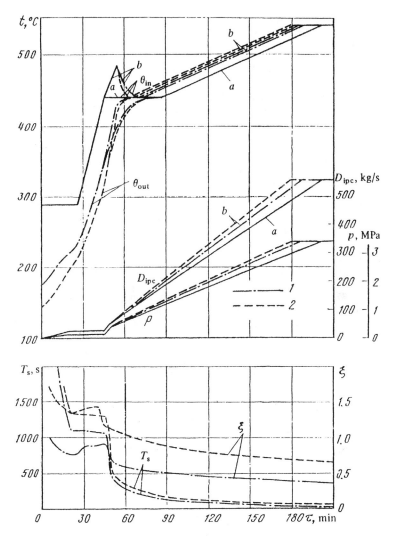

Figure 10-8. Calculated diagrams of optimal raising the reheat steam temperature at a warm start-up of a 800-MW supercritical-pressure power unit without (a) and with (b) allowance for the influence of the reheat steam-lines. *Source: A. Sh. Leizerovich and A.D. Melamed[11]*
1 and 2 – for the units with two and four hot reheat steam-lines, respectively
D_{IPC} – steam flow amount through the IP cylinder, p – steam pressure in the reheat system, θ_{in} and θ_{out} – steam temperatures at the hot reheat steam-line system inlet and outlet (after the boiler and before the IP cylinder), respectively.

4. *Standard Instruction for Starting up from Diverse Initial Thermal Conditions and Shutting down 500-MW Power Units with Boilers P-57 and Steam Turbines K-500-240-2* [in Russian]. *Moscow: SPO ORGRES, 1986.*

5. *Standard Instruction for Starting up from Diverse Initial Thermal Conditions and Shutting down 800-MW Power Units with Boilers TGMP-204 and Steam Turbines K-800-240-3* [in Russian]. *Moscow: SPO ORGRES, 1980.*

6. Avrutskii G.D., G.I. Doverman, R.M. Ostrovetskii, et al. "Starting Methods for Gas- and Oil-Fired Boiler-Turbine Unit 800-MW in Capacity at Uglegorsk Power Plant," *Thermal Engineering* 28, no. 9 (1981): 527-530.

7. Leyzerovich, A. Sh. *Technological Fundamentals for Steam Turbine Start-up Automation* [in Russian], Moscow: Energoatomizdat, 1983.

8. Leyzerovich A. *Large Power Steam Turbines: Design & Operation*, 2 vols., Tulsa (OK): PennWell Books, 1997.

9. Ryzhkov V.K., V.S. Shargorodsky, A.N. Kovalenko, and Y.E. Regentov. "A Nomogram of Starts and Changes in Load of a Steam Turbine Unit," *Thermal Engineering* 35, no. 4 (1988): 234-239.

10. Zaviska O. and H. Reichel. "Siemens Optimizes Start-up Sequences in a Power Plant Unit—New Unit Start-up Simulator Reduces Costs," *VGB PowerTech* 81, no. 7 (2001): 36-38.

11. Leizerovich A.Sh. and A.D. Melamed. "Calculation of the Influence of the Boiler-Turbine Steam-Lines on the Increase in Steam Temperature with Automated Start-ups of Generating Units," *Thermal Engineering* 29, no. 10 (1982): 542-546.

12. Leizerovich A.Sh., E.R. Plotkin, N.I. Davydov, et al. "Mathematical Modeling of Automated Turbine Start-ups," *Thermal Engineering* 40, no. 2 (1993): 107-112.

Chapter 11

Scheduled and Unscheduled Load Changes within and Beyond the Governed Range

There are two ways for power units to participate substantially in covering long-term power consumption ebbs in the power system: by means of either shutting down the unit with the subsequent start-up when power consumption increases or unloading with the subsequent rise of the load. The power plant personnel, if they have a choice, usually prefer the latter. This largely ensues not from any economic comparisons of fuel expenditures for each of these options but rather goes from certain apprehensions of equipment malfunctioning at start-ups. Undoubtedly, changing the load is a much easier and comfortable technological operation compared with start-ups. As a result, the scheduled unloading for night and weekend power consumption ebbs in power systems has become typical for many fossil-fuel power units, including those of the largest single capacity and supercritical (and even USC) steam conditions. Many power units are also deeply unloaded for the lunchtime of weekdays. So the number of scheduled deep load changes can reach 300-500 per year. A characteristic week load diagram for two power plants with 300-MW supercritical-pressure steam-turbine units is shown in Figure 11-1.

In addition, many units participate in covering occasional load changes in the power system under action of their automatic governing systems and devices to keep the system frequency and admissible power flows through the power transmission lines. As a rule, these load changes are less deep than the scheduled ones, but their number, depending on the unit's place and role in the system, can be quite significant.[1]

The recorded data of the routine output variations under action of an automatic governing system for the coal-fired 800-MW subcritical-pressure unit characteristic for some German fossil-fuel power plants is

295

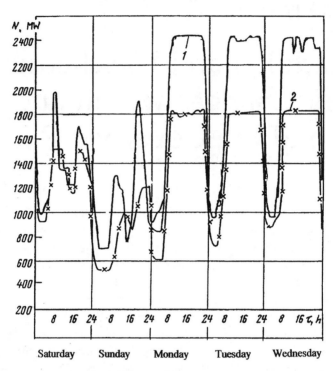

Figure 11-1. A week load diagram for two power plants with eight (1) and six (2) 300-MW supercritical-pressure steam-turbine units.

presented in Figure 11-2. The output demand varied in a range of 200 MW with a period of about 10 min with the ramp rate of about 5%/min, or approximately 40 MW/min. The actual turbine output traces the target value with a slight lag due to the boiler's furnace output changes by +28%/−24% of the full load.[2]

Since the reheat steam temperature and, hence, the heating steam temperatures for the IP critical elements do not vary materially with the turbine load, it is only the HP design elements that limit the load change rate. For modern steam turbines, this primarily concerns the HP rotors. The upper admissible values of the unsteady thermal stresses in the turbine's HP rotor at different kinds of load changes depend on the fact of how frequently these load changes repeat. The total number of scheduled load changes for covering the medium (intermediate) part of the power consumption graphs is usually estimated as not exceeding about 5×10^4 for the turbine lifetime. On the other hand, the number

Scheduled and Unscheduled Load Changes

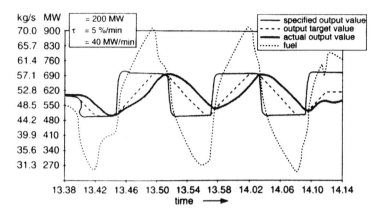

Figure 11-2. Unscheduled load variations under action of an automatic system for the German coal-fired 800-MW Heyden Unit 4 with the steam conditions of 18.2 MPa, 530/530°C. *Source: L. Lehman et al.*[2]

of occasional load changes under action of automatic governors is not restricted, and, consequently, the admissible thermal stress for these conditions should be estimated from their unlimited number. Therefore, the admissible thermal-stress value for the scheduled and unscheduled (occasional) load changes have to be set differently, and the admissible effective radial temperature differences for the HP rotors applied to the scheduled and unscheduled load changes commonly diverge from one another by a factor of 2-2.6 for steam turbines of different types with different stress concentrators on the HP rotor surface. In turn, for start-ups the upper admissible value of these temperature differences for the HP rotors of the same turbines are commonly set about twice greater than that for the scheduled load changes, taking into account that the number of start-ups for the turbine lifetime is still essentially smaller. These limits, different for start-ups, as well as for the scheduled and unscheduled load changes, are put into the operation instructions and automatic systems of governing the turbine load.[1,3]

The load changes under constant throttle steam pressure are carried out by decreasing the steam flow amount from the boiler with closing simultaneously the turbine's control valves. Even if the main steam temperature remains invariable, these load changes are accompanied with perceptible variations of the heating steam temperature in the HP cylinder—Figure 11-3a. This, in turn, causes unsteady-state temperature differences and thermal stresses in the HP rotor. This circumstance requires the load changes be distinctly controlled to prevent excessive thermal fatigue in the rotor.

As a rule, instructions of turbine manufacturers imply that the turbine load varies linearly in time. It is obvious that such a schedule is not optimal. If, for example, the load of a 300-MW supercritical-pressure turbine changes within the governed range under the nominal pressure with the ramp rate of 3.75 MW/min, the largest entire radial temperature difference in the HP rotor makes up to 32°C (58°F), but this value is reached only at the end of the process—Figure 11-4. Along with this, the same load change can be carried out with 30-35% smaller temperature differences due to an initial jump-like change of the load. Such an improved schedule is presented in the same diagram of Figure 11-4.

Due to a high level of heat transfer conditions to the HP rotor surface, after the steam temperature jump the radial temperature difference in the HP rotor quickly (in 20-30 seconds) rises to its maximum value and then slowly decreases under the law close to exponential with a time constant T_0. In order to keep further the thermal stress in the rotor at the upper admissible level, the initial load jump should be supplemented with a linear change of the turbine load. In doing so, it is supposed that

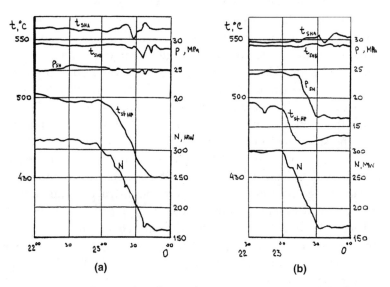

Figure 11-3. Evening unloading of a 300-MW supercritical-pressure unit under constant steam pressure (a) and combined governing program (b).

Scheduled and Unscheduled Load Changes

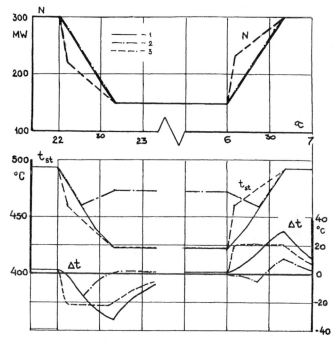

Figure 11-4. Variations in the heating steam temperature and radial temperature difference in the HP rotor of a 300-MW supercritical-pressure steam turbine at load changes from 100% to 50% MCR and back.

the heating-steam temperature for the HP rotor and the turbine load are linearly dependent. In addition, the heat transfer coefficients from steam to the metal surface and metal thermal conduction coefficients for the limited range of the load changes are supposed approximately invariable.

The structure of the optimal load changes (jump-like and subsequent linear change) can be correlated with the model for the radial effective temperature difference in the rotor $\overline{\Delta t}$ described by the transfer function

$$W_s = kT_0 s / (1 + T_0 s), \qquad (11.1)$$

where s is the Laplace operator. Such a model "predicts" the change of the thermal stress in the rotor. Therefore, like any prognostic model, it can be more approximate than diagnostic models needed to reflect the

300   *Steam Turbines for Modern Fossil-Fuel Power Plants*

actual conditions. (Mathematical models applied to monitor on-line the thermal-state state of the rotors are described in the next part of this book.) The values of factor $k = \overline{\Delta t} / \Delta \theta$ and time constant T_0 for each rotor type can be found analytically or from computations. For typical HP rotors of modern steam turbines, the maximum radial effective temperature difference makes up about 0.72-0.73 of the jump-like steam temperature changes, and the time constant, T_0, depends on the rotor size and thermal conduction of the metal, a:

$$T_0 = R_0^2 (1-\xi^2)/2a, \qquad\qquad (11.2)$$

where R_0 is the rotor body's radius, and ξ is the ratio of radii for the bore and outer surface. For HP rotors of modern large power steam turbines, this time constant commonly makes up about 1,200-1,900 s.

The admissible rate of the linear load change after the initial jump can be estimated as equal to

$$[W_N] = ([\overline{\Delta t}]/kT_0)/(\partial \theta/ \partial N), \qquad\qquad (11.3)$$

where $\partial \theta/ \partial N$ is the gradient of the heating steam temperature dependence on the turbine load. Such optimized load changes for a supercritical-pressure steam turbine are shown in Figure 11-4, lines 3.

If the load changes linearly without an initial jump, its rate can be significantly more than that at the quasi-stationary process $[W_N]$ after the admissible load jump $[\Delta N]$. However, in this case, the range of changing the load $[\delta N]$ with this rate value, W_N, is to be limited. According to the model described by (11.1), it follows that

$$\frac{[\delta N]}{[\Delta N]} = -\frac{W_N}{[W_N]} \times \ln \left(1 - \frac{[W_N]}{W_N}\right). \qquad\qquad (11.4)$$

Variations of the heating steam temperatures for the HP rotor with the turbine load significantly decrease if the turbine load changes are going on with sliding steam pressure. In this case, the same rate of the load changes causes essentially smaller temperature differences in the HP rotor—see Figure 11-4, lines 2. In turn, the turbine permits considerably greater load changes without inadmissible temperature differences and thermal stresses in the HP rotors.

Scheduled and Unscheduled Load Changes 301

The actual output change dynamics depend on the boiler-turbine control mode. In the *boiler-following mode*, the operator, or remote load dispatcher, or an automatic governor, control the turbine valves directly, with the power unit's control system reacting to keep the throttle steam pressure at the set level by adjusting the boiler's firing rate. In the *turbine-following mode*, it is the boiler firing rate that is controlled directly, with the control system keeping the throttle steam pressure by means of acting upon the turbine valves; this provides more stable boiler operation at a penalty of slower load response. So, most control system are constructed on the basis of the boiler-following mode that provides faster load responses, because in this case the turbine as if "borrows" temporarily the energy accumulated in the boiler and steam-line metal or, to the contrary, "store" it in the boiler.

Nowadays, many modern power units are operated with a "hybrid," or "combined," program of governing the steam pressure and position of the control valves. So, in the range of 100-75% nominal load, the HP control valves keep the nominal throttle pressure upstream of themselves, providing possible fast responses of the unit to power deficit in the system by opening the control valves while the boiler's load does not have time to increase.[4] At the lower load, the control valve position is fixed, and the unit operates under sliding pressure until the minimum acceptable pressure level, when the valves again return to keeping the steam pressure upstream of themselves. Comparative variations of the steam temperature at the first HP stage at the load changes under constant pressure and combined governing for actual operating conditions are obviously seen in Figure 11-3. It is worth noting that under sliding (variable) steam pressure the steam temperature at the HP first stage can even slightly rise when the turbine load decreases.

When the turbine operates under sliding pressure or combined governing, it becomes possible to optimize further the load change processes decreasing their duration. The diagram in Figure 11-5 depicts the results of calculated optimization for the scheduled decrease of the turbine load from the nominal value to the lower boundary of the governed range as applied to supercritical-pressure 300-MW units with the combined program of controlling the turbine valves; the upper rate of the load changes is limited by the value of 7%/min according to the boiler's hydraulic conditions. This pattern of the load changes was put into an advanced version of the standard system for automatic governing the frequency and load developed in VTI for the 300-MW units.[5]

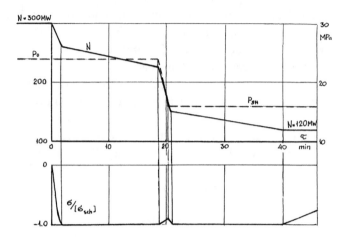

Figure 11-5. Calculated optimization of load changes from the rated output to the lower boundary of the governed range for a 300-MW supercritical-pressure power unit with a combined program of governing the turbine valves and limited maximum load change rate.
$[\sigma_{sch}]$—upper admissible value of thermal stress, σ, in the HP rotor at scheduled load changes

There also exist some other hybrid programs, as, for example, that called "free pressure mode." It is developed especially for turbines with nozzle-group steam admission control. In this case, with decreasing the turbine output, the power unit's control system successively drives the turbine valves toward their predetermined positions, maintaining the set throttle steam pressure, then allows it to slide with the invariable valve position until the newly set steam pressure value, then closes the valve, and passes to acting on the next one. Such a control mode was implemented at 500-MW subcritical-pressure power units of the Seminole power plant in Oklahoma. According to the authors,[6] this control mode promises the lowest heat rate performances of the turbine in the middle and low load ranges. Changes of the load, steam pressure, and valve position in the process of evening unloading of the turbine with such a control mode are shown in Figure 11-6. It is understandable that the admissible load change rates for each specific case of applying similar programs should be set with regard to the dependence between the turbine output and steam temperature drop at the HP entrance.

Scheduled and Unscheduled Load Changes

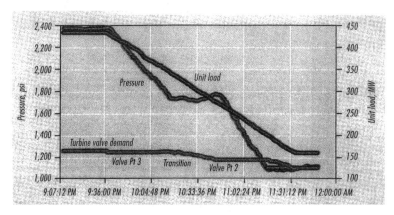

Figure 11-6. Evening unloading of a 500-MW subcritical-pressure turbine with a "free pressure control mode" at the U.S. Seminole power plant. Source: E.C. Hutchings et al.[6]

References

1. Leizerovich A. Sh., E.R. Plotkin, A.S. Sozaev, and Z.F. Goncharenko. "Limitations to Variations in Turbine Load with Participation of a Generating Unit in Control of Frequency and Power of a Power Supply System," *Thermal Engineering* 29, no. 6 (1982): 309-312.
2. Lehman L., H. Teigelake, and E. Wittchow. "Unit 4 of the Heyden Power Plant. Reference Plant for Modern 1000-MW Hard Coal-Fired Power Plants," *VGB Kraftwerkstechnik* 76, no. 2 (1996): 73-81.
3. Leyzerovich A. *Large Power Steam Turbines: Design & Operation*, 2 vols., Tulsa (OK): PennWell Books, 1997.
4. Vitalis B.P. "Constant and sliding-pressure options for new supercritical plants," *Power* 150, January/February 2006: 40-47.
5. Davydov N.I. and Melamed A.D., editors. *Automatic Control of Load for Fossil-Fuel and Nuclear Power Plants* [in Russian]. Moscow: Energoatomizdat, 1990.
6. Hutchings E.G., J.T. Coffman, and G. Cavano. "OG&E's Seminole Plant Operates under Free Pressure," *Power Engineering* 101, no. 12 (1997): 34-39.

Chapter 12

Cycling Operation and Its Influence on Turbine Performances

DIFFERENT APPROACHES TO COVERING
THE VARIABLE PART OF POWER CONSUMPTION

Because of substantial unevenness in the day and week power consumption, fossil-fuel power plants, which do not bear the base load, have to be operated in a so-called cycling manner, that is, with regular shut-downs for stand-by and subsequent start-ups of individual units, especially after night and weekend load ebbs, and/or frequent and deep load changes. More often than not, some of the power plant's units are shut down for weekends, whereas the others or some of them are deeply unloaded (up to the lower boundary of the governed load change range or even to the minimum continuous load), and all or most part of the power plant's units are unloaded for nights. In addition, steam-turbine units are also involved in automatic governing of the load and frequency of power systems and submitted to their emergency control. Such units have the annual capability factor of about 60-70% and the average annual start-up number at the level of about 20-30, most of them being warm start-ups (after outages for 36-60 hours, that is, mainly for weekends). They are considered operated in a so called *medium-load mode*.[1] In particular, this kind of operation has been typical for many power plants of the former Soviet Union with oil/gas-fired supercritical-pressure steam-turbine units, as it can be seen, for example, in Figure 11-1. A characteristic week load diagram of an oil/gas-fired 300-MW super-critical-pressure power unit at the Kostroma power plant (8x300 + 1,200 MW) is presented in Figure 12-1, and two fragments of regular diagrams with recorded results of temperature monitoring for the HP and IP rotors of a similar unit's LMZ turbine at the Kashira power plant (6x300 MW)

305

in the course of a warm start-up (after a weekend outage) and daily load changes are shown in Figure 12-2; another warm start-up of a similar turbine at the same power plant can be seen in Figure 9-4.

In many cases, power units are not shut downs for stand-by, even for weekends, but, if necessary, only deeply unloaded, as well as for nights. As a rule, such a mode of operation is rather characteristic for coal-fired power units and power units of a large single capacity. In this sense, the month load diagram for the 1,300-MW coal-fired supercritical-pressure unit of the U.S. power plant Gavin shown in Figure 12-3 is quite typical.

Along with this, some countries, especially Germany, the UK, and Japan, for many years have widely practiced regular shut-downs of fossil-fuel power units not only for weekends but also for nights, instead of unloading deeply, whereas the basis load has been mainly brought by nuclear and hydraulic power plants, as it can be seen in Figure 12-4. In this case, fossil-fuel power plants occurred to be operated in *medium-* or even *semi-peak-load zones* of operation with the average annual load factor of about 50% and less, and the annual number of start-ups counted in hundreds, most of which are hot start-ups after night outages.

Such a kind of cycling operation is commonly called two-shift or daily start/stop (DSS) operation and mainly put to subcritical-pressure steam-turbine units and CC power plants. Typical month and day load diagrams for DSS operation is presented in Figures 12-5 and 12-6 as applied to the German coal-fired Heyden Unit 4 with the gross output of 800 MW and steam conditions at the turbine inlet of 18.2 MPa, 530/530°C (2,640 psi, 986/986°F). The unit has been operated in a medium-load mode for 4,000-5,000 full-load hours with roughly about 100 start-ups per year. In 1987-1994, the annual number of hot and cold-and-warm start-ups varied from 33 and 16 to 103 and 29, and the total number of such start-ups for that period accounted to 535 and 200, respectively.[4]

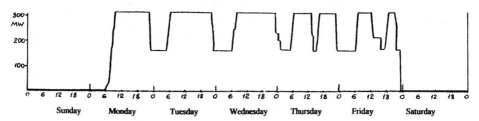

Figure 12-1. Week load diagram of a 300-MW supercritical-pressure power unit characteristic for cycling operation.

Cycling Operation and Its Influence on Turbine Performances 307

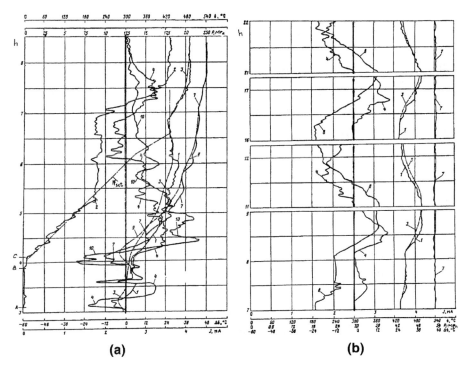

Figure 12-2. Fragments of regular diagrams with recorded results of temperature monitoring for the HP and IP rotors of a 300-MW supercritical-pressure steam turbine of LMZ at the Kashira power plant at a warm start-up (a) and daily load changes (b). *Source: A. Leyzerovich et al.[2]*

Recorded variables: 1 and 2 – steam temperature and pressure at the HP control stage, 7 and 8 – steam temperature and pressure in the JP steam admission chamber, 4 and 10 – effective radial temperature differences in the HP and IP rotors in the most thermally stressed sections, 3 and 9 – average integral metal temperatures of the HP and IP rotors in the same sections, respectively; N_{set} – the turbine load increase according to the power plant instruction

Characteristic start-up instants: A – giving steam into the HP cylinder with raising the turbine rotation speed up to 900 rpm, B – end of warming the hot-reheat steam-lines and giving steam into the IP cylinder with raising gradually the rotation speed up to 3,000 rpm, C – switching on the generator to the grid with closing the turbine bypass and accepting the initial load

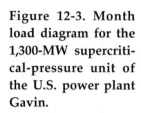
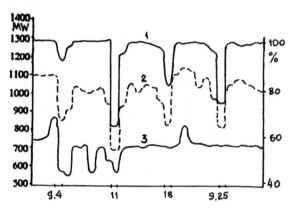

Figure 12-3. Month load diagram for the 1,300-MW supercritical-pressure unit of the U.S. power plant Gavin.
1, 2, and 3 – maximum, average, and minimum day load curves

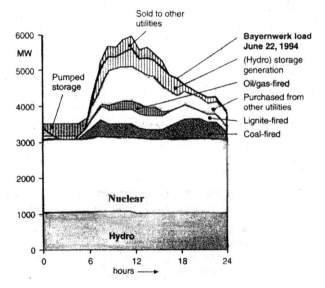

Figure 12-4. Power production diagram for the Bavarian electric utility (Bayernwerk) on a typical summer day. *Source: H.G. Houser*[3]

A similar operation mode has been also characteristic for some other coal-fired large power units of Germany both of subcritical pressure (as, for example, the 740-MW Sholven Unit "F," 720-MW Wilhelmshaven, 700-MW Mehrum, and many others of a smaller capacity[1,3]) and supercritical pressure as, for example, the 550-MW power unit Rostock, which was considered as a model, or "exemplary," power unit for cycling operation in a medium-load manner.[5] Since its initial syn-

Cycling Operation and Its Influence on Turbine Performances 309

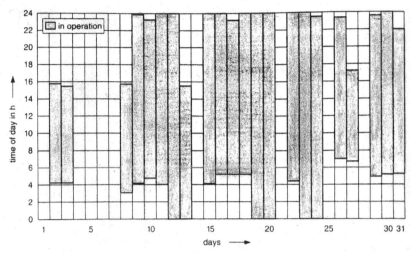

Figure 12-5. A month load diagram for the coal-fired 800-MW Heyden Unit 4 (May 1995). *Source: L. Lehman et al.*[4]

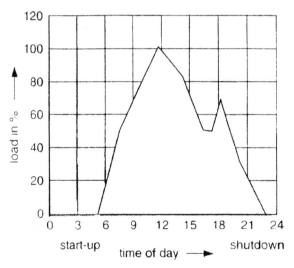

Figure 12-6. A typical summer-day load diagram for the coal-fired 800-MW Heyden Unit 4. *Source: L. Lehman et al.*[4]

chronization in April 1994 to July 1996, during its first 12,000 operating hours, this unit had passed 422 start-ups, including 395 hot start-ups. A possibility of operation in a cycling and even semi-peak mode was also proved for the 550-MW Staudinger Unit 5 with the steam conditions of 25 MPa, 540/560°C (3,620 psi, 1,004/1,040°F).[6] The instruction start-up diagram for this unit is presented in Figure 12-7.

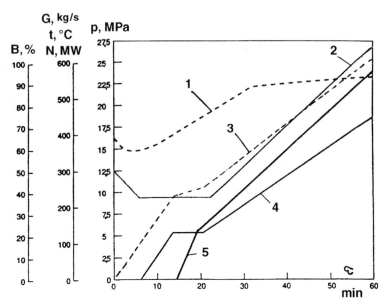

Figure 12-7. Instruction diagram of hot start-ups (after eight-hour outages) for the coal-fired 550-MW supercritical-pressure power unit 5 of the German power plant Staudinger. *Source: B. Stellbrink[6]*
1 and 2 – main steam temperature and pressure, 3 and 4 – fuel and steam flow amounts, 5 – turbine load

Another typical diagram of a hot start-up after a night outage is presented in Figure 12-8 for the 660-MW turbine of GEC at the Britain power plant Littlebrook "D" near London. Through its lifetime of 200,000 hours, the turbine having been operated in a two-shift mode was assumed to undergo up to 5,000 hot start-ups, 1,000 warm start-ups with the average outage duration equal to 36 hours, and about 200 cold start-ups. In addition, it is expected to have about 10,000 load change cycles between full and half load with the load change amp rate of about 10% MCR per minute. Such an operation mode is characteristic for a great number of Britain coal- and coal-and-oil-fired 500-, 600-, and 660-MW subcritical-pressure power units with drum-type boilers.[7] The total average annual number of start-ups for such power plants as the Fawley and Pembroke, with four 500-MW power units each, had varied for years between 200 and 600, and typically all the four power plant units have been shut down for night and started up in the morning. In so doing, the units are started up consecutively, and since the early 1980s,

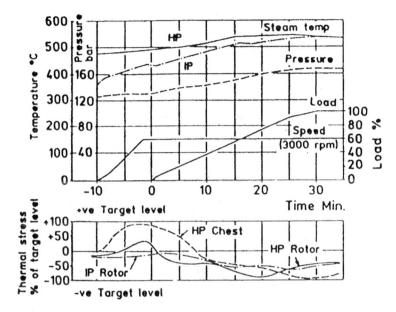

Figure 12-8. An actual hot start-up after a night outage of a 660-MW GEC turbine at the Britain power plant Littlebrook "D."

when four-unit two-shift practice has become routine, the time periods between these start-ups were reduced to one-half hour. Such a postponement is considered to be necessary to pay due attention to start-ups of each power unit. Start-up activity at the unit commences at 90 minutes before the scheduled synchronization. The initial turbine load, accepted immediately as the generator is switched on to the grid, makes up about 100 MW, and the unit is further loaded to the nominal output of 500 MW with the average rate of about 10 MW/min. Since the same early 1980s, at the similar power plant Didcot "A" (4x500 MW) the total annual start-up number has varied from over 600 to 200; in 2000 it neared to 750, and it 2001 accounted to 462.[8] A similar mode of operation is employed for such a flagship of the Britain power industry as the power plant Drax with eight 500-MW coal-fired power units. By 2000, the accumulated start-up number for individual 500-, 600-, and 660-MW units at power plants of the UK varied in the range between 800 and 2,000.[7]

In Japan, almost all new large supercritical-pressure fossil-fuel power units, up to the most up-to-date USC ones, have been designed with a capability of operation in a cycling mode, including weekly and daily stops-and-starts. Along with this, most of serviced power units, originally designed for bearing the base load and actually operated in this manner, were also refurbished for cycling operation, with extending the governed load range and reducing the minimum continuous load level, lowering the heat rate at partial loads, shortening the start-up duration, decreasing start-up energy expenditures, and facilitating the operational personnel labor due to wide automation of the transients, including start-ups.[1] A characteristic hot start-up diagram for a typical 600-MW LNG-fired supercritical-pressure power unit of a new generation at one of Japanese power plants is shown in Figure 12-9. These units, from the beginning of their operation, were intended for DSS operation with about 5,500 weekday restarts for the turbine lifetime. After the night outage, the unit can accept the full load in 90 minutes after igniting the boiler, with inclusion of approximately one hour of loading. After weekend outages, the unit is started up for 180 minutes, and the total number of such start-ups is estimated as 1,130 for the lifetime. Coal-fired units are started up more slowly. So, for example, hot start-ups of a 700-MW unit at the power plant Hekinan take 155 minutes from igniting the boiler up to the full load, with 99 minutes of loading. The load changes in the 50% range (from 350 MW up to 700 MW and back) are handled with the scheduled rate of 5 %/min. The mentioned figures are also close to the flexibility characteristics of 1,000-MW and 1,100-MW LNG-fired supercritical-pressure units. So, for example, at the Higashi Ogijima Units 1 and 2, the load change rate makes up to 50 MW/min, and total duration of hot start-ups after night outages does not exceed 170 minutes. Actual diagrams of a hot start-up and load changes for the USC-pressure double-reheat 700-MW unit of the above mentioned power plant Kawagoe gotten in the course of field start-up tests are shown in Figure 12-10. Hot start-up of the 1,000-MW supercritical-pressure Matsuura Unit 2 is presented in Figure 9-5, and characteristic diagrams of hot start-ups after night outages for Japanese CC power units are shown in Figures 9-8 and 9-9.

Even though in the 1970s-1980s some U.S. steam-turbine-based power plants experienced a two-shift operation mode,[1] presently there are no such units in the USA known to be operated in this manner on a regular basis.[7] Covering power consumption variations in the power systems is mainly achieved by operating the fossil-fuel units with partial loads—see Figure 12-3. Most units can operate comfortably at a minimum load of 25% MCR, and some are able to maintain stable operation at loads down to 10%. Individual units are shut down over weekends.

Cycling Operation and Its Influence on Turbine Performances 313

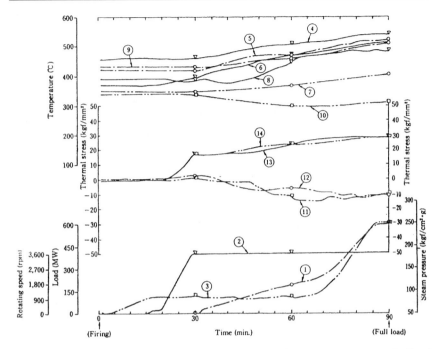

Figure 12-9. Hot start-up after night outage for a 600-MW LNG-fired supercritical-pressure unit of the Japanese power plant Gobo.
1 - generated output, 2 - rotation speed, 3 and 4 - main steam pressure and temperature, 5 - reheat steam temperature, 6 and 7 - metal temperature near the internal surface and on the external surface of the HP control-valve steam-chest wall, 8 - metal temperature near the internal surface of the HP inner casing in the first stage's zone, 9 - metal temperature near the internal surface of the IP valve steam-chest, 10- steam temperature at the IP cylinder outlet, 11 and 12 - thermal stresses on the surface of the HP and IP rotors in the zones of their first stages, 13 and 14 - the same on the surface of the central bores. *Source: By courtesy of Hitachi.*

In France, nuclear power plants generate about 80% of the annual demand, and hydroelectric plants, which are very flexible and relatively cheap in operation, account for approximately 15%, with only the balance of 5% generated by fossil-fuel power plants in a peak mode. There are some twenty fossil-fuel power plants with an average capability factor of less than 15%. Each of them makes about 70 start-ups per year, including cold start-ups. Three 600-MW subcritical-pressure power units at the Cordemais and Le Havre plants generate about 20% of this power generation amount, with the rests commonly operating for less than 1,000 hours per year.[7]

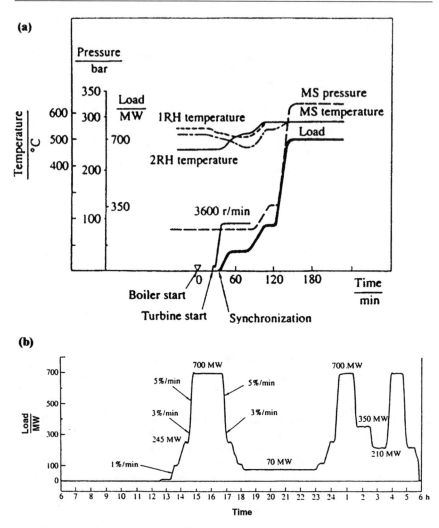

Figure 12-10. Hot start-up (a) and load changes (b) according to field test results for the 700-MW USC-pressure double-reheat Kawagoe Unit 1. *Source: K. Iwanaga et al.*[9]

Serious concern about cycling operation of fossil-fuel power plants also exists or arises in Australia, Canada, China (including Hong Kong), Denmark, India, Italy, Korea, South Africa, and many other industrially developed and developing countries. It should be said that since the mid-1990s, even more steam-turbine power plant operators in different countries throughout the world released that their mode of operation

Cycling Operation and Its Influence on Turbine Performances 315

has been likely to change to cycling due to deregulation processes in the power generation industry, as well as mass implementation of highly efficient low-emission gas-fired CC power units intended primarily for operation in a base-load manner. True, in the subsequent years, with the increase of gas prices, there already arose a demand of shifting this kind of power equipment into the medium- and semi-peak-load zones, too. All these changes resulted in a higher level of interest in the costs associated with cycling operation and optimizing its scale.

COST OF CYCLING OPERATION AND WAYS TO REDUCE THIS COST

Power units operated in a cycling mode unavoidably lose in their operational efficiency compared to base-load operation owing to an increase in the heat rate while operating under partial loads, as well as because of additional fuel and energy expenditures for start-ups. These energy losses are quite obvious, can be relatively easily assessed, and could and should be minimized by improving the employed start-up systems and technologies and excepting any undue delays in the course of transients. As mentioned before, one delay minute in the process of start-up loading increases the specific start-up energy expenditures by about 0.08-0.1 tons of equivalent fuel per 100 MW of the unit's rated capacity. Experience shows that, with improving start-up technologies and optimizing start-up diagrams, it becomes possible to decrease the start-up time expenditures by at least 15-20 minutes; sometimes, the saved time reaches 1.5-2 hours and even more. While operating under constant steam pressure, with unloading the turbine to 30% MCR, the turbine heat rate increases by 2-6%, depending on the steam admission control mode, but with changing over to sliding steam pressure these figures could be decreased to 0.5-4%.[1] Speaking of other required expenditures for cycling, it is also obvious that cycling power units need more complicated and sophisticated automated and computerized data acquisition and control systems (DACSs) and require extra labor of the power plant's operational personnel, increasing the O&M and capital costs. Cycling power plants also unavoidably demand for greater stores of treated water to compensate flow amounts of spoiled steam discharged at the beginning start-up stages.

Along with this, there exist some less obvious but perhaps much more substantial losses tied with or caused by cycling operation. According to Aptech Engineering Services, specializing in such investigations, apart from the fuel and auxiliary power needed for start-ups, as well as a higher heat rate caused by operation at less favorable operating conditions, accounting the real cost of cycling operation should also take into consideration the following factors:

- increased forced-outage rates of power equipment, with the resultant higher usage of less economical generating units, including the purchase of additional short-term capacities, and additional maintenance and recurring capital costs associated with overhauls,

- additional capital expenditures for new capacities because of the shortened lifetime of the cycling units,

- and, finally, long-term changes in the operating efficiency of power equipment caused by its deterioration in the process of cycling, including, for example, worn out seals.[10,11]

Among some other negative consequences of power plant cycling, it would also be pertinent to mention such effect as increased emissions of nitrogen oxides (NOx) and carbon monoxide (CO) into the atmosphere at the transients and even at subsequent stationary operating conditions while stabilizing the combustion processes. In particular, this is especially important as applied to CC power units.[12] For steam-turbine fossil-fuel power units, their frequent start-ups also feature increased emissions into the atmosphere.

Aptech developed a software program, called COSTCOM, for processing the power plant data and with the use of this program has analyzed and surveyed the "real costs" of cycling operation for more than 250 fossil-fuel power units with the single capacity of up to 1,300 MW.[13] According to these studies, it is the first of the above mentioned factors, that is, the increase in the maintenance scope and overhaul costs, that mainly contributes to the additional expenditures caused by cycling operation.[10] Figure 12-11, based on the data of Aptech, shows an increase in years of the annual maintenance costs for a model U.S. base-load 600-MW coal-fired power unit plus additional expenditures caused by its cycling operation with the average annual number of hot start-ups equal to 76.

Cycling Operation and Its Influence on Turbine Performances 317

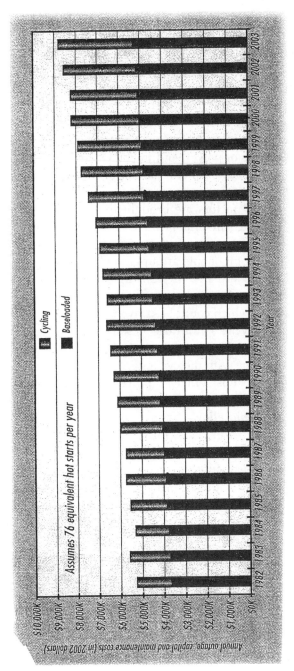

Figure 12-11. An increase of annual maintenance costs for a model base-load 600-MW coal-fired power unit with additional expenditures caused by its transition to cycling operation, according to Aptech Engineering Service. *Source: D. Smith*[13]

According to Aptech, as applied to steam turbines, possible damages and deterioration caused by cyclic operation are the following: water induction into the turbine, damages of bearings, fatigue cracking and plastic distortion of valve steam-chests, nozzle boxes, cylinder casings, and bolt joints, solid particle erosion of nozzle and rotating blades, an increase in rotor stresses and growth of rotor flaws, wear and destruction of gland seals, fatigue of blade attachments, fatigue and cracking of disc bores, silica and copper depositions in the steam path, contamination of oil in the lubrication and hydraulic control systems, etc.[11] Similar lists are composed for other kinds of power equipment. Of course, these lists can be paraphrased, somewhat modified, and supplemented, but this does not change the matter.

All the enumerated factors are brought to the equivalent forced-outage rate (EFOR), which invariably increases when a power unit is pressed out into cycling service. This results in shortening the unit's lifetime, unless its key components are upgraded for cycling, as shown in Figure 12-12. Along with this, as from the power plant operator's standpoint, the capital and maintenance costs evidently increase with the intensity of cycling, which could be conventionally characterized by the annual number of cycles (shut-downs/start-ups and/or deep load changes), from the view of the power system's dispatcher the overall system expenditures decrease with the cycling intensity because it becomes possible to support the most efficient power generation distribution between different producers (see Figure 12-4). Comparison of both these counteracting trends would allows finding an optimal solution with an economic optimum number of cycles for specific power units, minimizing the overall system and plant operating costs—Figure 12-13. That is, the increased capital and maintenance cost for the power plant should be counterbalanced by reduction of overall fuel consumption and power production cost in the power system.

In a qualitative sense, all these considerations are absolutely correct. However, some of their quantitative aspects require some notes. First, it is very difficult to assess trustworthily the influence of cycling operation of an individual power unit or power plant on the total power production cost and fuel saving in the power system because of a multiple alternative options. Nevertheless, in many cases, the resulted system effect turns out to be of a significantly larger scale than the power plant's detriment. As a result, the optimal intensity of cycling (average number of cycles per year) shifts in the diagram of Figure 12-13 far rightward,

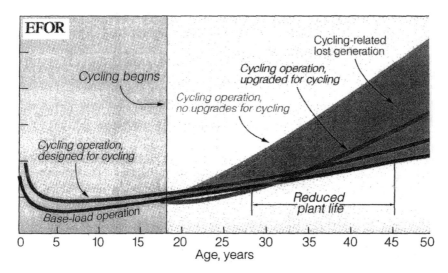

Figure 12-12. Growth of equivalent forced-outage rate with cycling operation as applied to a typical 600-MW coal-fired unit. *Source: S.A. Lefton et al.[11]*

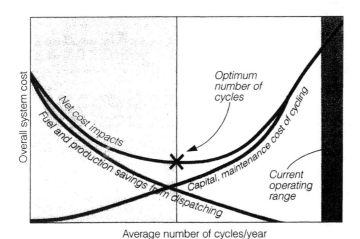

Figure 12-13. Search of an optimal cycling operation mode (the annual number of cycles) as a balance between fuel and production saving for the power system and capital and maintenance cost for the power plant. *Source: S.A. Lefton et al.[11]*

320 *Steam Turbines for Modern Fossil-Fuel Power Plants*

that is, just to the "current operating range." And second but perhaps the first in significance reason is that the detriment from cycling operation for the power plant substantially depends on not only design features of power equipment, start-up system, automation control degree and information support for the operational personnel, that is, all the "material" factors, which are really extremely important, but also on the qualification and skills of the operational personnel which determine the quality of operation, especially as applied to cycling operation conditions. In other words, comprehension and estimation of the cycling costs are certainly important, but it is even more important to search for possibilities to decrease these costs if though only qualitatively, without a possibility to assess precisely the quantitative effect.

For power units well prepared to cycling operation, including special teaching and training the operational personnel, the power plant detriment caused by cycling occurs not so great as it can be expected looking, for example, at Figure 12-11. Different level of power plant readiness to cycling operation is one of the causes why different assessments of the cycling operation costs turn out to differ very much. In addition, there exist some substantial differences in methodologies used for estimating the overall costs of the transients. As a result, different assessments of overall O&M and capital costs for the same-class power units can significantly differ, even in the magnitude order.

According to evidence from Aptech Engineering Services, there exists clear correlation between the annual number of shut-downs/start-ups (or other cycles, as for example, load changes) and the equivalent forced-outage rate (EFOR)—see Figure 12-12. Nevertheless, in many cases the presence of such correlation is not corroborated. Thus, for example, special statistic analysis of forced-outage rates for three similar oil-fire power units (two operated in a cycling manner with shut-downs for weekends for a number of years and the third one operated in a purely base-load manner) did not give any difference in their forced-outage rates (FOR): for the turbines they were equal to 0.27, 0.82, and 0.55, respectively, whereas the average FOR level attributed for large steam turbines is assessed as low as about 0.5; similar results were obtained for other power equipment types of these units. This led to a conclusion, made by the European Technology Development in their special report on power plant damages due to cycling, that "there is no clear relationship between cycling and FOR."[7] A similar conclusion was done on the basis of special comparative surveys of the operating reliability for a few 300-MW supercritical-pressure units

Cycling Operation and Its Influence on Turbine Performances 321

at different power plants of the former Soviet Union, operated in both the base-load and cycling modes.[14-16] More precisely, it should be said that such a conclusion about the absence of correlation between the cycling intensity and FOR is true for power units well prepared for cycling and with due quality of operation and, to the contrary, such an evident correlation arises if these conditions are not met.

For cycling power units, there are two factors involved in deterioration of their operating performances: 1) a mere increase in the number of actions whose effects are accumulated with time, the scale of these effects depending on the power equipment preparation, readiness for cycling, and quality of operation, and 2) an increase in the probability of unfavorable operating conditions with one-time consequences which manifest themselves at once without accumulation in time, whereas the probability level and the effect scale depend on the power equipment preparation, readiness for cycling, and quality of operation. It seems reasonable to consider the above mentioned possible steam turbine damages and deterioration associated with or caused by cyclic operation just from such a standpoint, with some modification of the aforesaid factor list for the sake of convenience.

TURBINE DAMAGES CAUSED BY OR ASSOCIATED WITH THE TRANSIENTS

Thermal Fatigue of Thick-wall Turbine Design Elements

All the thick-wall turbine design elements, such as valve steam-chests, nozzle-boxes, casings, casing rings, rotors, and others, are prone to thermal fatigue caused by unsteady-state alternating-sign thermal stresses arising at the transients. In turn, these unsteady-state thermal stresses are predominantly caused by non-stationary temperature differences across the thickness of the considered elements (radial temperature differences) supplemented by axial and tangential (circumferential) temperature differences, as well as temperature unevenness of adjoined components of the considered elements and their different rigidity (see, for example, Figure 8-13). Since the temperature differences and thermal stresses in different design elements depend on the same external actions, that is, changes of the main (for the HP turbine sections) or reheat (for the IP and LP sections) steam temperatures, as well as changes in the HP valve position (for the HP sections) and the steam flow amount through the turbine (the turbine rotation speed or load), it may be thought that if the above mentioned leading indices of the turbine's temperature and thermal-stress states (that is, the temperature differences in the critical elements) do not exceed their admissible values, the thermal stresses in other turbine elements (not only in critical ones) do not exceed their admissible values and do not cause premature fatigue cracking in all these design elements. By contrast, long-term systematic violations of the set limits for the leading indications in the course of cycling operation testify of the possibility of fatigue cracking not only for the critical elements but for other thick-wall elements, too. The upper and lower (positive and negative) admissible limits for the leading indices should be set different for different types of the transients, taking into consideration different frequency of their repetition. The maximum alternating-sign values of the leading indices at actual transient operating conditions and the number of alternating cycles within individual transients characterize the quality of handling them. Without continuous temperature monitoring for the critical elements evaluation of the operation quality at the transients cannot be trustworthy.

As it was mentioned before, the critical elements for modern steam turbines are predominantly their HP and IP rotors; the effective radial temperature differences in their steam admission sections should be used as the leading indices of the turbine's temperature and thermal-stress states, and continuous monitoring of these temperature differences should be arranged by means of mathematical modeling of the rotor heating based on the measured heating steam temperatures for the considered rotor sections. Long-term regular continuous monitoring of these temperature differences at different power plants shows that, depending on the turbine preparation for the transients and the quality of operation, the maximum values of the monitored temperature differences and number of sub-cycles within individual cycles of transients can differ from one another by several times, and the resulted specific fatigue damages for typical cycles of transients can differ tens times. Some results of this temperature monitoring for 300-MW supercritical-pressure steam turbines at different power plants are presented in Figures 7-8, 9-4, and 12-2; a thermal stress-strain diagram for the HP rotor of such a turbine at its typical transients is presented in Figure 12-14, and the ranges of variations for the specific thermal fatigue values for the HP and IP rotors of a few 300-MW turbines based on results of temperature monitoring for their rotors at characteristic

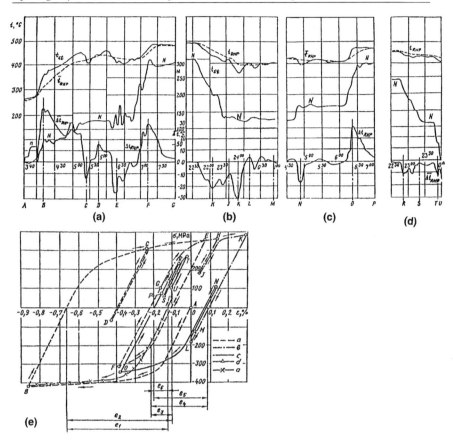

Figure 12-14. Temperature state variations for the HP rotor of a 300-MW supercritical-pressure steam turbine of LMZ (a-d) and cyclic thermal stress-strain diagram (e) at typical transients. *Source: A.Sh. Leizerovich et al.[17]*

a – warm start-up after weekend outage, b – unloading for a weekday night, c – loading in a weekday morning, d –shut-down for stand-by; t_{st} – steam temperature at the control stage, $\bar{t}_{RHP}$ and $\Delta \bar{t}_{RHP}$ - average integral metal temperature of the HP rotor and the effective radial temperature difference in its most stressed section

cycles of transients are brought into Table 12-1.

It might be well to note that cold start-ups are often considered to be the transients most damaging to the turbine in terms of its thermal fatigue.[10] It may be correct as applied to specific damages, that is, for individual start-ups, since the range of increasing the steam temperatures

324 *Steam Turbines for Modern Fossil-Fuel Power Plants*

Table 12-1. Variations of specific low-cycle (thermal) fatigue values for the HP and IP rotors of a few 300-MW supercritical-pressure turbines at their characteristic cycles of transients.

Cycle of Transients	HP Rotor	IP Rotor
Shut-down for weekend and subsequent warm start-up after two-day outage:		
• without mandatory temperature monitoring of the rotors,	0.065 – 0.13	0.2 – 0.81
• with mandatory temperature monitoring of the rotors.	0.0025 – 0.005	0.06 – 0.1
Shut-down for night and subsequent hot start-up after six-hour outage:		
• without mandatory temperature monitoring of the rotors,	0.1– 0.5	0.2– 0.35
• with mandatory temperature monitoring of the rotors.	0.0025 – 0.005	0.05 – 0.08
Load variations within the governed range for covering the night power consumption ebb:		
• under sliding main steam pressure,	~ 0.002	~ 0.0025
• under constant main steam pressure.	~ 0.01	~ 0.0025

Source: A. Sh. Leyzerovich et al.[18]

is maximal just for cold start-ups, and possibilities of steam temperature excursions and other operational errors during these start-ups are significantly greater than those for other start-up types. However, it is not completely correct because the number of such start-ups for the turbine lifetime is rather limited and their total contribution to thermal fatigue of the turbine's design elements is commonly inferior to other types of transients (warm and hot start-ups and load changes within the governed range).

Intense accumulation of low-cycle (thermal) fatigue of the rotor metal in the vicinity of stress concentrators on the rotor surface can result in appearance of fatigue cracks in grooves at the entrance of the front and central gland seals, fillets of the first stage's wheel discs (for impulse-type turbines), and root grooves of the first stages' rotating blades (for reaction-type turbines). It is obvious that the hazard of these damages grows with the accumulated number of different cycles, but the quality of running the transients plays a decisive role—one "bad" start-up can contribute to thermal fatigue of the turbine metal as much as several tens of "good" start-ups do. To decrease the rate of thermal fatigue accumulation, it is also important to diminish the stress concentrator factors on the rotor surface in the above mentioned most dangerous zones by means of increasing the radii of curvature for the considered concentrators. It can be done even as applied to the turbines in service in the process of their overhauls. In these cases, improvement of the rotor surface shape is combined with skin-peeling the rotors. This implies removal of the metal layers near the surface with maximal accumulation of thermal-fatigue damages.[1,14,19] (This subject is considered in more detail in Chapter 17 and Figure 17-9.)

Plastic Distortion of Casings and Other Stator Elements

Great temperature unevenness arising in casings, casing rings, and diaphragms of the high-temperature (HP and IP) cylinders at the transients (predominantly start-ups) can also lead to plastic deformation of these elements, apart from contribution to thermal fatigue cracking. Of importance is that in addition to other negative consequences, this type of damages can lead to permanent steam leakages through the split surfaces for the distorted joint components, with resulted energy losses and a decrease in the turbine efficiency at all the subsequent operating conditions, including the stationary ones. Special experiments obviously corroborated the notion that even a single bad start-up (for example, the first cold start-up after an overhaul) with great temperature differences in the considered elements (across the flange width, between the casing wall and flange, between the flange and stud-bolts, along the radius of the casing ring or diaphragm) can make them distort.[1,15] This type of damages does or does not take place at steam turbines no matter how frequently they are started up—the clue is the quality of running the transients. It can look like a paradox, but if the turbine is started-up frequently, the quality of these operations commonly occurs higher and a probability of resulted distortion is less that for turbines with rare, occasional start-ups.

Along with unsteady temperature differences, plastic distortion, with all the mentioned consequences, can also be caused by metal creep with stationary temperature differences in the considered elements under conditions of high metal temperatures (over 450°C, or 840°F).

Brittle Fracture of Rotors

Most incidents with brittle fracture and burst rupture of turbine rotors, which took place at power plants of different countries in the recent decades, were matter-of-factly tied with turbine transients—as a rule, they happened during start-ups (especially, cold start-ups) and were caused by a gradual growth of initial hidden flaws in the rotor metal under combined action of the tensile centrifugal and thermal stresses in the metal depth.[1] These combined stresses are especially dangerous for large IP and LP rotors, with great centrifugal forces, at the beginning stages of cold start-ups, when the metal temperatures are lower than the fracture appearance transition temperature (FATT).

One of the most "famous" failures of this type happened in 1987 at the four-cylinder Siemens (KWU) 330-MW turbine at the German power plant Irshing with the rated steam conditions of 17.8 MPa, 530/530°C (2,580 psi, 986/986°F).[1,20] By the time of failure, the turbine had been in service for 15 years with about 58,000 operating hours and experienced 838 start-ups, including 110 cold ones. The failure came during an ordinary start-up after ten-day outage. The turbine was held at the nominal rotation speed (3,000 rpm), and the generator was prepared for synchronization, when one of two forged solid (without a central bore) LP rotors suddenly burst into 34 pieces. Subsequent investigations showed the failure started because of initial flaws near the rotor axis. The largest of these flaws was schematized in its section as an ellipse with the semi-axes of 65 and 130 mm (2.5 and 5 in), and the stress intensity factor, K_1, was estimated as equal to approximately 75 $N/m^{3/2}$ which is over the fracture toughness, K_{1c}, for 2%Ni-Cr-Mo-V steel at the temperature below 80°C (176°F). As a matter of fact, the rotor forging would have had to be manufactured from the steel with 3-3.5% nickel with higher strength properties.

For modern rotor steels, their FATT does not exceed 100-120°C (210-250°F). With modern improved start-up technologies, comprising the pre-start warming up of the rotors and additional warming them at intermediate rotation speeds, as well as temperature monitoring of the IP and, if necessary, LP rotors to prevent them from inadmissible unsteady thermal stresses at start-ups, a hazard of this kind of turbine failures should not be associated with cycling.

Among incidents with rupture of turbine rotors, noteworthy are also those caused by transient torsional oscillations of the turbine-generator shaft.[1,21] Apparently the latest of such incidents took place at one of the 300-MW supercritical-pressure steam turbines of LMZ at the

Kashira power plant in the vicinity of Moscow in October 2002.[22] The turbine, its electric generator, condenser, and numerous auxiliaries were completely destroyed; the turbine-generator foundation and the turbine hall columns were damaged severely, and the roof collapsed across four spans of the building. The turbine's shaft-line was broken in eight sections—Figure 12-15, and the generator's rotor was cracked into a few large pieces, too. Until the failure, the turbine had been in operation for about 228,000 hours with 190 start-ups and 14 overhauls, whereas the turbine's individual lifetime was assigned equal to 250,000 hours. At the failure instant, the turbine was in operation for 11 days after the regular overhaul and bore load of about 235 MW; all the steam parameters, vibration level, and other monitored operating indices were absolutely normal, and nothing pointed out to the subsequent failure. Thorough subsequent investigations showed that the failure happened due to the superposition of several causes, the key of which was destruction of a shroud ring on the generator rotor after appearance of a surface fatigue crack and its development to the critical size. The crack had appeared and propagated as a result of transient torsional oscillations. The failure could be prevented if the turbine-generator shaft had been carefully tuned out from the torsional resonances at the nominal and doubled rotation frequency and abnormal conditions causing torsional oscillations were not allowed.

Numerous researches associated with a few preceding failures of this type in different countries showed that the above mentioned torsional oscillations are excited by several kinds of torque disturbances caused by dynamic interaction between the turbine's generator and electric grid. The specific fatigue of the rotor metal because of these torsional oscillations adds up to several percents of the turbine's lifetime and in the most serious cases it reaches even several tens of percent for each event. The most dangerous events of this type are: subsynchronous resonance, out-of-phase synchronization of the generator, and both successful and unsuccessful high-speed reclosing (HSR) after short-circuits.[1,21] This kind of failures can be hardly associated with cycling operation excepting the cases of switching on the ill-synchronized generator to the grid at start-ups, which must be obligatory prevented.

Solid-particle Erosion

Solid-particle erosion (SPE) of turbine blades, nozzles, and radial seals in the first HP and IP stages, as well as HP and IP control valves,

328 Steam Turbines for Modern Fossil-Fuel Power Plants

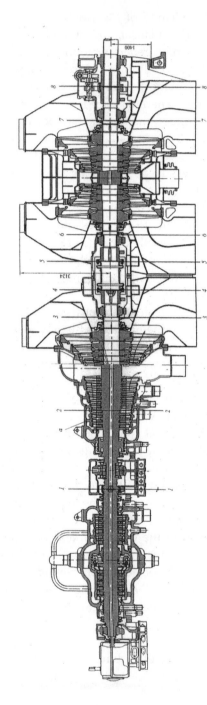

Figure 12-15. The shaft breakage sections (1-8) of the 300-MW supercritical-pressure turbine of LMZ in the failure of October 2002 at the Kashira power plant. Source: I.Sh. Zagretdinov et al.[22]

has been a problem of concern for power plant operators in many countries for many years.[1,23] This kind of erosion is caused by oxide scale that exfoliates from high-temperature boiler surfaces, including superheater and reheater (SH/RH) tubes, outlet headers, main and reheat steam-lines. This scale is brought by the steam flow to the turbine steam path causing its erosion. The rates of growth and exfoliation of oxide in SH/RH circuits depend on the tube material and operating conditions. Among other effects, SPE leads to wear and distortion of blade and nozzle profiles, an increase of their surface roughness, and greater radial seal and spill strip clearances, resulting in a significant reduce in the turbine efficiency, an increase in the repair and maintenance costs, and even can lengthen the scheduled turbine overhaul or carry out an additional, unscheduled one.

According to a survey of the effect of SPE on U.S. electric utilities conducted by EPRI in 1992 and embraced 33 utilities representing 559 fossil-fuel power units with the total installed capacity of 174,986 MW, or about 40.5% of all the NERC fossil-fuel power plants, 76% of the covered units experienced SPE-related problems and losses. For these units, the performance losses were found to range from 0.2% to 6.5%, or 84-211 kJ/kWh (80-200 Btu/kWh). The average annual heat rate penalty was found to be \$457.000 per unit, based on the heat rate degradation of 147 kJ/kWh (140 Btu/kWh), the fuel cost of \$1.85/10^6 Btu, and the capacity factor of 65%. About 50% of the total loss in the HP and IP cylinder efficiency between overhauls was attributed to SPE. The average additional maintenance cost penalty for an average power unit was assessed ranged between \$20,000 and \$40,000 per year. The average annual cost of decreased availability related to extended or more frequent overhauls was found equal to \$261,000 per unit, and the total annual estimated costs amounted to \$758,000 per unit.[23]

Fossil-fuel power units with once-through boilers, of both supercritical and subcritical pressure, were found most commonly affected by SPE, whereas power units with drum-type boilers were typically less prone to SPE, as well as HRSGs of CC units. Reheater tubes appeared to be more likely to exfoliate than superheater ones, explaining that SPE often trends to be more severe in the first IP stages compared to HP ones. Main SPE zones for the first IP stages for a supercritical-pressure cogeneration steam turbine of TMZ with the electric output of 250/300 MW are shown in Figure 12-16.

Since the presence of free oxygen is mandatory for the exfoliation

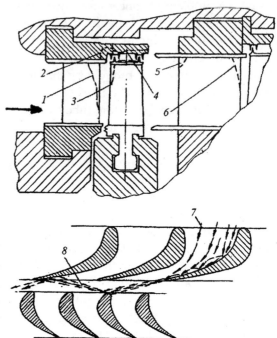

Figure 12-16. Main SPE zones for the first IP stages (1-6) and movement of solid particles in the first stage (7, 8). Source: V.A. Khaimov et al.[24]

process, it predominantly takes place at the boiler's start-ups when the tubes are not cooled by steam, whereas the external side of the tubes is already swept by high-temperature flue-gas. The turbine's hot start-ups, when the boiler is led to a high fuel load level from the very beginning of ignition, are considered especially unfavorable from the standpoint of SPE. As a result, in general, the SPE damage scope was accepted typically increasing in proportion to the number of start-up cycles.[23] Noteworthy also is that hot start-ups often feature steam temperature excursions, with the SH/RH circuit metal temperatures increasing by up to 50°C (90°F), which could even double the exfoliation rate. Along with this, it would be wrong to extend directly the aforesaid conclusion about the dependence of the SPE scope to the SPE expense, because the additional maintenance cost penalty and overhaul duration are not in proportion to the damage scope. In addition, it seems likely that although transition to cycling operation with more frequent start-ups, up to two-shift operation, does cause enlarged initial difficulties with SPE, once the inner loose oxide layer shed, the problem rather disappears.[7]

It is known that steam turbines of different power plants experience

Cycling Operation and Its Influence on Turbine Performances 331

SPE of different degrees. In particular, most of European power plants do not have so serious problems with SPE as do, for example, most of U.S. power plants.[7,23,25] This is partially explained by the more widespread use of more oxidation-resistant or exfoliation-resistant steels (like 12% Cr one) in the SH/RH circuits and steam-lines. Along with this, of importance are the shortages of start-up schemes, their bypass systems. It is significant that most of Western-European steam-turbine power units are started up with the use of two-bypass systems, with equal steam flow amounts through both the SH and RH circuits, and most of steam-turbine power units of the former Soviet Union and Eastern Europe, with one-bypass start-up systems, have the outlet SH/RH surfaces heated by flue gases of a relatively low temperature.

Noteworthy is that the SPE intensity significantly depends on the velocity of steam entering the turbine. For the first HP stage, this velocity is maximum with partial steam admission (nozzle-group control) and substantially lower with throttle steam admission control. In the last case, the nozzle erosion factor occurs to be approximately eight times lower (equal to about 0.25 instead of 2.00).[23] So, if the turbine control system allows, it is advisable to start-up the turbine with fully open HP control valves.

For power units in service experiencing intense SPE, periodic chemical cleaning of the SH/RH circuits, including steam-lines, can have significant benefit in controlling SPE. In addition, it seems very effective to protect the susceptible turbine surfaces with plasma coatings or diffusing alloying.[1,23,26] Such coatings with a thickness of 250 to 400 mkm are also proposed to protect high-temperature boiler tubes from the scale exfoliation. Cladding by Inconel 625 is considered the most common solution. There also exist numerous design developments of solid particle catchers settled in the steam-lines, valve steam-chests, and directly before and within the first HP and IP stages. [24,25]

Water Induction into Turbines

With regard to special countermeasures worked out to prevent water induction into turbines from their steam extraction systems, the most frequent sources of water induction are the main and reheat steam-lines and their spray attemperators.[1,23] Water induction can also descend from drainage lines of the main steam-lines, crossover pipes, and cylinder casings themselves. Sometimes, water and/or cold steam enter turbines through the end gland seals. More often than not, all these situations are associated with turbine start-ups and shut-downs, even though in any case they are caused by gross deviations from the start-up technology, violence of the operation instructions. A probability of such situations can be accounted increasing with the start-up quantity. However, it hardly exist some comparative data to trustworthily assess such a probability for different power units operated in different modes.

When water enters a rotating turbine, this can cause significant, often brutal, mechanical damages for blades, nozzles, and the thrust bearing. In addition, entering water or cold steam causes thermally induced stresses in turbine elements swept by these fluids. The damage extent depends on the water amount, the point where it enters the turbine, turbine rotation speed at this instance, and initial temperature of the exposed metal surfaces. If the water amount is significant enough to retard the turbine rotation, this often lead to complete destruction of the turbine steam path. Otherwise, water induction can cause local damages and deformations which, however, can be so serious that the damaged elements, even as large as a rotor as a whole or the thrust bearing, have to be replaced.

At the same time, many start-ups are accompanied with insignificant inductions into the turbine of water or cold steam from insufficiently pre-heated steam-lines or crossover pipes. These incidents do not cause any serious mechanical damages, do not tell on the turbine vibration readings, and more often than not remain unnoticed by the operational personnel if the turbine is not equipped with sufficiently quick-response steam temperature measurements in the steam admission zone. If such measurements exist, the considered events of induction manifest themselves by very short-term, practically impulse, decreases of the measured steam temperatures. Just that event can be seen in Figure 9-4 at the instant when steam is given into the HP cylinder through the ill heated crossover pipes. A similar situation took place when the turbine was put under load and the main steam attemperator was switched on. These events cause appearance of tensile thermal stresses on the heated surfaces significantly contributing to thermal fatigue accumulation in the metal of both rotors and casings. In particular, such en effect was recorded experimentally at the actual 200-MW steam turbine with the HP casing furnished with special experimental strain gauges.[1,15] Such "temperature shocks" were found to be one of the causes of fatigue cracks revealed in the HP casing's steam admission zone of these turbines, commencing from the internal surface. True, these cracks were revealed in the casing metal of steam turbines operated in both cycling and base-load manners.[27]

Wear out of Gland Seals

Commonly, in the operation process the turbine's internal efficiency noticeably comes down what in a great degree occurs to be associated with the transients, primarily start-ups. In particular, this process is seen in Figure 12-17 demonstrating the relative changes with time for the internal efficiency of two 300-MW supercritical-pressure steam turbines of LMZ at the Russian oil-fired power plant Kirishi according to periodical heat-rate performance field tests. These tests were carried out as a part of a complex research for investigating how cycling operation of power units affects their reliability and efficiency.[14] Some of this plant's (6x300 MW) power units were shut down regularly for weekends and sometimes for weekday nights, whereas others were unloaded up to the lower minimum load. The fulfilled investigations did not reveal any remarkable difference in FOR and availability factor for the power plant's units operated in a cycling manner and other power units of the same type operated in the base-load mode. At the same time, the steam turbine's internal efficiency of the considered units substantially decreases in the course of operation, even though without any obvious dependence on the start-up number.

Not mentioning breakage of blading, entering foreign bodies into the steam path, and other rather abnormal incidents, the decrease in

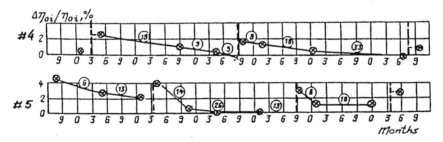

Figure 12-17. Relative changes with time for internal efficiency of two 300-MW supercritical-pressure steam turbines of LMZ at the Kirishi power plant (Russia) according to periodical heat-rate performance field tests. *Source: E.R. Plotkin and B.I. Shmukler[14]*
The dotted vertical lines correspond to turbine inspections and repairs; the circled figures indicate the number of start-ups between the tests, designated by the signs of ⊗

the internal efficiency of turbines in the operation process is commonly caused by such factors as SPE of blading in the first HP and IP stages and water drop erosion (WDE) of blading in the LP cylinders, depositions on the blading surfaces (see below) throughout the steam path, and wear out of the turbine gland seals. As distinct from the other events, whose effect gradually increases in time of operation, the last factor manifests itself at once. The wear or even complete destruction of the gland seals can be caused, for example, by water induction into the turbine with resulted sharp axial displacement of the turbine rotor and axial rubbing between the stationary and rotating elements of the steam path. Water induction from the bleeding steam-lines or turbine drainage lines adjoined to the high-temperature cylinder's bottom is also accompanied with radial rubs. Therewith, the bottom part of the casing sharply cools, causing a "humping" effect because of a great temperature difference "top-bottom." As a result, the lower clearances between the rotor and stator components come to zero with wearing out or crushing the over-shroud and diaphragm (or under-shroud) seals.

However, more often than not, destruction or wear of the turbine gland seal take place at start-ups conducted with some violations of start-up instructions. It can be, for example, rolling up the turbine with excessive "top-bottom" temperature differences in the high-temperature cylinders. Therewith, if the rotor is somewhat bowed, rubbing goes on its one-side, and the pattern of rubbing is dramatized by further bowing of the rotor because of its uneven heating up to the appearance of plastic strains. It might be useful to note that the top-bottom temperature differences for the outer casing also cause turning the inner casing hanged inside the outer one. This results in somewhat additional decrease of the lower radial clearances in the steam path.[1,15] For impulse-type turbines, lower radial clearances in gland seals of individual stages can also remarkably decrease at start-ups because of radial temperature unevenness of the diaphragms that results in some "opening" of their bottom half. [1,15,28]

Along with this, the most typical situation is a decrease to zero of axial clearances in the turbine steam path in the start-up process because of an excessive increase or decrease of the RRE in individual cylinders. This results in brushing and rubbing in both the end and central (if exists) gland seals and seals of individual stages with their resultant wear. Of importance can also be the absence of single-valued correlations between variations in the monitored RRE and actual axial clearances in

Cycling Operation and Its Influence on Turbine Performances   335

the gland seals. So in the case of a local temperature unevenness, some axial clearances can come to zero even with the RRE varying within the admissible range.[1,15] It might be well also to note that newly developed and implemented type of turbine seals (see Part 1) are less vulnerable to brushing and rubbing.

Copper Depositions on Blading Surfaces

Copper and its oxides have appreciable solubility in superheated, high-pressure steam. For units with the main steam pressure over 16 MPa (2,300 psi) and feedwater heaters (FWHs) made with the use of both copper-alloy and carbon-steel tubes, the deposition of copper and its oxides in the inlet HP stages can be a significant problem, resulting in a rapid loss of the turbine output and efficiency and requiring chemical or mechanical cleaning of the turbine steam path.[23] Historically, the problem used to be fairly topical in supercritical-pressure units, but has now disappeared with the replacement of copper alloys in the feedwater trains. At the same time, it remains significant for many subcritical-pressure units in service. Unfortunately, it is difficult, and sometimes impossible, to control corrosion of both carbon steel and copper alloys in the pre-boiler circuits, including FWHs. The dilemma is that the optimum pH for corrosion control of copper alloys is lower (8.5-9.0) that the optimum pH for carbon and low-alloy steels (> 9.5). During operation, copper alloys corrode as a result of interaction with ammonium hydroxide and oxygen, ammonium carbonate, and other acids, as well as high purity feed water. Transport of copper into the turbine occurs both during start-ups and stationary operating conditions. Corrosion of the copper alloys in the feedwater can occur during inadequately protected (oxidizing) shut-down and start-up periods, resulting in large amounts of copper corrosion products being transported around the cycle during start-ups. Another mode of copper transport into the turbine is by attemperating sprays. With a steam pressure decrease, copper and its oxides lose their solubility and deposit in the first HP stages. The impact of depositions on the IP steam path tends to be much lower.

The presence of copper in the steam is known to bring about a rapid (in worst cases after only few hundred operation hours) and in many cases very significant degradation of the HP steam path efficiency. Some turbine producers, including ALSTOM, have reported this efficiency decreases in extreme cases in excess of 3% after relatively short operation periods.[29] An example of this performance degradation over a year of operation after refurbishing the HP and IP steam paths with replacement their blading is shown in Figure 18-19. After a year of operation, the HP section's efficiency dropped by approximately 1.2%. Scraping the blading surfaces showed the copper contents equal to 39% in the scraping composition for the last HP blades and 21% for the last IP ones. Subsequent performance tests, carried out immediately after a cold re-start, showed the HP section's efficiency recovered to the level of only 0.8% down of the guarantee tests carried out 18 months earlier. Such recovery after a restart when the turbine has been allowed to cool is considered to be typical for efficiency drops caused by this kind of depositions. When the turbine cools down, adhesion between the deposits and blade surfaces is weakened; in the subsequent re-start some of deposits breaks away, and the steam path efficiency recovers slightly, although further depositions then occur and efficiency degrades back relatively quickly.

On the other hand, during start-ups, copper, iron, and dissolved oxygen levels can rise to tens or even hundreds of their values during normal operation. Further, it is not unusual for these elevated levels to last for up to 6-8 hour after synchronization. Some modifications of the operating procedures allow minimizing the transport of copper oxides into the turbine.

The copper deposits are successfully removed from the HP steam path by chemical cleaning. But, in doing so, the choice of solvent for cleaning is critical and substantially depends on the deposit composition.[23]

Stress Corrosion Cracking of the LP Wheel Discs

Typical locations of stress-corrosion cracks found in turbine wheel discs are shown in Figure 12-18. As a rule, these damages are evident in the phase transition zone of the LP cylinders, in a so called Wilson zone of saturated steam.[23,30] Cracks in the blade attachment zone (Figure 12-19) can entail rupture of the disc rim and blade liberation with all the potential consequences of avalanche-shaped damages in the steam path, jump-like rise of the shaft vibration, damages of bearings, and so on. On the other hand, cracks on the bore surface can be equally or even more dangerous, because their propagation can cause fracture of the disc as a whole with subsequent destruction of the turbine.

The most general explanation of the stress-corrosion cracking (SCC) phenomenon is given by its electrochemical theory. According to this concept, the main factor causing crack creation and propagation is an-

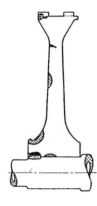

Figure 12-18. Typical locations of SCC- related damages found in LP turbine wheel discs. *Source: F.F. Lyle and H.C. Burghard.*[30]

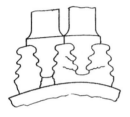 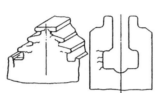

Figure 12-19. Stress corrosion cracks at a disc rim in the blade attachment zone.

odal dissolving of metal in the crack root. It is known that under certain circumstances there appears a protective film on the metal surface (a so-called *passivation* phenomenon). If the passive layer is broken under action of, for example, mechanical stresses, the crack root is brought into circumstances of chemical activity, and metal begins to dissolve actively with continuous depassivation. This process is accompanied with hydrogenization of metal on the juvenile surface at the crack root and embrittlement of metal, as a result. There exist some concepts that this process is intensified if the turbine operates under conditions of frequent load changes with resulted shifts of the phase transition zone and variations of the steam pressure at the turbine stages working in this zone. In particular, this also means that the Wilson region at transient operating conditions embraces a greater number of stages, and more discs turn out to be involved in this process.

Along with the steam/water chemistry, intensity of stress corrosion to a great degree depends on composition and heat treatment of the steel applied to the discs and their design features. So, stress corrosion of the discs has been almost absent not only at power plants of some countries with traditionally high culture of steam/water chemistry, but also at steam turbines of such producers as ABB and Siemens, even if their machines operate at the plants without high-quality water treatment. By contrast, steam turbines manufactured by the US producers were prone to stress corrosion to a large degree.

Corrosion cracks most frequently originated in places where liquid concentrates of steam contaminants could gather and be stored for a long time. Impurities are consolidated in various slots, recesses, grooves, and other design crevices, or occasional fissures or pores in the metal, which simultaneously become places of stress concentration. For SCC to occur, three conditions must be satisfied: a sufficiently high tensile stress must be applied to a susceptible kind of steel in a corrosive environment. Combination of all three factors creates a real menace of the SCC appearance and propagation.

If a turbine operates under conditions of aggressive steam (with undesirable impurities) and the disc steel with regard to its heat treatment is prone to stress corrosion, the probability of stress corrosion damages for the wheel discs in Wilson region nears 100%—Figure 12-20. These statistic data for steam turbines of Westinghouse and GE are very close to similar data for turbines of some European and Soviet turbine prodicers.[31,32] On the basis of field measurements for 40 damaged wheel discs, Westinghouse researchers derived an empirical equation for crack growth which can be recast as follows:

$$\ln(v) = -16.829 - \frac{4057}{t} + 0.04 \times \sigma_{0.2}, \tag{12.1}$$

where v is the crack growth rate in m/s; t is the metal temperature in °C, and $\sigma_{0.2}$ is the conventional yield limit in MPa.

According to investigations of EPRI, the crack growth rate mostly depends on the steel composition and tempering temperature, yield limit, and disc metal temperature. For standard NiCrMoV steels used for LP wheel discs and rotors, the most influential components of the steel composition are manganese, vanadium, nickel, and sulfur. If the sulfur content rises to 0.01%, the crack growth rate increases, but a further increase of the sulfur content is accompanied by some decrease in the crack grow rate. It is assumed that the manganese content (Mn in %) is the most representative, and the crack growth rate is proposed to be foreseen by means of the empirical equation with regard to this value and tempering temperature, t_t:

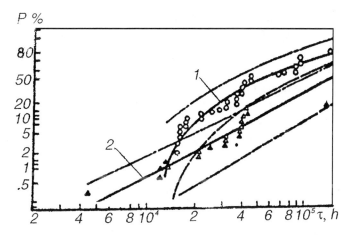

Figure 12-20. Probability of damages in the key slot (1) and on other surfaces (2) of wheel discs in the Wilson region. *Source: O.A. Povarov and E.V. Velichko*[31]

$$\ln(v) = -4.82 - \frac{5150}{t+273} + 0.0048\sigma_{0.2} + 4.53Mn - 0.0127(t_t+273). \quad (12.2)$$

As can be seen from the aforesaid equations, the crack growth velocity rises appreciably with the disc metal temperature, and steels with higher strength properties (yield limit) are more sensitive to stress corrosion.

All these factors are taken into consideration by turbine manufacturers when they choose the material for the wheel discs of designed and retrofitted turbines. So, presently there are some steel kinds applied to the LP discs and rotors with extremely high resistance to stress corrosion. These steels are used in newly designed and retrofitted turbines, as well as for substituting the damaged wheel discs. At the same time, it is the plant's responsibility to keep due steam-water chemistry to prevent stress-corrosion cracking.

Although SCC damages are primarily driven by steady-state stresses (by definition), there exists an effect of dynamic stresses on the number of cycles to initiation of cracking in rotor materials. Nevertheless, it is beyond reasons to say of a greater turbine vulnerability to SCC in the case of cycling operation.

Overheating of Blading Caused by Windage

Overheating of blading because of windage occurs when a turbine is operated with some high rotation speed (mainly, synchronous or subsynchronous) with low or none steam flow amounts through the steam path. Great circumferential speed of blades rotating in a relatively dense stagnant steam makes them overheat because of the effects of fan and friction. This phenomenon is also referred to as the ventilation effect. It results in thermal elongation of rotating blades, possible brushing and rubbing, and cracking in the blade buckets and attachment roots.[23]

Such situations can arise as applied to the HP stages of turbines operated with two-bypass start-up systems if the HP section is not separated from the cold reheat steam-lines and is not furnished with a blowing down line. In this case, the time length of operating the turbine under the rotation speed over the set value with the closed HP control valves is strictly limited. It could be also possible that the overheating effect in the HP steam path appears after turbine trips or load discharges because of insufficient flow capacity of the cold reheat steam-line drainage lines. As a result, the steam pressure at the HP outlet decreases too slowly, and the HP rotating blades rotates in rather dense steam for minutes. In this case, two options are available—faster reduction of the cold reheat steam pressure by increasing the cold reheat drainage's flow capacity or arranging a special steam-line to pass steam from the HP casing into the condenser.

For LP steam paths with long LSBs, the exhaust steam temperature can rise dramatically at low or no loads and high backpressure. To prevent this, the LP exhaust hoods are commonly equipped with spray systems which are activated whenever the measured temperature in the steam exhaust hood increases to the set level or the turbine load falls below the set value.

Water Drop Erosion of LP Blading

While steam expands through the last few stages of the LP steam path, it becomes wet, and the water droplets can cause significant blade erosion. Water drop erosion (WDE) is an expected form of damage to the turbine allowed for a certain degree to provide the maximum available energy with the taken main and reheat steam parameters. For modern steam turbines, with their long LSBs, the maximum permissible end wetness for the steam expansion process does not commonly exceed about 10% (see Figure 2-1), even though for steam turbines with shorter LSBs

Cycling Operation and Its Influence on Turbine Performances 341

some higher wetness degrees (up to 12-13%) are allowed. It is understandable that the greatest steam wetness takes place in the LSBs, and modern turbines are designed with maximum protection of the LSBs against WDE. So, the next-to-last or other intermediate LP stages operating in wet-steam environment can suffer from WDE to a greater degree than the LSB, even though at the lower steam wetness.

Typical locations of WDE for turbine rotating blades are shown in Figure 12-21. It mainly concerns the upper part of the leading edge (zone A), but can also involve some other parts shown in the picture. To mitigate the most dangerous WDE forms, the upper third of the leading edge and adjacent back surface, that is, the most susceptible zone, of steel LSBs are shielded with Stellite laminas brazed to the blade surface or manufactured with a profiled welded bar-nose made of erosion-resistant insert material with a weld filler (see Figure 5-17). As to titanium LSBs, as a rule, they are not protected against WDE owing to sufficient erosion resistance of titanium alloys (see Figure 5-20). Along with this, some turbine producers do protect the tip sections and shroud of the longest modern titanium LSBs with special shields—Figure 5-19. To reduce the steam wetness and diminish in this way the WDE severity, diverse design methods of water removal from the LP steam paths are applied: circumferential water catchers, suction slots in hollow stationary blades, increased axial spacing between the stationary and rotating blades, and so on—see Figures 5-21—5-25, and 5-27.

Evaluation of the WDE severity level for the leading edge of blades is presented in Figure 12-22.

There exist different approaches to determining a so-called WDE criterion, E, as well as different empirical equations having been developed to estimate its value. They are applicable for both wet-steam turbines of nuclear power plants[32] and LP steam paths of superheated-steam turbines for fossil-fuel power plants. The simplest and most obvious equation was developed and employed by Hitachi:

$$E = 4.3 \times (0.01u_\mathrm{p} - 2.44)^2 \times y_1^{0.8}, \tag{12.3}$$

where u_p is the circumferential speed of the blades (in m/sec), and y_1 is the steam wetness in the gap between the nozzle and blade rows, in %. The blade longevity is supposed to be assured if $E < 2$; if the value of E ranges from 2 to 4, the blades are regarded as being exposed to moderate corrosion, and if $E > 4$, this condition is treated as inadmissible.

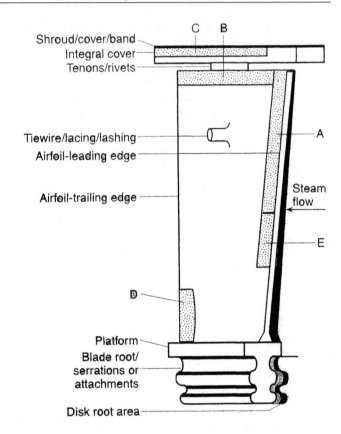

Figure 12-21. Typical locations of WDE for turbine blades of last LP stages.
Source: W.P. Sanders[33]

According to another approach, the rotating blade longevity is assumed assured if $E < 1.0$, with calculating this value in the following way:

$$E = \frac{y_0}{\psi} \times \frac{c_{2a}}{200\nu_2} \times \left(\frac{u_p}{100}\right)^4 = \frac{y_0}{\psi} \times \frac{G}{200F} \times \left(\frac{u_p}{100}\right)^4. \tag{12.4}$$

Here, y_0 is the steam wetness at the stage inlet; ψ is the stage's moisture separation factor varying in the range between 0.8 and 1.0; v_2 is the specific steam volume at the stage outlet (in m³/kg); c_{2a} is the axial component of the exit steam flow velocity (m/s); G is the steam flow amount through the stage (kg/s), and F is the annular area of the stage blades (m²).

Turbine producers and research institutions in the power industry have proposed many other expressions that are more complex and take into account more factors, such as the blade length-to-mean-diameter

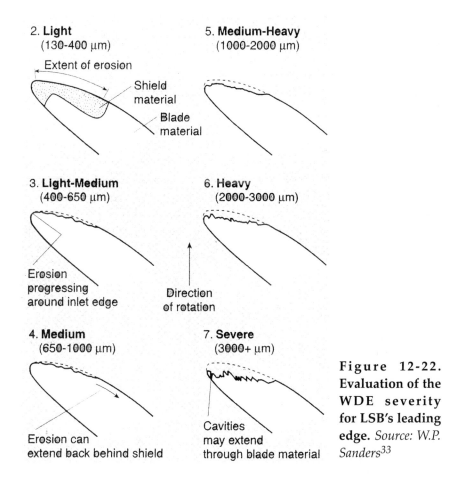

Figure 12-22. Evaluation of the WDE severity for LSB's leading edge. Source: W.P. Sanders[33]

ratio, axial clearance between the nozzle and blade rows, share of coarse-grain drops in the total moisture content, stage reactivity at the blade tip, outlet edge thickness of the nozzles, the stage's enthalpy drop, and so on.[32] The admissible value of the erosion criterion E calculated with the use of these equations also varies.

With lowering the turbine load, since the reheat steam pressure declines, if the reheat steam temperature does not substantially change, the end steam wetness decreases, too. Along with this, during start-ups, at their initial stages, with lower reheat steam temperatures, it is possible that the end steam wetness reaches its rated value and even exceeds it. Most often than not, such events are rather of a temporarily nature being caused by errors in controlling the reheat steam temperatures. At the transients, with variations of the turbine load and reheat steam conditions, the Wilson (phase transition) zone in the LP steam path can move back and forth, and as a result WDE can arise in the stages not originally protected or insufficiently protected. What is even more important, when the turbine is operated under low-flow steam conditions and/or higher backpressure in the condenser, there often appear reverse flows and vortices in the last and next-to-last stages, causing WDE of their blades in unprotected parts, such as the trailing edge near the blade root (zone D in Figure 12-21). In addition, such operating conditions are often accompanied with switching on the water sprays into the turbine exhaust hood to prevent overheating because of windage (see above) or into the condenser neck to cool the steam flows discharged through the turbine bypasses. The reverse steam motion in the exhaust and LSB (see Figure 5-8) can catch this water and carry it out to the LSB, intensifying its erosion.

The WDE rate can be cut by approximately a third if the wet steam flowing through the LP steam path contains microdoses of surface-acting fluids (surfactants). Such an amines-based surfactant, octadecylamine $C18H37NH2$, also called ODA (brand name ODACON—that is, ODA conditioned), was developed by Russian (of VNIIAM and MEI) and German (of REICON) specialists. They also developed and have widely implemented an effective technology of its usage for fossil-fuel and nuclear power plants.[34] This reagent is presently industrially produced in Germany, has a European certificate of quality, and the technology of its usage completely meets European ecological requirements. The WDE rate diminishes in the presence of surfactants in the two-phase flow due to a better dispersion, that is, reduction of water drop sizes. This effect is obvious from the images presented in Figure 12-23, which compares a water drop impact against a solid surface for "pure" wet steam, without any ODA additions, and with ODA added to the steam (about 4 mg per litre).

Added to the working fluid, ODA also effectively promotes protection of the rotating blades, wheel discs, and other LP steam path elements against SCC by means washing out chlorides and other aggressive deposits from the surfaces swept by wet steam. Technically, ODA can be injected into feedwater after the BFP. Along with fossil-fuel power plants, ODA has also been successively used at wet-steam turbines of nuclear power plants.[32]

Surfactants are also used to protect the steam/water paths from so

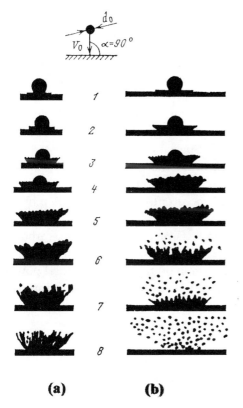

Figure 12-23. The effect of adding microdoses of amines-based surfactant in wet-steam flow on interaction between water droplets and solid surface. *Source: G.A. Filippov et al.*[34]
a – without ODA, b – with ODA concentration of 4 mg/l; instants after impact: 1 – 0.35×10^{-6} s, 2 – 0.85×10^{-6} s, 3 – 1.35×10^{-6} s, 4 – 1.85×10^{-6} s, 5 – 2.35×10^{-6} s, 6 – 2.85×10^{-6} s, 7 – 3.35×10^{-6} s, 8 – 3.85×10^{-6} s; mean droplet size $d_0 \approx 200$ mkm; two-phase steam flow velocity $w_0 \approx 50$ m/s.

called *downtime corrosion*—that is, atmospheric corrosion under action of ambient air in the process of relatively long outages. The air influence destructs a protection oxide film on the inner surfaces of equipment, resulting in their pitting. Furthermore, when a power unit is started up after the outage, its working fluid turns out to be contaminated by various impurities, including the products of downtime corrosion, and the steam/water path should be "washed out" for a long time period with withdrawing the contaminated water from the cycle until the water-chemistry indices reach their normal values. Downtime corrosion can seriously affect the entire steam/water circuit, including the turbine itself and its auxiliaries, steam-lines, and feedwater pipelines. So, if a turbine is shut down for more than a week, some national rules and standards for operation and maintenance of power plants advise preservation of the steam/water path.

Methods of preservation used in different countries can be roughly subdivided in two groups. The first one is based on creating and maintaining an anticorrosive environment in the internal spaces of equipment to be protected. The second group involves forming protective films on the inner surfaces of equipment. As applied to steam turbines, the most widely applied approaches propose blowing hot and dry air through the steam path or filling up the turbine with nitrogen or volatile inhibitors. Preservation of condensers and feedwater heaters is most frequently accomplished by washing them out with a hydrazine-ammonia solution or filling them up with nitrogen. Special conservation lubricants or glycerin are used for valves. The use of microdoses of surfactants added to the working fluid before the outage allows preservation of the entire steam/water path much easier and more effectively.[35,36] ODA inhibits downtime corrosion by forming a molecular hydrophobic film on internal equipment surfaces. This film protects metal from oxygen, carbonic acid, and other corrosive and aggressive substances contained in air. Experiments showed that ODA adsorption on the main structural power equipment materials (of both the turbine and its auxiliaries) is approximately equal for carbon and austenitic steels, copper, and brass.

References

1. Leyzerovich A. *Large Power Steam Turbines: Design & Operation*, 2 vols., PennWell, Tulsa (OK): 1997.
2. Leyzerovich A.Sh., K.N. Tikhonova, A.F. Kuz'min, et al. "Experience of Long-Term Regular Monitoring of the Rotor Temperature State of LMZ K-300-240 Turbines" [in Russian], *Elektricheskie Stantsii*, no 3 (1988): 30-35.
3. Houzer H.G. "Retrofit of the C&I of Two Coal-Fired 150 MW Units with Fully Automatic Start-Up and Shutdown," *VGB Kraftwerkstechnik* 77, no. 5 (1997): 356-361.
4. Lehman L., H. Teigelake, and E. Wittchow. "Unit 4 of the Heyden Power Plant. Reference Plant for Modern 1000-MW Hard Coal-Fired Power Plants," *VGB Kraftwerkstechnik* 76, no. 2 (1996): 85-94.
5. Hojczyk B., W. Hühne, and H.G. Thierfeleder. "The Rostock Hard Coal-Fired Power Plant," *VGB Kraftwerkstechnik* 77, no. 4 (1997): 297-303.
6. Stellbrink B. "Initial Successes with an Innovative Technique at Staudinger 5 Power Station," *VGB Kraftwerkstechnik* 74, no. 4 (1994): 322-326.
7. Starr F., J. Gostling, and A. Shibi. *Damage to Power Plant Due to Cycling*. ETD Report No. 1002-IIP-1001, London: European Technology Development, 2000.
8. "Didcot station adapts to competitive British market," *Power* 146, September 2002: 36-40.
9. Iwanaga K., A. Ohji, and H. Haneda. "The construction of 700-MW units with advanced steam conditions," *Proc. of the Institute of Mechanical Engineers* 205, no. A4 (1991), 249-252.
10. Lefton S.A., P.M. Besuner, and G.P. Grimsrud. "Understand what it really costs to cycle fossil-fired units," *Power* 141, March/April 1997: 41-46.
11. Lefton S.A., P.M. Besuner, and G.P. Grimsrud. "What you don't know will hurt

you," *Power* 146, November/December 2002: 29-34.

12. Parkinson G. "Facing the challenge of cycling," *Turbomachinery International* 45, no. 6 (2004): 27-28.

13. Smith D.J. "Market Competitiveness Requires Constant Control of Plant Operations and Maintenance Cost," *Power Engineering* 107, no. 7 (2003): 35-40.

14. Plotkin E.R. and B.I. Shmukler, editors. *Flexibility of Large Fossil-Fuel Power Units* [in Russian]. *Proc. of VTI*, issue 14, Moscow: Energiya, 1978.

15. Plotkin E.R. and A.Sh. Leyzerovich. *Start-ups of Power Unit Steam Turbines* [in Russian], Moscow: Energiya, 1980.

16. Prokopenko A.G. and I.S. Mysak. *Stationary, Unsteady, and Start-up Operating Conditions of Fossil-Fuel Power Plant Units* [in Russian]. Moscow: Energoatomizdat, 1990.

17. Leizerovich A.Sh., A.D. Trukhny, and A.A. Kochetov. "Influence of Quality of Controlling the Turbine Transients on Low-Cycle Fatigue of Rotor Metal" [in Russian], *Teploenergetika* 30, no. 6 (1983): 13-18.

18. Leyzerovich A.Sh., A.D. Trukhny, V.G. Grak, et al. "Low-Cycle Thermal Fatigue of Steam Turbine Rotors at Transients" [in Russian], *Elektricheskie Stantsii*, no. 11 (1989): 61-67.

19. Hirota V., V. Kadoya, T. Goto, et al. "Changes of Material Properties and Life Management of Steam Turbine Components under Long Term Service," *Technical Review—Mitsubishi Heavy Industries* 19, no. 10 (1982): 202-215.

20. Merz A. and R. Reinfenhäuser. "Failure of the Turbine at Power Plant Irshing" [in German], *VGB Kratwerkstechnik* 69, no. 3 (1989): 255-259.

21. Leyzerovich A.Sh. "Accumulation of Fatigue Damage in the Steam Turbine Rotors because of Torsional Oscillations due to Interaction with the Power System" [in Russian], *Energokhozyajstvo za Rubezhom*, no 4 (1982): 12-17.

22. Zagretdinov I.Sh., A.G. Kostyuk, A.D. Trukhnii, and P.R. Dolzhanskii. "Destruction of the 300-MW Turbine-Generator Unit at the Kashira District Power Station: Causes, Consequences, and Conclusions," *Thermal Engineering* 51, no. 5 (2004): 345-355.

23. McCloskey T.H., R.B. Dooley, and W.P. McNaughton *Turbine Steam Path Damage: Theory and Practice*, 2 vols., Palo Alto (CA): EPRI, 1999.

24. Khaimov V.A., Yu.Ya. Kachuriner, and Yu. A. Voropaev. "Erosion Wear by Solid Particles of the IP-1 Steam Path of T-250/300-240 Turbines" [in Russian], *Elektricheskie Stantsii*, no. 4 (2004): 14-20.

25. Holmes A. "Solid Particle Erosion in the USA vs. European Experience," presented at the Power Steam Turbine Retrofit Conference, San Francisco, 2003.

26. Schülein R.W., M. Born, and J. Korb. "Thermal Spray Coatings to Reduce Corrosion and Erosion," *VGB PowerTech* 86, no. 7 (2006): 58-64.

27. Gladshtein V.I., G.I. Moseev, and V.P. Plotnikov. "Investigation of the effect of variable operating conditions on the growth of creep cracks in the metal of cast stator parts in turbines," *Thermal Engineering* 39, no. 2 (1992): 77-81.

28. Orlik V.G. "The Off-Centering of Split Parts to a Radial Temperature Gradient," *Soviet Energy Technology*, no. 10 (1989): 55-58.

29. Hogg. S. and D. Stephen. "ALSTOM-CPS San Antonio Retrofit of JK Spruce Unit 1 HP-IP Turbine—an Example of an Advanced Steam Turbine Upgrade for Improved Performance by a non-OEM Supplier," presented at the 2005 International Joint Power Generation Conference, Chicago, 2005: PWR2005-500227.

30. Lyle Jr. F.F., A. McMinn, and G.R. Leverant. "Low-pressure steam turbine disc cracking—an update," *Proc. of the Institute of Mechanical Engineers* 199, no. A1 (1985): 59-67.

31. Povarov, O.A. and E.V. Velichko. "Corrosion Cracking in the Metal of Steam Turbines," *Soviet Energy Technology*, no. 4 (1989): 10-18.
32. Leyzerovich A. *Wet-Steam Turbines for Nuclear Power Plants.* Tulsa (OK): PennWell, 2005.
33. Sanders W. P. *Turbine Steam Path. Maintenance & Repair,* in 2 vols., Tulsa: PennWell Publishing Company, 2001.
34. Filippov G.A., A.N. Kukushkin, G.A. Saltanov, et al. "Protection, Cleaning and Preservation of Steam/Water Paths of Fossil-Fuel and Nuclear Power Plants Using Film-Forming Amine-Based Microadditives," *Proc. of the American Power Conference* 59, Chicago, 1997: 936-940.
35. Filippov G.A., O.I. Martynova, A.N. Kukushkin, and others "Preservation of the Equipment of Thermal and Nuclear Power Stations by Means of Film-Forming Amines," *Thermal Engineering* 46, no. 4 (1999): 307-311.
36. Filippov G.A., A.N. Kukushkin, G.A. Saltanov, et al. "Preservation of Thermal Power Equipment Using Reagents Based on Film-Forming Amines," *Thermal Engineering* 46, no. 9 (1999): 789-794.

PART III

DIAGNOSTIC MONITORING AND INFORMATIVE SUPPORT FOR TURBINE OPERATORS

Chapter 13

Automated Data Acquisition and Control Systems for Modern Power Plants

RECENT YEARS' EVOLUTION OF
AUTOMATED CONTROL AND SUPERVISORY SYSTEMS

With growing a single capacity and steam parameters of newly designed and constructed fossil-fuel power units and heightening requirements to their reliability, availability, efficiency, and environmental safety, they need more sophisticated control and monitoring systems (CMS) to ensure a proper quality of operation with minimum deviations from the set operating conditions. This also refers to aging power units in service to keep them in operation with due quality as far as necessary. From this consideration, it may occur reasonable to refurbish or replace their existent CMSs with up-to-date ones in parallel with or even without refurbishing main power equipment. Demands to the aforesaid systems become more complicated with transference of power units to a cycling mode of operation. And all these demands become even more topical and serious with regard to the situation with a shortage of the experienced operational personnel for power plants. All these circumstances raise a necessity to make the CMSs more complex and complicated, more automated and automatic, turning them into highly computerized automated data acquisition and control systems (DACSs) and substantially extending the scale of their functions. This is possible only with the use of modern achievements of advanced information technology (IT), including computing technique. In turn, appearance of newly developed means for getting, processing, and presenting information, suitable for the use at power plants, and improved capabilities of these means make

351

it possible to extend further the scope of functions fulfilled by modern DACSs, refine their functions, and heighten the quality of fulfillment. A combination of modern improvements in power equipment and IT makes power plants more competitive, with remarkably greater productivity and better operation quality from both the serviced equipment and servicing people.[1-3]

Modern automated CMSs or DACSs for power plants, also called Supervisory and Control Systems (SCSs), have passed in their evolution several stages.[4] The first one was characterized with the use of decentralized systems, where every automated function was performed by a specialized device. This provided a high degree of the system's vitality since a failure of individual devices did not lead to any significant loss in the total capability for control or monitoring, while a simultaneous failure of several devices was rather improbable because of their independence and territorial distribution. Among evident disadvantages of such systems were their unwieldiness, great number of components, and, as a result, great laboriousness for their service and maintenance. In addition, the scope of functions for such a system occurred to be unavoidably limited.

In the early 1960s, there began the use of computers at power plants, which enabled the entire DACSs to be build as highly centralized systems. However, even the early steps in this direction met serious problems. To a great degree, this was due to insufficient reliability of computer facilities and the lack of standard software. The difficulties grew because of the desire of system developers to load the computers with all the possible functions, even though from the present-day point-of-view these computers had very low operating speeds and insignificant memory capacities. Such central controlling and supervising computers needed specialized software, which was exceptionally time-consuming and expensive in development and debugging. The low reliability of those computers forced the developers to duplicate their functions, making the systems excessively redundant.

The appearance of microprocessor technologies and achievements in network communications created prerequisites for changing over from rigidly centralized systems to distributed control systems (DCSs), where information was processed and control was executed for the most part in a decentralized way with the use of local computer networks. Such systems have had high productivity because information is processed in parallel and high vitality because of decentralization. A decrease in

Automated Data Acquisition and Control Systems 353

the cost of microprocessor facilities themselves allows developers to create the systems that could resolve at the lower hierarchical levels many problems that used to be solved in the central computer of centralized systems. Software is an important constituent of DCSs. It also passed through several evolution stages—from individual algorithms and programs, which were developed entirely anew for every new system (as well as for initial centralized systems), to software packages constructed in accordance with a modular principle.

All these circumstances allowed the modern highly computerized automated DACSs to become absolutely indispensable in power plant operation and successfully respond to new operational requirements. Of significance has become the CRT-based arrangement of operational control and monitoring. This not only leads to significant reduction in the size of the control desk but also makes it possible to provide the operational personnel with the relevant and timely informative support that represents a part of new functions that can be charged on the DACS. Many of newly proposed and traditional control and informative functions are fulfilled at a new elevated level with the use of advanced approaches up to the application of expert systems, neural nets, and artificial intelligence (AI). So, for example, the self-adjusting governing system with the use of AI technologies provide substantially lower deviations of the governed variables at any transients and disturbances which is very important for power units with the elevated steam conditions, especially as these units are to be operated in a cycling manner. Of importance also is a possibility of long-term storage of operation data and their retrieval for post-operating analysis. In brief, corresponding to new requirements, functions of computer-based automated DACSs expand and become more comprehensive. Some features of this process, according to the vision of Hitachi, are shown in Figure 13-1. Total configuration of the DACS, or SCS, developed by Hitachi in the early 1990s and called "Hitachi Integrated Autonomic Control System (HIACS-5000)" is presented in Figure 13-2.

These systems were adapted to specific power plant features and requirements and implemented in the mid-1990s at several objects of the Japanese power industry, including, for example, the coal-fired 1,000-MW Matsuura Unit 2 with the steam conditions of 24.1 MPa, 593/593°C (3,495 psi, 1,100/1,100°F) and new CC units like those of the Himeji-I Unit 1, Yokohama Unit 8, and Kawagoe Unit 3.[7] The feature of new systems was the use of projector-type displays with 70-in (178 cm) or

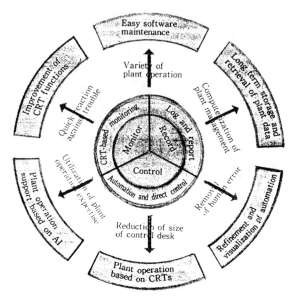

Figure 13-1. Advantages of a computer-based automated supervisory and control system for a power plant unit. *Source: A. Kaji and K. Aoki[5]*

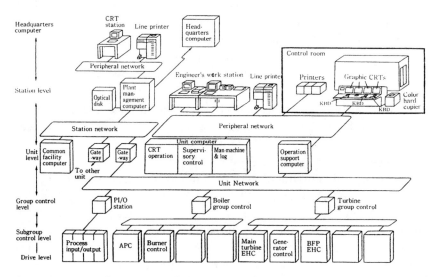

Figure 13-2. Configuration of the power plant's supervisory and control system of Hitachi HIACS-5000. *Source: A. Ito et al.[6]*

110-in (279 cm) large screens, along with common CRTs. HIACS-5000 was followed by SCSs of a new generation HIACS-7000 with a horizontally distributed structure, enhanced reliability and maintenance owing to standardization of hardware and software and economic human interface.[8] In particular, these systems were implemented at the 1,000-MW Haramachi Unit 2 with the steam conditions of 24.5, 600/600°C (3,550 psi, 1,112/1,112°F) —Figure 13-3—and the CC Unit 1 of the Shin-Oita Power Station Group 3.[9] This development was also employed to refurbish the SCSs at some power units in service, such as the Oi Unit 2 (after 27 years of operation), Anegasaki Unit 4, and Fuutsu CC Group 2.[10] Somewhat later, such a system was also implemented at the advanced combined cycle (ACC) Fuutsu Group 3, consisting of four ACC units with the output of 380 MW each.[11] Based on the HIACS, Hitachi has also developed and deployed a wide range of subsidiary control systems subordinated to HIACS, including the new steam turbine control system HITASS-2000.[12]

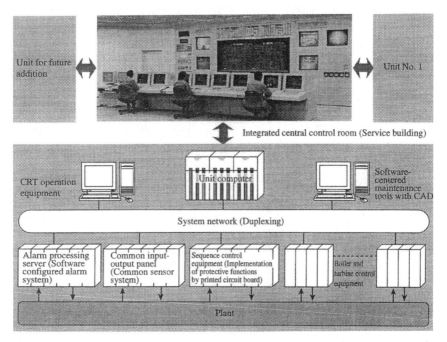

Figure 13-3. Integrated central control room and SCS (HIACS-7000) configuration for the coal-fired 1,000-MW Haramachi Unit 2. *Source: A. Takita et al.*[9]

Similar developments were fulfilled by some U.S. and European designers and producers and implemented at newly constructed or refurbished power plants based on the distributed control concept and friendly human interface with the combined use of CRTs and large screens. So, such an approach was used in the process of retrofitting the SCSs of several Britain coal-fired 500-MW power units operated in a cycling DSS manner, including the above mentioned power plants Drax (8x500 MW) and Didcot (4x500 MW).[13] For the retrofitted systems, continuous, or analogous, control (governing) was built on the basis of DCSs, while discrete (logical) control was automated with the use of programmable logical controllers (PLCs), and main operating information was presented at three video panels (67-in large screens) and nine 21-in monitors for each unit.

In this regard, one of the most well-known power units of the recent years was the 555-MW Unit 5 of the German power plant Staudinger with the steam conditions of 26.2 MPa, 545/562°C (3,800 psi, 1,013/1,044°C), which was put into commercial service in 1992. This unit has been often used as a reference point in comparative analysis of the efficiency and other operating qualities. It was also renowned by its science-based cockpit-type control room developed by ABB with highly automated control and almost full data presentation by means of videoterminals instead of individual and group instruments.[14,15] This resulted in significant reduction in the control board size—Figure 13-4, as well as in handling the unit by one operator per shift even at start-ups. Its concept was further developed as applied to the lignite-fired power plant Schkopau comprising two 450-MW units with the steam conditions of 28,5 MPa, 545/560°C (4,130 psi, 1,013/1,040°F) completely commissioned in 1996. An advanced visualization system allows the plant to be handled by two operators per shift from the common central control room exclusively via large-screen display terminals and mouse-controlled interactive communications. Each one of two operators has at their disposal three operating displays and two overview screens. Fully networked process management system in its redundant configuration is shown in Figure 13-5. The units are automatically controlled on the basis of ABB's Procontrol P system. Similar highly automated up-to-date computerized DACSs were installed at a number of units at German power plants under refurbishment, including two 250-MW units at the same Staudinger power plant, two 500-MW units at the Boxberg, and some others.[17]

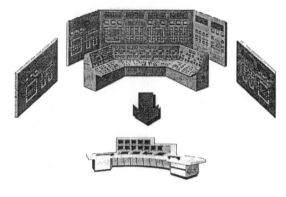

Figure 13-4. Reduction in size for the control board of the Staudinger Unit 5 as compared with an ordinary power plant. *Source: K.-H. Kraushaar*[15]

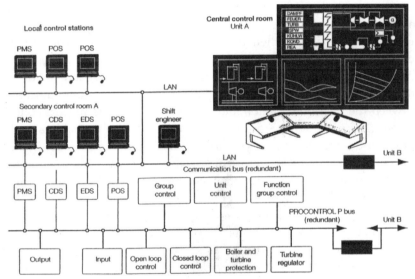

Figure 13-5. Fully networked process management system with redundant configuration for the Schkopau power plant. *Source: L. Herbst and W. Rieger*[16]

Among other developments, it is also pertinent to note a microprocessor-based DCS for the power industry developed by Siemens: TELEPERM M and its modifications TELEPERM ME, MI, and others. For years, they have been extensively used in different industry branches, including automation of power plants. In parallel, another department of Siemens has engaged in developing a line of industrial controllers and systems on their basis—SIMATIC. Since the mid-1990s, its new versions,

SIMATIC S5 and SIMATIC S7, had appeared on the market, as well as a new modification of TELEPERM TELEPERM XP, which was declared to be optimally adapted to special demands of power plant operation and enabling a greater degree of user-friendliness and a clear control room concept. Its individual modules complement each other in forming a complete I&C system. The system's TQM (Total Quality Management) ensures that the systems are tailored to the customer's specific requirements and comprises individual steps taken to draw up it up to service.[18]

The structure of the TELEPERM XP system and its division into subsystems is shown in Figure 13-6. The automation systems AS 620 perform the task of the group and individual control levels, acquire measured data from the process, execute open-loop and closed-loop control functions, and then transmit the resultant commands to the process. Three individual subsystems, denoted B, F, and T, are used for different equipment groups with regard to different demands made for them. The bus system SINEC provides a network for communication between the distributed autonomous components of the process control system. The operational terminals connected by the terminal bus in the control room area have an access to the bus system via sub-coordinated processing units (PU). The terminal bus also serves as a link to other on-site computing devices. The power plant's operational personnel fulfill their functions for operating and monitoring via the process control and information system OM 650. The system is modular and has an open structure; the number of terminals can be varied depending on the process requirements. The system is operated with a mouse or trackball. Large-screen displays can be used instead of monitors in the control room to replace conventional panels. The engineering system ES 680 is used to configure all the system components.

In particular, this system was employed for the largest lignite-fired German power plant Schwarze Pumpe 2x800 MW with the steam conditions of 26.8 MPa, 547/565°C (3,885 psi, 1,017/1,049°C, put into operation in 1997-8.[19,20] Diverse TELEPERM modifications have been also used for refurbishing the I&C of other German power plants, especially those operated in a cycling DSS manner.[21,22] It is emphasized that the resulted optimization of the start-up process and enhancement of the unit's flexibility provides a short payback time.

Along with this, noteworthy is the IT software tool PROFIT developed by Siemens Power Generation Instrumentation and Control.[2,23]

Automated Data Acquisition and Control Systems

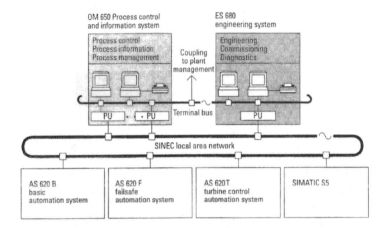

Figure 13-6. Schematic of the Process Control and Information System of Siemens TELEPERM XP. *Source: U. Sill and W. Zörner*[18]

The PROFIT is a decision-making tool that brings together information from the power generation process and the power industry market to improve profitability. This tool permits continual access to all technically and economically significant data on the cost and profit situation referring to the energy trade market. Decisions can then be made under consideration of the current costs per kWh and costs per kWh at the set time during the week, overview of energy resources and availability during the week, and other relevant variables. In this way, the plant operator knows how profitably his power generation equipment is operating at any given instant and what he must do to achieve the best results. A possible data presentation form is given in Figure 13-7.

The modular I&C concept is also applied to the Monitoring, Control, and Protection System for a steam-turbine-generator set produced by Siemens. With its modular structure, the digital I&C facility offers a tailor-made solution adapted to the specific output, steam parameters, operating conditions, and requirements of the object. Figure 13-8 shows the basic structure of this system, including circuits for protection, closed-loop and open-loop control, monitoring, and diagnostics. It is suitable for both the newly constructed and refurbished, upgraded power plants.

In the TELEPERM XP system, as well as many others, the operating and engineering workstations are made PC-based with the use of international software standards, such as UNIX, Windows, and others. Along with this, in the last years there has arisen a trend of constructing

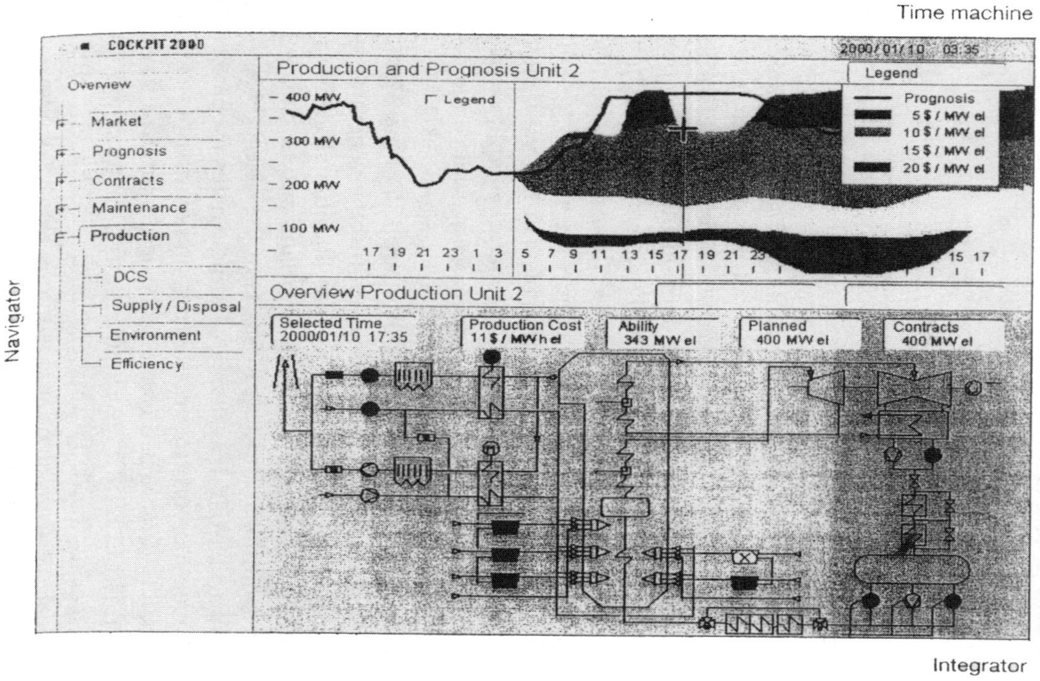

Figure 13-7. Data presentation form of Siemens' PROFIT – an IT decision-making tool for the free energy market. *Source: E. Dubslaff and K. Riedle*[2]

Automated Data Acquisition and Control Systems

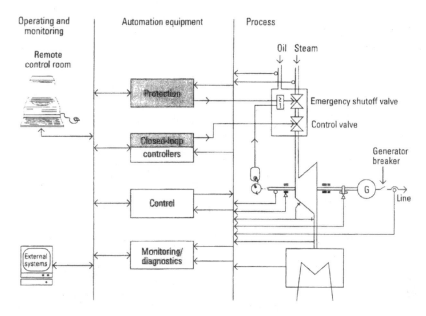

Figure 13-8. Basic structure of Siemens' Monitoring, Control, and Protection System for a steam-turbine-generator set. *Source: U. Sill and W. Zörner*[19]

the DACSs based on the DCS technology with handing over the functions of processing units and controllers to personal computers (PC). In particular, such a fully PC-based control system for fossil-fuel power plants, called the Ovation system, was developed by Westinghouse Process Control Inc., becoming a subsidiary of Emerson Electric Corporation. The system features Pentium controllers and provides rapid and smooth upgrading of the DCSs featuring microprocessor-based WPDF controllers developed and manufactured by Westinghouse. In 2000, such mass transition from WPDF to the Ovation was fulfilled at the entire 1,500-MW Midland Cogeneration power plant in Michigan during a two-day power plant outage.[24] The Ovation system was also chosen for newly constructed coal-fired units of 800 MW for the Yonghungdo power plant of Korea Electric Power Corporation and 600 MW for Tuoketuo of Beijing Datang Power Generation Co.[25] Hardware for the Yonghungdo system will include more than 18,000 I/O devices, over 36 remote controllers, several servers, and multiple operating and engineering workstations.

362 *Steam Turbines for Modern Fossil-Fuel Power Plants*

Even though many skeptics still doubt if the PC-based hardware and software are robust enough for the use in control process systems of power plants, it is beyond of question that they can be successfully used at elder power plants for additional informative support of the operational personnel even without serious reconstruction of the control desk and changes in the existent data acquisition contours. Such a partial refurbishment seems especially advisable for power units with non-computerized DACSs or with rather obsolete computers of limited capabilities. With this notion in mind, as early as the early 1990s, the PC-based Local Subsystem of Diagnostic Monitoring of the Turbine (LSDM-T) was developed for a 250/300 MW supercritical-pressure cogeneration steam turbine of TMZ and implemented at one of Moscow cogeneration plants.[26,27] A diagram showing the information flows and the set of algorithmic modules of the LSDM-T is presented in Figure 13-9. The developed LSDM-T was supposed to be supplemented with a PC-based subsystem for the turbine's vibrodiagnostics,[28] as well as newly designed additional PC-based LSDMs for the boiler, generator, and the unit's steam/water chemistry, however these plans have not been materialized. Another PC-based monitoring subsystem was developed for another Moscow cogeneration plant with non-reheat 110-MW steam turbines fed from the common steam header.[29] As distinct from the LSDM-T, this subsystem—Figure 13-10—was intended for informative support of the operational personnel as applied to both the turbines and boilers, controlled from the common control room, but only at the transients (mainly start-ups). Taking into consideration a shortage of space at the existent control room and since the turbines and boilers are never started up simultaneously, the developers restricted themselves by the use of a single PC-workstation for the group of three turbines and four boilers. The number of measurements input from each turbine and each boiler made up about 20 and 50 items, respectively. Experience showed that such PC-based subsystems can be mounted, connected up to the regular measurements, tested, and handed over to the plant personnel at the operated equipment without interruptions in current operation.

DIAGNOSTIC MONITORING OF POWER EQUIPMENT

Speaking of DACSs for modern power plants, it is necessary to dwell specifically on *technical diagnostics* of power equipment that in the

Automated Data Acquisition and Control Systems 363

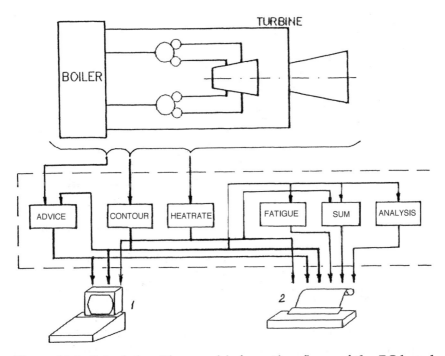

Figure 13-9. Set of algorithms and information flows of the PC-based Local Subsystem of Diagnostic Monitoring of the Turbine (LSDM-T).
Intention of the algorithmic modules: "ADVICE" – shaping recommendations about necessary or possible discrete control actions at the turbine's transients and comparing the actual course of the transients with the relevant instruction diagrams, "CONTUR" – monitoring the temperature and thermal-stress state of the turbine and determining admissible changes for the main operating parameters, "HEATRATE" – calculating the efficiency indices for the turbine as a whole and its individual components, "FATIGUE" – estimating low-cycle thermal fatigue for metal of the turbine's critical elements, "SUM" – statistic processing data on the turbine's operating conditions and their deviations from the assigned admissible values, "ANALYSIS" - shaping hypothesis about the causes why the leading indications of the turbine state exceeded their admissible values. Data presentation: 1- on-line, for the operational personnel, 2 – off-line, for post-operating analysis

recent years has become one of inalienable functions of such systems to the extent that automated systems of technical diagnostics (ASTD) are considered to be special subsystems of the data acquisition and processing systems. Technical diagnostics for equipment of power plants has some distinguishing features compared with its notion for other

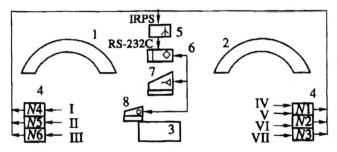

Figure 13-10. Schematic of the PC-based system for informative support of the operational personnel for the non-unit cogeneration plant.
1 – control desks for turbine no. 9 and boilers nos. 14 and 15, 2 – the same for turbines nos. 7 and 8 and boilers nos. 12 and 13, 3 – workstation of the shift engineer, 4 – analogous-digital converters, 5 – data controller and concentrator, 6 – PC, 7 - monitor and keyboard, 8 – printer; inputs: I – from turbine no. 9, II and III – from boiler nos. 14 and 15, IV and V – from turbines no. 8 and 7, VI and VII – from boilers nos. 12 and 13.

applications. For this reason, some traditional definitions in technical diagnostics need to be refined referring to the power industry.[31,32] The main goal of technical diagnostics is formulated as detection of equipment elements which are defective or faulty from the standpoint of not only reliability but also efficiency. In addition, technical diagnostics at power plants could be aimed at determination of origins of the revealed malfunctions of faulty equipment. It is also necessary to subdivide technical diagnostics of power equipment in service into *operative diagnostics* carried out at running equipment and *maintenance diagnostics* at non-operating equipment during its overhauls, repairs, and inspections. Operative diagnostics mostly utilizes methods and approaches of so-called *functional diagnostics* based on measuring and processing data from operating equipment obtained without any "provoking" outside influences. Along with this, operative diagnostics does not only come to functional one, but can also use some methods of *testing diagnostics*, based on inflicting some definite test signals and analyzing the responses to them. By contrast, maintenance diagnostics is carried out off-line, by definition, and mainly based on methods of testing diagnostics. Apart from visual inspections, including remote ones with the use of various intrascopes (endoscopes and boroscopes), one of the most potential means of maintenance diagnostics is the non-destructive testing (NDT) of equipment metal.

Operative diagnostics is mainly carried out on-line during the operating process, continually or periodically. Some functions of operative diagnostics can also be fulfilled occasionally (in particular, off-line) upon the operator request based on the data continually acquired and stored during the process. Matter-of-factly, operative diagnostics should be considered an advanced version of common monitoring, and it is difficult to draw a definite boundary between these two concepts. This being so, operative diagnostics can be also called *diagnostic monitoring*.

Even though diagnostics, by definition, refers to equipment, some experts coin the term of *situation diagnostics*, meaning revelation of the origins of the observed abnormal deviations of the monitored variables and advice for the operator on how to remedy these situations. However, to say strictly, if these deviations are not associated with equipment's damages or defects, solution of such a problem is not referred to diagnostics. It seems more correct to attribute this function to *operative surveillance* or *informative support* of the operator. At the same time, the latter is a somewhat more extensive category, including the more general help for the operator in optimizing the stationary and transient operating conditions of power equipment.

Of importance is that both of the mentioned kinds of technical diagnostics (operative and maintenance) are intended for different groups of the plant's professionals: operational personnel and maintenance staff, and results of these diagnostic procedures should be presented and used in different manners. Some of these results should be stored and forwarded to outside diagnostic systems for further processing. A general chart of technical diagnostics arrangement for a power unit is shown in Figure 13-11.

For complex objects, such as power equipment, it seems reasonable to regard their diagnostics as a totality of individual *diagnostic functions*. By these functions are meant diagnosis as applied to either individual equipment elements or individual factors determining the general state of the diagnosed equipment. Each diagnostic function can be indicated by the specific faults to be detected, the applied type of diagnostics, and specific diagnostic parameters. Different diagnostic functions intended for pursuing different diagnostic tasks essentially differ from each other in their diagnostic approaches, methods, parameters, and means. As a result, at many power units and plants, only individual diagnostic functions are implemented, and equipment is furnished with corresponding specialized instrumentation and diagnostic means, or autonomous diagnostic subsystems. As applied to different power equipment types, it is

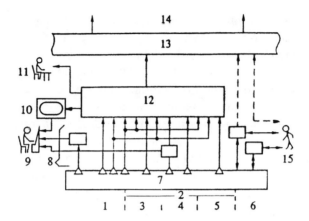

Figure 13-11. Technical diagnostics for power equipment as a portion of a general task for automated data acquisition and processing
1 - conventional on-line monitoring, 2 – operative diagnostics, or diagnostic monitoring, 3 – diagnosis for efficiency, 4 – diagnosis for reliability, 5 – situation diagnosis, 6 – maintenance diagnostics, 7 – power unit with its measurements, 8 – specialized measurements, 9 – operational personnel, 10 – operator's workstation(s), 11 – plant engineers, 12 – automated systems of technical diagnostic as a subsystem of power unit's data acquisition and processing system, 13 – the power plant's data acquisition and processing system, 14 – outside computerized data processing system, 15 – maintenance staff

possible to compose the lists of most expedient and effective diagnostic functions that can be primarily recommended for power plants. It is understandable that these lists vary depending on the equipment features and conditions of operation. Along with this, the advisable scope, completeness, and depth of technical diagnostics considerably depend on the equipment readiness to diagnostics, that is, its *diagnosibility*. In many cases, power equipment occurs badly adjusted to diagnosis—for example, is not furnished with necessary embedded measurements and does not allow installing them at site.

The first attempt of arranging general requirements on diagnosibility of thermal equipment for fossil-fuel steam-turbine-based power plants (their steam turbines, boilers, auxiliaries, and steam-lines) was undertaken in the early 1990s in Russia following recommendations of the International Electric Committee.[32,33] In particular, the developed document comprised a list of the most advisable diagnostic functions for steam turbines, including the followings: vibrational diagnostic monitoring of the turbine's shaft-line and bearings, vibrational diagnostics of the LSBs, diagnostic monitoring of the steam path regarding its flow capacity and heat rate performance of individual sections, diagnostic monitoring of temperature and thermal-stress states and assessment of thermal fatigue damage accumulation for the most thermally stressed design elements, diagnosis of thermal insulation for the high-temperature cylinders, valve steam-chests, and steam-lines, and some others. For different applications, depending on specific conditions, this list varies.

Along with this, for example, Siemens also gives its list of preferable diagnostic functions materialized in the turbine's diagnostic system DIGEST.[18] Of importance is that in this case the developers also strictly distinguish on-line and off-line diagnostics. The former is considered to be a computer-assisted assessment of parameter readings (during operation) for the purposes of the early recognition of incipient faults, identification of their causes, and finding suitable countermeasures. And the latter enables processed, stored, and measured data to be analyzed at any time. The off-line diagnostics may be performed by inspections of certain components without opening the turbine during its outages and by measurements and tests performed on the turbine during standstills, that is, it is a typical maintenance diagnostics. The principal parts of the system concern the following directions:

- turbine vibration,
- life expectancy in reference to thermal fatigue, creep damage, and damage accumulation because of torsional oscillations, and
- "thermodynamic monitoring" covering the turbine itself and its individual sections, the turbine's condenser and feedwater heaters.

The system has modular architecture that enables it to be developed with a high degree of flexibility for different application conditions.

Complex experimental ASTDs were developed and put into service in the former Soviet Union at the Ukrainian 300-MW unit of the Zujev-2 power plant and 800-MW unit of the Zaporozhye power plant.[34-36] The list of their diagnostic functions happened to be fairly close to the set materialized at the EPRI's Diagnostic Demonstration Center at the Eddystone power plant.[36,37]

Some of the above mentioned ASTDs can be considered a part of traditional, stationary and local, automated DACSs of the specific power units. In the same traditional way, some tasks of maintenance diagnostics at the power plant can also be fulfilled with the use of specialized mobile devices and intelligent computerized system—for example, for NDT of metal. The use of similar specialized mobile systems as applied to certain functions of operative diagnostics, as it is sometimes proposed, for example, as applied to vibrational diagnostics of turbines based on the use of expert systems and AI, hardly can be convenient and helpful in practice: independently of such a system, the turbine should be furnished with some stationary sensors and facility for preliminary diagnostics, and in the case of any alarming symptoms it would take time to deliver and connect a proposed more sophisticated system with its indispensable additional sources of information.

Along with this, if the turbine or other power equipment is already equipped with the necessary measurements, an advanced diagnostic support with the help of the most progressive technologies and experienced specialists can be fulfilled if the power plant is connected to a remote monitoring-and-diagnostic (M&D) center. Such a center was first arranged in Orlando, Florida by Westinghouse for servicing a few adjacent power plants with its steam turbines and generators.[36] The output data terminals of these power plants were connected to the center's input terminals via the phone network. The center's hardware and software were designed to be capable of diagnosing more than 50 individual units simultaneously on a continuous basis. By the early 1990s, this diagnostics center had covered three main directions: monitoring and analysis of steam/water cycle chemistry, diagnostics of generators, and diagnostics of steam turbines in their vibrational state and heat-rate performances. The center was staffed by skilled experts in power equipment and its operation, responsible for supporting and expanding its activity.[38] Presently Siemens Power Generation and Siemens Westinghouse Power Corporation as its subsidiary have two M&D centers: one in Orlando and the other in Erlangen, Germany that together service more than 130 power units worldwide, including steam-turbine based, gas-turbine, and CC power plants.[39,40] In the recent years, there appeared a few of other web-based M&D centers in the USA and other countries mainly intended for servicing gas turbines and steam/water chemistry for steam-turbine power plants.[39-42]

Remote M&D centers indeed can use more sophisticated technologies of processing the measured variables and diagnostics of serviced equipment than facilities available for individual power plants and even power utilities. However, if the information change between the center and power plant goes in real time, for the power plant operator does not matter if some portion of information comes to the control desk from the power unit's DACS or a certain remote center.

HUMAN-MACHINE INTERFACE AND
INFORMATIVE SUPPORT FOR THE OPERATOR

Even in the recent past, it was normal that a large power unit was operated and serviced by a shift team of three and more operators, including the roving personnel and operators of local control boards. Gathering the whole information acquired from the unit's equipment at the single control desk, computerized processing of these data, and their more proficient presentation with the use of a limited number of displays, as well as wider automation of control functions, allow operating large modern power units by as few operators as possible. At some power plants, two units operated from the same control room are serviced by a shift team of two operators.

It is understandable that with the rise of the single capacity and steam parameters of power units, as well as appearance of new control and monitoring methods and, as a result, new tasks for power plant control and monitoring systems (including technical diagnostics), the quantity of measured variables to be monitored and processed permanently grows. Proliferation of information flowing to the control room strengthens the need to make it more meaningful, readable, and comprehendible for the operators. It has been repeatedly emphasized that the information redundancy overwhelms the operator and occurs to be rather useless for them.[43]

The operator errors and non-optimal control actions too frequently affect the unit's operating reliability, efficiency, and flexibility. Apart from emergencies, this especially applies to transient operating conditions. Running the units non-optimally (in particularly, deviating from instruction diagrams) slows down the transients, that is, decreases the load change rates to cover the power demands and increases the fuel expenses. In addition, it usually results in substantial abnormal accumulation of thermal fatigue in metal of the most stressed equipment elements and can potentially lead to plastic deformation or brittle fractures. The operator problems with timely perception and comprehension of coming information can aggravate an equipment state fraught with consequences and even lead to failures. Of principal significance is that at the transients and under pre-emergency and, especially, emergency situations, the operators act in circumstances of time shortage, and all the problems with the information perception are exacerbated. Even at the most serene, normal stationary operating conditions under load, the operator's lack of skills and qualification, incapability of comprehending the available information often cause some deviations from the optimal operation parameters and, as a result, remarkable additional fuel expenses.

The operational personnel do bear main responsibility for the power unit control, even for automated power units. The control quality significantly depends on the personnel's qualification, knowledge, and skills, and their shortage often shows itself very drastically. It is well known that the "human factor" turns out to be the main cause of most failures at power plants, and just the same factor to a great degree affects the power plant performances.[44] As sounded at one recent international power generation conference, "We have met the enemy, and this enemy is us!" Aging of experienced personnel (one can sometimes hear, "Too much grey hair at control rooms!" which seems hardly fair) and their retirement have made the situation much worse, as well as the lack and insufficient skills of younger successors. It can be said that in most cases, while examining how an accident at the power plant and the ensuing power generation loss could be prevented, it occurs that the answer would lie in an improved competencies of the power plant's operational personnel and maintenance staff. [45]

At the same time, it should be admitted that frequent claims to the operational personnel are at least not always correct. Matter-of-factly, for many power plants the operators are not physically able to catch all necessary information, apprehend, process, and apply it properly and in timely fashion. It has been true for older power units with their numerous rather primitive instruments at the control board, but introduction of modern computer-based DACSs, unfortunately, has not solved this problem at all, as it had been assumed before, but sometimes even hampered its solution, flooding the operators with enormous flows of additional information and making the forms of its presentation too intricate.[46] As emphasized by Christopher Ganz of ABB as applied to steam turbines,[47] "The Human Machine Interface is where all the measured signals of a turbine are available for analysis. Detailed information about the turbine should be read here. Being a window to the process, the interface is the place to go in case of operation irregularities. But often little attention is given to the operator interface. It is seen only as a necessary add-on to the controllers providing the means to start and stop the turbine, and modify the load set. But there is more to interface whose key advantages are often overlooked."

Developments in the microprocessor application technology, network communication, and knowledge engineering have enabled the computerized DACSs and DASs to penetrate almost all the constituents of operation for fossil-fuel power plants, including both monitoring and control of the unit's equipment. This has resulted in changing entirely the control desk concept with removal of almost all the switches and auto/manual stations and almost complete abandonment of individual instruments for controlled operating parameters and multi-point recorders for indices of the equipment state. Instead, practically all the data used by the operator turns out to be exclusively concentrated on a few monitors and large-screen displays (see Figures 13-3—13-5) with appearance of the necessary windows at the operator request.

With all the advantages of such state-of-the-art systems, noteworthy is that significant scope of various, ill differenced information output at the displays makes it difficulty to perceive and "stomach" this information on-line. In addition, in many cases, especially for elder power units with not so fast computers in their DACSs and dated displays, there takes place noticeable time delays for shifting the windows after the request. As a result, in reality more often than not the operational personnel never use most of the data presentation options and information bulks potentially presented at the control panels, confining themselves to the most important, convenient, and habitual windows that bear not so much information. Moreover, with a huge amount of initial unprocessed data, the operator often lacks the most necessary information characterizing the current and forecasted state of equipment, as well as the current operating conditions of the unit and admissible or necessary operating actions.

A golden rule of a thumb is that any irrelevant, excessive information at the control panel, if it is not necessary, is harmful for the operator because it diverts their attention. However, at many modern power plants, the operators staying in front of display terminals find themselves choosing among a huge number of data presentation options and trying to comprehend numerous many-colored figures, graphs, diagrams, tables, and so on. With diverse colorful captions and figures, graphs, and technological charts, the display screen often looks like a Christmas tree. Such an abundance of information is very advantageous and worthwhile for researchers but overwhelms and irritates the operator having to evaluate the current situation and make decisions under conditions of shortage of time.

372 *Steam Turbines for Modern Fossil-Fuel Power Plants*

It is also pertinent to note that the information images presented at the control desk displays are mainly intended for the use under stationary operating conditions, and there lack special windows for different kinds of transients or they present necessary information in rather rough, unprocessed forms. So, at many power plants, one can see the operators calculating in their head or on their fingers some temperature differences or loading rate or tracing the start-up course with a shabby piece of paper beside a powerful computer complex. Calculating capabilities of computers are not practically used for calculating more sophisticated, multi-factor indices of the equipment state or analyzing operating conditions. Similar vices are often also inherent in arranging alarm information, when numerous beacons begin simultaneously flash-lighting, hiding the initial signal amidst those marking its secondary consequences, and only the operator's skills and experience can prompt them to take necessary actions.

Well-grounded control of modern power equipment often requires the use of mathematical models to monitor those important variables that cannot be measured directly but should be calculated on the basis of measured ones. A good example is temperature differences in the turbine rotors that should be employed for running the transients. Similar models are worth applying to the monitoring of the thermally stressed conditions of such design elements as, for example, HP valve steam-chests. Mathematical models would also be needed to foresee the consequences of possible control actions at both stationary and transient operating conditions, their influence on the unit's operating efficiency and reliability and to determine the admissible or desirable values of the possible variations in the current operating conditions. Unfortunately, the DACS software for most modern power units does not have these functions.

To overcome the mentioned negative phenomena, the power unit DACSs should provide special informative support for the operator and present all the necessary information on-line at request with the use of special windows. Possible content of this informative support in some of its aspects as applied to stationary and transient operating conditions is presented in the next chapters of this part. Some of the tasks of informative support for the personal as applied to the turbine are called in Figure 13-9 as functions of the LSDM-T.

Of extreme important is that effective informative support for the operator demands developing and implementing a special enclosure to the power unit's operating instruction. Such an enclosure should order

Automated Data Acquisition and Control Systems 373

the obligatory use by the operator of the corresponding windows and attention to certain equipment state indices in the process of different operating conditions. Development and implementation of such an enclosure or special section in the power unit's operating instruction seems extremely advisable whether the operator is supplied with additional informative support or not.

As a general rule, the operator should be provided with all information that is necessary for running the operating conditions, and nothing but this information should be brought on the control panels, being presented in the most complete, convenient, and obvious graphic forms, being duly grouped, hierarchically arranged, and easily available. In particular, at emergencies, all the input signals must be ranked and filtered, and the operator should be primarily supplied with those requiring direct actions. All the windows should be grouped by the equipment systems and the data presentation type with thought out transition from one system and type to another—Figure 13-12. It is implied that different types of windows are selected and used for different operating tasks. So, mimic diagrams, being most widespread, are intended for general surveillance and control of the equipment state and the course of technological processes. In doing so, the operator should have a possibility to see windows of the equipment components with not only the values of directly measured variables, but also indices to be calculated on the basis of these variables in comparison with their admissible or targeted values. At request, the operator should also have a possibility to see the changes of the controlled operating parameters with time in the form of trend graphs. In many cases, it can be advisable to plot these graphs against the instruction diagrams precisely for the current case. Along with this, as applied to logical operations of discrete control, the operator not only needs common information about the state of the objects of this control (open/closed, switched on/off, and so on) but may be glad to find additionally some pieces of advice about possible or necessary changes in their state. For processes of analogous control, for many cases it can be helpful to show the resulted changes in correlation between diverse variables in the diagram form as compared with the targeted dependence. The effects of different factors on the equipment state can be conveniently compared with the use of bar charts, and so on.

It should be also said about the necessity of storing and recording the main operating parameters and indices of the equipment state. As a rule, these data are automatically recorded only for the emergency

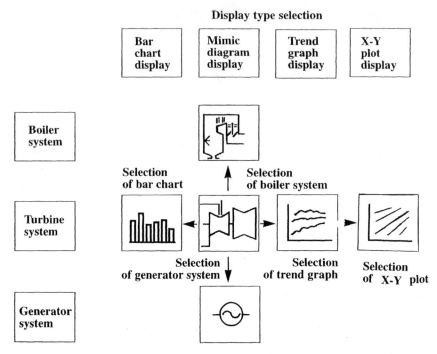

Figure 3-12. Main types and groups of on-line data presentation windows for a fossil-fuel power unit. *Source: After A. Kaji and K. Aoki*[5]

events and a certain time periods preceding these events. In addition, the measured and calculated data could be stored and recorded for arbitrary time periods at the operator's requests. Unfortunately, at many power plants this possibility is not used for transient operating conditions, (or, at least, the data are recorded in the unprocessed form). If so, possible errors committed by the operator in the course of transients remain unnoticed by both the personnel themselves and power plant engineers or supervisors, and the same typical errors will repeat again and again in the future. Under these circumstances, it is difficult to speak of teaching and training the personnel and improving their qualification. In addition, this makes practically impossible to assess accumulation of the metal damages (creep and fatigue) with regard to the actual operation quality. On this basis, it is strictly advisable to arrange obligatory fulfillment of the mentioned function as applied to all the transients and all other operating conditions with inadmissible deviations of the monitored variables with their necessary statistic processing. Among other things, in this case the informative support algorithms should comprise a sub-module for automatic identification of the operating conditions. Even if the data of transients are recorded and stored, power plant engineer often merely do not have time to study and analyze them. Along with this, sometimes plant engineers do not have sufficient qualification for such analysis. In this connection, automatic post-operating analysis of the past operating conditions would be very useful.

All the mentioned functions for DACSs providing informative support for the operational personnel can considerably contribute to improvement of their qualification and raising the operation quality. Under consideration below are a few of such functions that to a greatest degree combine the tasks of operative technical diagnostics (or diagnostic monitoring) for steam turbines and informative support for turbine operators. Therewith, this consideration will not touch on such specific problems as vibrational diagnostics of turbines, monitoring of steam/water chemistry, diagnostic monitoring of the turbine condenser and other auxiliaries, and so on.

References:

1. Smith D. "Integrating power plant control increases availability and reduces costs," *Power Engineering* 98, no. 9 (1994): 25-29.
2. Dubslaff E. and K. Riedle. "Innovation and Competition in the Power Plant Sector," *VGB PowerTech* 82, no. 7 (2002): 32-37.
3. Cooper P. "IT: the key to competitiveness," *Modern Power Systems* 22, no. 10 (2002), 27-28.
4. Sergievskaya E.N. "Trends in Development of Automated Process Control Systems for Thermal Power Plants," *Thermal Engineering* 47, no. 11 (2000): 1025-1030.
5. Kaji A. and K. Aoki. "Advanced Features of a Computer System for Thermal Power Plants," *Hitachi Review* 37, no. 4 (1988): 237-244.
6. Ito A., Y. Tennichi, and H. Hanaoka. "Thermal Power Generation Supervisory/ Control System," *Hitachi Review* 44, no. 1 (1995): 25-30.
7. Takahashi S., K. Shimizu, M. Kikuchi, and T. Ikematsu. "Advanced Information and Control System for Thermal Power Stations," *Hitachi Review* 46, no. 3 (1997): 143-146.
8. Ito A., K. Furudate, and M. Fukai. "High-Reliability, Next-Generation Supervisory and Control System for Power Stations," *Hitachi Review* 47, no. 5 (1998): 214-218.
9. Takita A., S. Takahashi, M. Fukai, and T. Takei. "Recent Technical Developments in Thermal Power Station Supervisory and Control Systems," *Hitachi Review* 48, no. 5 (1999): 267-272.
10. T. Kimura, S. Takahashi, and M. Fukai. "Latest High-Reliability Supervisory and Control Systems for Thermal Power Stations," *Hitachi Review* 49, no. 2 (2000): 53-60.
11. Iijima T., K. Ouchi, Y. Maruyama, and S. Nemoto. "Hitachi's Latest Supervisory and Control System for Advanced Combined Cycle Power Plants," *Hitachi Review* 51, no. 5 (2002): 154-157.

12. Kami T., T. Tomura, and Y. Kato. "Latest Power Plant Control System," *Hitachi Review* 52, no. 2 (2003): 101-105.
13. "The name of the game is a flexible friend," *Modern Power Systems* 18, no. 10 (1998): 39-41.
14. Neupert D., M. Schlee, and E. Simon. "Science-Based Process Control at the Staudinger Power Plant—Status, Operating Experience, and Prospects" [in German], *VGB Kraftwerkstechnik* 74, no. 3 (1994): 225-229.
15. Kraushaar K.-H. "The Cockpit-Type Control Room for the Staudinger Unit 5— Technical Concept, Philosophy, and Operating Experience to Date" [in German], *VGB Kraftwerkstechnik* 74, no. 7 (1994): 581-585.
16. Herbst L. and W. Rieger. "Unique process visualization system for Schkopau power plant," *ABB Review*, no. 1 (1997): 13-18.
17. Leyzerovich A.Sh. "New ABB Developments for Steam-Turbine Power Plants of Germany" [in Russian], *Elektricheskie Stantsii*, no. 12 (1999): 57-61.
18. Sill U. and W. Zörner. *Steam Turbine Generators Process Control and Diagnostics*, Erlangen: Publicis MCD Verlag, 1996.
19. "Schwarze Pumpe: a new era in lignite fired power generation," *Modern Power Systems* 17, no. 9 (1997): 27-36.
20. Schroeter S. "Schwarze Pumpe switched to the grid" [in German], *Standpunkt* 11, no. 3 (1998): 34-38.
21. Houzer H.G. "Retrofit of the C&I of Two Coal-Fired 150 MW Units with Fully Automatic Start-Up and Shutdown," *VGB Kraftwerkstechnik* 77, no. 5 (1997): 356-361.
22. Leibbrandt S. and B. Meerbeck. "New Automation Concepts Increase the Profitability of Power Plants. Concepts and Experiences from Four Plants Including 'Consumption Controlled Start-up'," *VGB PowerTech* 84, no. 12 (2004): 40-43.
23. Abbe M., A. Sesser, and G.K. Lausterer. "Cockpit takes off," *Modern Power Systems* 22, no. 6 (2002): 31-33.
24. "Automated tool cuts I&C migration time to days," *Modern Power Systems* 20, no. 12 (2000): 21-25.
25. "PC-based controls reach global markets," *Power* 145, September/October 2001: 10.
26. Leyzerovich A.Sh., B.R. Beyzerman, N.F. Komarov, et al. "First Experience of Applying the PC-Based Local Subsystem of Diagnostic Monitoring for Steam Turbine" [in Russian], *Elektricheskie Stantsii*, no. 4 (1993): 18-22.
27. Leyzerovich A. "PC-Based Operative Information Support at Transients and Their Postoperative Analysis in Reference to Steam Turbine," *Proc. of the American Power Conference* 56, Chicago, Part 1, 1994: 36-40.
28. Zile A.Z., V.V. Komarov, M.N. Rudenko, and S.V. Tomashevskii. "Vibration Monitoring and Diagnostics of Turbine Units in the Power Industry," *Thermal Engineering* 51, no. 4 (2004): 262-267.
29. Leyzerovich A.Sh, D.I. Bukhny, Yu.A. Radin, et al. "PC-Based System of Information Support for Fossil-Fired Non-Unit Power Plant's Group Board Operator" [in Russian], *Elektricheskie Stantsii*, no. 7 (1994): 27-31.
30. Leizerovich A.Sh. "Information Support for the Operating Personnel of a Thermal Power Station during Transients and Their Post-Operative Analysis," *Thermal Engineering* 43, no. 10 (1996): 817-823.
31. Leyzerovich A.Sh. "The Concept of Operative Technical Diagnostics for Thermal Engineering Equipment of Power Plants" [in Russian], *Elektricheskie Stantsii*, no. 7 (1991): 28-31.
32. Leizerovich A.Sh. and G.K. Sorokin. "Draft Standard on Diagnosibility of Heat-Generating and Mechanical Equipment of Power Plants," *Thermal Engineering* 40,

no. 5 (1993): 404-406.

33. Leyzerovich A. "General Requirements on Diagnosibility of Power Unit Equipment—First Approach to Standardization," *Proc. of the American Power Conference* 58. Part 2. Chicago, 1996. 1463-1467.

34. Leyzerovich A.Sh., L.P. Safonov, A.V. Antonovich, et al. "Developing and Mastering Automated Systems for Diagnostic Monitoring of Power-Generating Units at Thermal Power Plants," *Thermal Engineering* 42, no. 2 (1995): 154-160.

35. Leyzerovich A.Sh., L.P. Safonov, A.M. Zhuravel', et al. "Automated Diagnostic Systems and Subsystems for Fossil Power Plant Units: the Experience of Designing and Mastering in the Former USSR," *Proc. of the American Power Conference* 55. Part 1. Chicago, 1993. 881-885.

36. Leyzerovich A. *Large Power Steam Turbines: Design & Operation*, 2 vols., PennWell, Tulsa (OK): 1997.

37. Armor A.F., K.H. Sun, S.M. Divakuruni, et al. "Expert Systems for Power Plants: the Opportunities," *Proc. of the American Power Conference* 50. Chicago, 1988. 460-467.

38. Osborne R.L. "Centralized diagnostics uses artificial intelligence," *Modern Power Systems* 7, no. 2 (1987): 53-57.

39. Smith D. "Remote monitoring," *Power Engineering* 108, no. 3 (2004): 24-30.

40. "Long distance diagnostics," *Modern Power Systems* 22, no. 2 (2002): 23-25.

41. Baxter R. "Remote M&D leverages expertise, supports outsourcing," *Power* 146, March/April 2002: 44-52.

42. "Internet Delivery of Plant Monitoring and Diagnostics," *EPRI Annual Report*, 2000: 7.

43. Elliot T.C. "Control room instrumentation and equipment," *Power* 139, October 1995: 41-50.

44. Makansi J. "Powerplant training. Ensure your team's survival in the trenches," *Power* 139, January 1995: 23-27.

45. Wood W.A. "The Brave New World in Power Plant Operations," presented at the 2000 International Joint Power Generation Conference, Miami Beach, 2000, IJPGC2000-15026.

46. Leyzerovich A. "Information support for turbine operators," *Energy-Tech*, no. 4 (2001): 14-15.

47. Ganz C. "The importance of operator displays. All system and process views should be easily accessible," *Turbomachinery International* 46, no 1 (2005): 27-28.

Chapter 14

Diagnostic Monitoring of Turbine Heat-rate and Flow-capacity Performances

MONITORING OF HEAT-RATE PERFORMANCES AND
DEVIATIONS FROM THE RATED VALUES

The diagnostic function of on-line monitoring of turbine heat-rate performances, called sometimes "thermodynamic diagnostics," can be characterized as functional, mostly periodical or occasional, current and prognostic with the long-term forecast depth estimated in months. It is based on calculations of such performance indices as the gross heat rate and internal efficiency for the turbine as a whole, its cylinders, and individual sections, as well as the changes of these indices brought to the comparable conditions in time. These calculations can cover both the turbine itself and its auxiliaries, including the condenser, feedwater heating regenerative system, BFPs, and their turbine drives. Together with diagnostic monitoring of the heat-rate performances for the boiler and its auxiliaries, this function allows tracing the heat-rate state of the power unit as a whole.

More often than not, this thermodynamic diagnostics is based on simplified, or "alternative," versions of the standardized heat-rate test procedures used for acceptance tests of steam turbines.[1,2] As applied to the Performance Test Code for steam turbines (PTC-6) of the American National Standard Institute (ANSI) and American Society of Mechanical Engineers (ASME), there exist special regulations to qualify individuals and institutions as certified to have the capability of conducting correctly such tests.[3] These regulations are not formally extended on the heat-rate diagnostic monitoring, but some general approaches remain the same. As distinct from the acceptance tests even in their simplified, alternative versions, the heat-rate diagnostic monitoring is almost completely based on the use of common, regular measurements with a limited, small number of special measurements of a heightened accuracy. As a result,

379

the total errors and uncertainty of calculating the final performances are higher than those for the acceptance heat-rate tests and often occur to be comparable with the changes of calculation results to be analyzed. Naturally, this substantially hampers the diagnostics and limits its possibilities.

Even though the heat-rate monitoring, as a diagnostic function, would have to be intended for the revelation of changes in the equipment state that fraught with possible failures, in reality such a goal practically is not pursued, and the main purpose of this kind of diagnostics is to reveal the distinctions of the actual performances from the rated ones and traces the changes of these performances in time, in the course of operation. Just as in the case of periodic heat-rate tests (see Fig. 12-17), the performance trends are analyzed to reveal the influence of O&M on the turbine heat-rate—mainly its changes under action of different operating conditions, its restoration after repairs and overhauls, and so on. Knowledge of the actual heat-rate performances and their degradation in time opens possibilities of optimizing the turbine O&M up to transition to the predictive, or condition-oriented, maintenance.[4-8]

Typical results of such monitoring for a 420-MW power unit are shown in Figures 14-1 and 14-2. This function of the process efficiency monitoring and visualization of its results for the operators was materialized at the computerized data acquisition systems for several power units operated by the German electric power utility E.ON Kraftwerke GmbH (EKW).[9] The raw data measured with a periodicity of 1 to 10 sec and performance data calculated on their basis are compressed to mean values (per minute, hour, day, and so on) and stored. According to the authors,[10] visualization of the history process data trends give the operational personnel necessary tools for condition-oriented maintenance, as well as contribute to identification of possible disturbances and damages, not imminent but those that have already taken place. It is obvious that this information is needed for plant engineers to schedule the unit's repairs, but is unlikely usable for the operational personnel. Along with this, in the present forms the gotten information can hardly be used by the plant engineers, too, and it would need some additional processing.

In the USA, over the last two decades many fossil-fuel power plants have been equipped with a variety of computerized data acquisition systems with on-line diagnostic monitoring of the heat-rate performances. With time, many electric power utilities have brought data from these systems together into an integrated real-time heat-rate

Diagnostic Monitoring of Turbine Heat Rate and Flow Capacity 381

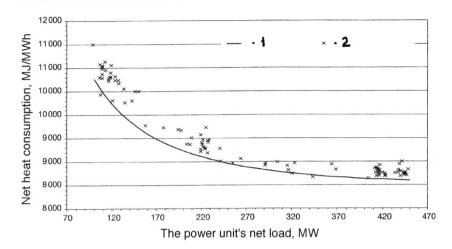

Figure 14-1. Comparison of the on-line rated (1) and calculated (2) data of the heat-rate dependence on the net load for the 420-MW EKW power unit. *Source: J. Mylonas et al.*[10]

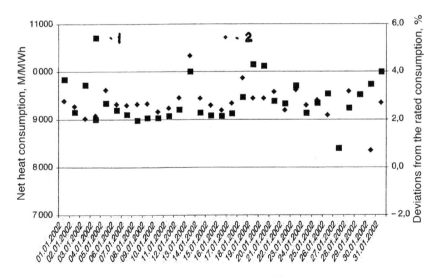

Figure 14-2. Changes from day to day of the on-line calculated heat-rate performances (1) and their deviations from the rated value (2) for the 420-MW EKW power unit. *Source: J. Mylonas et al.*[10]

monitoring system. Operators and power plant engineers can use this information to run their power equipment in a more efficient way. Of importance is that at many power plants their operational staff from the beginning has been actively involved in the selection and application of the heat-rate monitoring programs and discussions about visualization of their results. This has facilitated acceptance of these programs by the operators. In particular, such a program called Plant Performance Assessment was developed and implemented by EPRI at the 1,000-MW coal-fired power plant Merom. In the development and implementation process, eight seminars were conducted for the plant staff to make them more aware of the heat-rate diagnostic monitoring and the use of its results. Unfortunately, because of insufficient accuracy of the used regular measurements, the calculated data were not always reliable and had a significant spread which limited possibilities of their usage. Following to recommendations of EPRI, a number of key instruments were upgraded or replaced, including, for example, feedwater flowmeters and various temperature and pressure sensors. With improved measurement basis and awareness training, the Merom has reduced in a year its heat-rate performances to the lowest level since the plant went into operation in the early 1980s. According to EPRI, the plant's heat rate has been improved by 1.5%, the net capacity factor increased by 5.2%, and the power production cost reduced by 2.3%.[11]

Great possibilities are laid in the "thermodynamic module" for steam turbines of the diagnostic system DIGEST of Siemens.[12] This module's architecture can be seen in Figure 14-3. Its basis is the thermodynamic computation program KRAWAL calculating the essential performance data for both the turboset as a whole with its auxiliaries and for the turbine itself with its individual sections. Data presentation allows tracing historic changes of the calculated variables in time and trends for the performance data brought to the comparable values, as well as seeing current deviations of the steam/water cycle parameters from the rated values and their influence on the efficiency indices. A display image derived from a hardcopy with the main data characterizing the current operating conditions and heat-rate performances for a turboset is shown in Figure 14-4. It should be noted that in reality the actual number of values carried out on the screen happens to be much more than it is presented in Figure 14-4, including, for example, steam pressure and temperature values in the bleedings and feedwater temperatures after the FWHs, as well as the internal efficiency data for

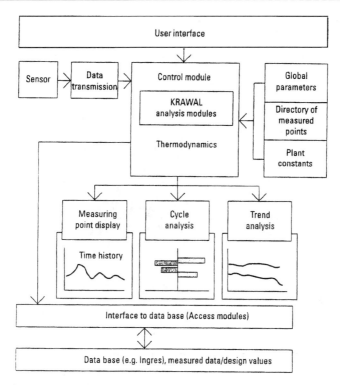

Figure 14-3. Architecture of the thermodynamic module for a steam turbine's diagnostic system DIGEST of Siemens. *Source: U. Sill and W. Zörner*[13]

individual turbine cylinders. Various means of visualization also allows the user to see in graphic forms the influence of individual factors on the turbine efficiency or directly compare operating conditions in terms of the turbine's heat rate and output—Figure 14-5.

When the turbine works under the set load, in order to provide its most efficient operation, all its steam conditions should be kept at the rated levels set for the given load, not exceeding these levels. Each deviation to smaller values results in a noticeable increase of the heat-rate or a decrease in the turbine output. Siemens' estimations of the costs of such deviations in terms of power underproduction for a medium 400-MW steam-turbine unit are given in Table 14-1.

While any lowering of the cycle parameters causes a decrease in the turbine efficiency and output, whereas heightening these parameters beyond the rated values is not admitted from the standpoint of long-term

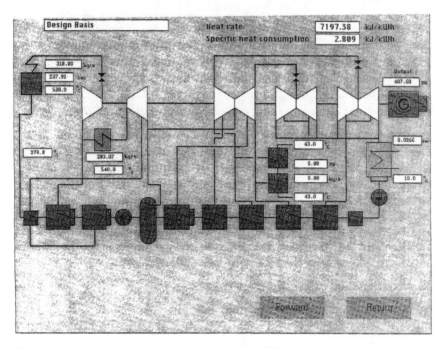

Figure 14-4. Main data on operating conditions and heat-rate performances of a turboset carried out on a display of a steam turbine's diagnostic system DIGEST of Siemens. *Source: U. Sill and W. Zörner*[13]

Table 14-1. Power underproduction with deviations of steam/water cycle parameters for a medium 400-MW steam-turbine power unit, according to Siemens

Steam/Water Cycle Parameter	Deviation from the Rated Value	Power Underproduction, MW
Main steam temperature	–10 °C	1.2
Main steam flow	–0.5%	2.0
Hot reheat steam temperature	–10 °C	0.9
Averaged FWH TTD	+ 1 °C	0.4
One FWH bypassed	-	2.4
Condenser TTD	+ 1 °C	0.6
Circulating water flow amount	–10%	0.6

Source: U. Sill and W. Zörner[13]

Diagnostic Monitoring of Turbine Heat Rate and Flow Capacity 385

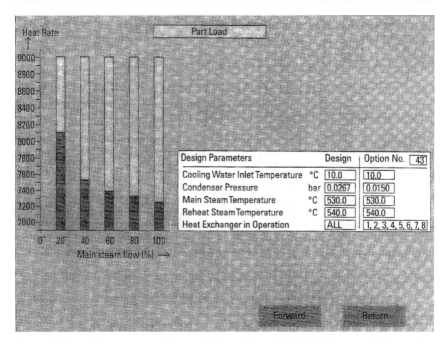

Figure 14-5. Heat-rate of the turbine related to design conditions for partial loads carried out on a display of a steam turbine's diagnostic system DIGEST of Siemens. *Source: U. Sill and W. Zörner[13]*

strength, the only advice which could be given to the operator is—to keep the cycle parameters as close to the rated values for the set turbine load as possible. In reality, this is a very common situation when one or several cycle parameters deviate from their rated values to less, and, unfortunately, too frequently these situations remain unnoticed or underestimated by the operator. That is why it seems very important not only to impassively record the happening deviations of the cycle parameters but also attract the operator's attention to these deviations and show on-line the cost of these deviations in terms of underproduction of power or additional fuel expenses up to their money equivalents. An example of a display image with such data for the above mentioned Germany EKW power plant is presented in Figure 14-6 (derived from a hardcopy).

Another example of displaying the losses caused by deviations of the cycle parameters is presented in Figure 14-7 as it is shown for the operator at the U.S. power plant John P. Madgett operated by Dairyland Energy Cooperative.[11] The heat-rate monitoring system was implemented

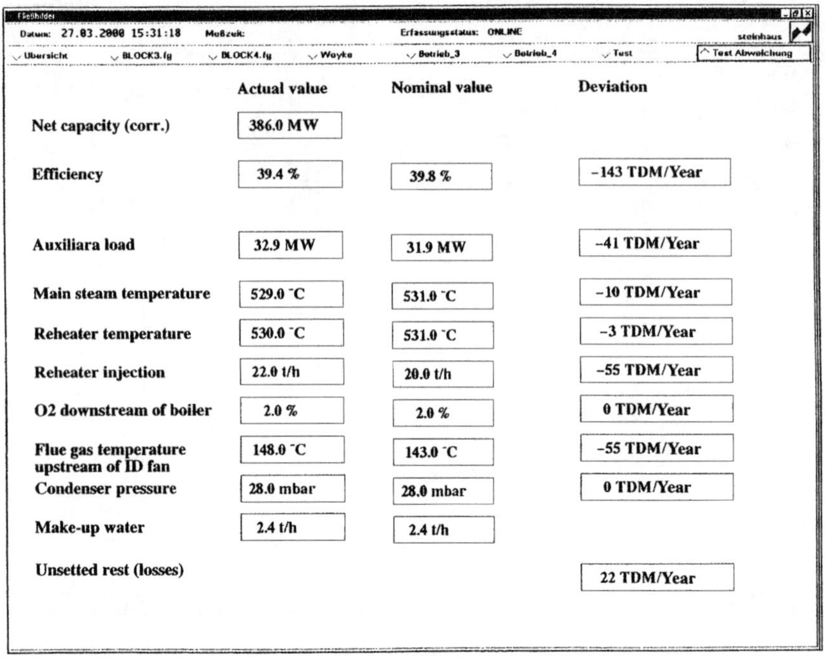

Figure 14-6. The cost of deviations of the cycle parameters from the rated values for the 420-MW EKW power unit, as presented to the operator. *Source: W. Woyke[9]*

at this power plant in 1998. According to the Dairyland Energy manager and performance administrator, over the last 10 years the heat-rate at the John P. Madgett Station has reduced by 6-7%, whereas at the other power plants it has improved only by 4%. It is emphasized that on-line heat-rate performance data allows the personnel to take notice of the efficiency degradation with time more quickly. At the same time, it is necessary to note that at the illustrated instant, the main and reheat steam temperatures substantially exceeded their "target" (rated) values without identifying these events and attracting the operator's attention to them. Therewith, the main and reheat steam temperature spray attemperators were switched on, but the spray flows were probably insufficient to reduce the controlled temperatures to the required levels. Otherwise, it can be definitely said that the current heat-rate performances were much worse. In addition, if such inadmissibly elevated steam temperatures were kept for long time, this would cause a significant detriment to the

John P. Madgett 13:38:43 November 30, 2001				Net Heat Rate: 10788 Net MW: 300.4 $/MW: 10.7	
Description	Actual	Target	Units	Heat Rate Dev. Btu's	Heat Rate Dev. Cost ($/hr)
Net Unit Heat Rate	10788	9976	Btu/kwh	878	241.30
Throttle Steam Temp.	968	955	°F	−98	−27
Throttle Steam Press.	1796.20	1800	psig	4	1.10
Reheat Temperature	980.50	955	°F	−195	−53.60
SH Spray Flow	5.40	0.0	klb/hr	1	0.10
Reheat Spray Flow	17.60	0.0	klb/hr	18	4.80
Final Feedwater Temp.	437.20	445.90	°F	35	9.50
Auxilliary Power	21877.10	23917.70	kw	−73	−20.10
Cond. Back Press.	1.77	1.27	In Hg	25	6.90
Excess 02	1.73	1.96	%	−6	−1.50
APH Gas Out Temp.	301.20	314.4	°F	−31	−8.60
LOI	1405.10		%		
		Total	Btu/kw	−325	−89.60

Figure 14-7. Displaying the losses caused by deviations of the cycle parameters at the John P. Madgett Station. *Source: D. Smith[11]*

long-term strength of the high-temperature turbine elements, main and reheat steam-lines and headers, and this detriment could be much more serious that the temporary heat-rate gain.

It can be supposed that with more well thought-out, convenient, and obvious graphic presentation of the gotten results, the operator would not commit the errors caused the noted shortcomings in running equipment. It is understandable that visualization of the gotten results in a table form, as in Figures 14-6 and 14-7, is far from optimal. It seems more convenient for operational perception to use some graphic forms like, for example, bar-charts (see, for example, in Figure 14-3—"Cycle analysis"). A good example of such presentation with the use of colored bar-charts with meaningful colors can be seen in Figure 14-8—with displaying the deviation effects in both the heat-rate changes and their cost in $/hr, as it was developed by Tennessee Valley Authority (TVA) and employed for their 700-MW Paradise Unit 2.[14]

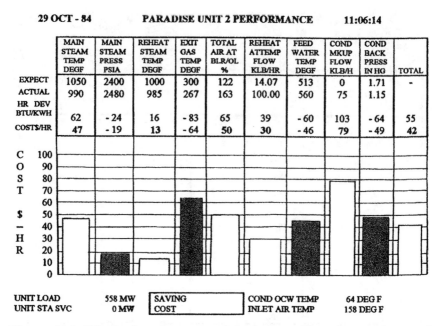

Figure 14-8. Displaying the costs of deviations from the rated steam/water cycle parameters as applied to TVA's 700-MW Paradise Unit 2. Source: J.T. Young[14]

As said before, any optimization of the turbine efficiency at the set load roughly speaking comes to keeping the steam/water cycle parameters as close to the rated values for the given load as possible. However, this golden rule can have two exceptions. The first one refers to operation with the constant or sliding throttle steam pressure, especially for steam turbines with nozzle-group steam admission control. The corresponding variations of the turbine's efficiency with the turbine load are single-valuedly determined by the turbine design, steam conditions, and the mode of steam admission control, presenting as if the turbine's characteristics—Figure 14-9. For turbines with nozzle-group steam control, at certain load values changing the control valve position and throttle steam pressure can give some noticeable gain in the turbine heat-rate. The optimization of this kind can be quite reasonable if the turbine is scheduled to work at the set partial load for long.

The second exception concerns optimization of the backpressure in the turbine condenser if the operator has a possibility to vary the cooling water flow amount by means of varying the number of cooling water

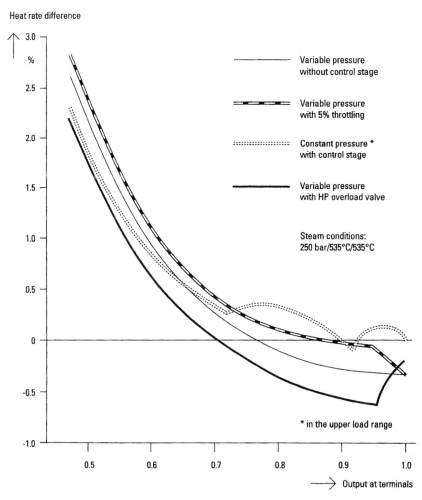

Figure 14-9. Heat-rate changes with the load for reheat steam turbines with different steam admission control modes. *Source: U. Sill and W. Zörner*[13]

pumps in service, turning their blades (for cooling water pumps with adjustable blades), or changing their rotation speed (for example, with the use of variable-frequency static converters). In the early 1980s, such a function of on-line optimization of the cooling water flow was developed and used at 800-MW power units of the Uglegorsk power plant in the former Soviet Union (now in Ukraine) in spite of pretty limited possibilities of varying the water flow by turning the blades of the cooling water pumps of that time.

It is usually assumed that the deeper vacuum in the condenser always means the more efficient turbine operation. Indeed, with lowering the turbine backpressure, the turbine output increases. However, first, if the LSB annular exhaust area is not sufficiently great, this effect sharply diminishes for too low backpressure values, and, second, sometimes the power underproduction because of lower vacuum can be completely compensated by reduction of power consumption by the cooling water pumps. The optimum cooling water flow amount can significantly vary depending on the water inlet temperature and steam flow into the condenser (that is, the turbine load). In particular, such calculations were conducted as applied to the 540-MW Unit 2 of the Dutch Maasvlakte power plant.[15] The unit is equipped with cooling water pumps with adjustable blades which allows varying the cooling water flow between 10 m^3/s and 20 m^3/s by adjusting the pumps' blade angles, and the cooling water temperature varies during a year in the range from 1°C to 25°C (34-77°F). The effect of varying the cooling water flow on the unit's net heat-rate at different inlet water temperatures and the turbine load equal to 500 MW is shown in Figure 14-10. With a decrease of the cooling water temperature from 20°C to 4°C, the optimum cooling water flow decreases from 18.8 m^3/s to 12.2 m^3/s. This water temperature fall itself decreases the unit's net heat-rate by approximately 118 kJ/kWh (about 1.4%), and reducing the water flow to the optimum gives an additional gain of about 44 kJ/kWh (that is, about 0.5%). The possible daily economy (in Netherlands guldens) with variations in the water flow is shown in Figure 14-11. At lower loads, the optimum cooling water flow shifts to the even lower values. Thus, if the turbine works with the load of 200 MW, with the water temperature fall from 20°C to 4°C, the optimum water flow decreases from 14.6 m^3/s to 11.4 m^3/s. A similar diagram can be displayed for the operator as the help for keeping the optimal "cold end" parameters. However, in doing so, these data should be presented for different values of the turbine load (or just for the current turbine output) as applied to the actual current cooling water temperature taken as an invariable value for each day.

Optimizing the cooling water flow amount depending on the cooling water temperature and steam flow amount to the condenser can be even more effective for cogeneration steam turbines, where the latter can substantially decrease with an increase of steam extractions to the

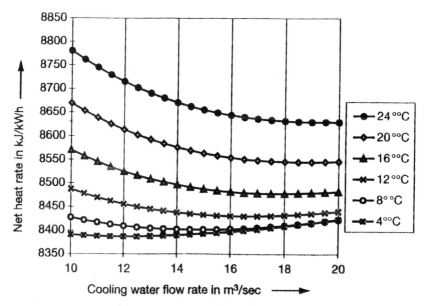

Figure 14-10. The effect of varying the cooling water flow on the net heat-rate at different cooling water inlet temperatures and invariable load of 500 MW for the 540-MW unit of the Dutch power plant Maasvlakte. *Source: J. Kromhout et al.*[15]

network-water heaters. In particular, such optimization was carried into effect at the Walsum cogeneration plant in Germany, with a 410-MW steam-turbine unit equipped with nine two-stage axial cooling water pumps with adjustable impellers.[16]

It is known that many modern large steam turbines with a number of LP cylinders are made with their condensers connected consecutively by cooling water with resultant different backpressure values at the LP exhausts. Such a scheme happens to be profitable under conditions of relatively warm cooling water, and this effect increases when the turbine is operated under partial loads.[17] For such turbines, it may occur advisable to pass to the "common," parallel connection of the condensers with falling the cooling water temperature (if, of course, such a possibility is laid in the project). Optimization of the "cold end" depending on the power output and ambient temperature conditions can be also highly effective for steam-turbine power units with so called "parallel" condensing systems (that is, furnished with both traditional water-cooled

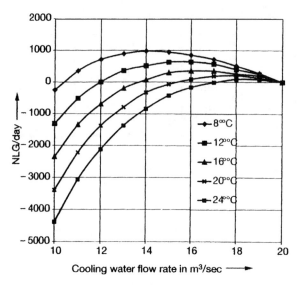

Figure 14-11. Economy with varying the cooling water flow at different cooling water inlet temperatures and invariable load of 500 MW for the 540-MW unit of the Dutch power plant Maasvlakte. *Source: J. Kromhout et al.*[15]

condenser and air condensers).[18]

It is understandable that such operating actions do need informative support for the operator to make optimal decisions. In all these cases, it is extremely desirable for the operator to have a possibility of predictive calculations and looking through "what if" scenarios to estimate the potential gain.[19]

It might be well to note that the actual data of power equipment efficiency and their changes in time, taken from the power plants' diagnostic systems, can be also very useful for planning the fuel consumption and power generation in the scale of individual power plants and electric power utilities upon the whole.[20] In particular, these data allow determining generic and individual corrections for the power equipment efficiency with aging. According to some U.S. electric power utilities, their managers mainly look at the heat-rate performance data on a weekly basis. According to the Santee Cooper Energy Control Center, dispatching the units relying on their actual continually refreshed performance data saves about $1,400 per day during a typical summer season.[11]

In France, as generally known, nuclear power plants generate about 80% of the entire electricity production, and 30 fossil-fuel units (3x125 MW, 18x250 MW, and nine of 600-MW and 700-MW units) are to be operated in semi-peak and peak manners, maintain the power grid's stability, participate in voltage control, and provide power reserve for the event of a potential drop in the nuclear power availability. Under these conditions, with a broad spectrum of possible options, it is especially desirable to know the actual operating efficiency of individual fossil-fuel power units with regard to its deterioration in time and variations with load. On this ground, Electricité de France (EdF) developed a special diagnostic module called "On-Line Economic Control" as applied to the existent computerized data acquisition and control systems of fossil-fuel power units.[21] This module uses the power unit's real time input database (about 50 measured variables and 8 to 10 manual inputs for fuel parameters), calculates the performance deviations from the rated performances for the stable operating conditions, and displayed the calculated results in a WINDOWS NT environment in the forms of a generic screen with the main process parameters and block diagrams with the deviations in the measured operating parameters and calculated performances. The prototype version of this module was installed in December 1998 on two 125-MW gas-fired units and one 600-MW coal-fired unit. The standard version, incorporating some improvements made after initial application months, was implemented in 1999 on the pilot sites and then was assumed to be transferred to all the rest units. According to EdF, thanks to the use of this module, in the first six months of its application the average heat-rate fell by approximately 300 kJ/kWh compared with the values of 1998.

It is supposed that diagnostic monitoring of the actual current heat-rate performances of power units integrated with financial information from the power industry market will allow the power plant operators to service their equipment in a really optimal manner quickly reacting to all the market changes. To attain this goal, it is supposed to connect the PC-based performance monitor servers operating in situ at the power plant via the Internet to a remote computing center possessing all the necessary real-time financial information. Such a system was developing by Sega Inc. as applied to the Kansas City Power & Light Company's Hawthorn power plant featuring with a mix of generating units: simple-cycle gas turbines, CC units, and traditional steam-turbine-based ones.[22]

REVELATION OF TURBINE STEAM PATH DAMAGES AND SOURCES OF HEAT-RATE PERFORMANCE DETERIORATION

Scrutinizing the turbine's heat-rate performances and their changes with time does often allow the researchers to reveal possible, hypothetical source(s) of the observed deterioration in the turbine efficiency, including supposed damages of the turbine, that is, to diagnose its state. However, this requires furnishing the turbine with a pretty great number of additional high-accuracy measurements. Hence, in this case the question is not about on-line heat-rate monitoring but rather about periodical, or repeated, heat-rate tests of the turbine. An obvious example of such an approach is a long-term investigation conducted at the Northport power station operating four 375-MW steam turbines that were installed between 1967 and 1977. The Northport, as well as many other power plants, has dealt with noticeable performance degradation as the turbines age. The power plant invested considerable time, labor, and money in performance evaluation of its turbines to identify the potential sources of the problem while the turbines were operating to eliminate them in the next overhaul. Such an investigation was conducted by TurboCare Inc. together with the power plant based on repeated heat-rate tests of the turbines.[23] The heat-rate test series fulfilled in summer 2004 at all the plant's four turbines showed the internal efficiency of the HP and IP sections ranged from 78.9% to 81.7% and from 84.6.1% to 92.3%, respectively. Detailed analysis of the suspect data and their changes as compared with results of the previous tests made it possible to suppose that the main source of the turbine efficiency deterioration is the increased leakage between the HP and IP sections, as well as some degradation of the HP and IP steam paths. These conclusions were conformed in the process of the nearest overhaul at the Unit 3. The expected decrease of the turbine heat-rate after the repairs owing to the steam path improvement, including the use of brush seals, was estimate by GE as equal to 271 kJ/kWh (257 Btu/kWh). The following heat-rate tests of the turbine after its overhaul showed that these expectations were even exceeded—the unit's heat-rate improved by 491 kJ/kWh (465 Btu/kWh), with 334 kJ/kWh attributed to the turbine (instead of above mentioned 271 kJ/kWh). The HP-IP leakage fell from 12% of the main steam flow to 5%, and the internal efficiency of the HP and IP sections improved by 2.6% and 6.6% up to 81.9% and 91.2%, respectively.

On-line diagnostic monitoring, mainly based on regular measurements of a less number and worse accuracy and stability, cannot provide such accuracy in heat-rate performance estimations. Figures 14-12 and 14-13 present some rather typical results of such on-line diagnostic monitoring for a Korean 500-MW power unit with the steam conditions of 16.7 MPa, 538.538°C (2,415 psi, 1,000/1,000°F): dependences on the main steam flow for the actual and design heat rate and changes with operating time of the actual and design efficiencies for the HP, IP, and LP sections.[24] Indeed, such data and their changes with time in the long-term scale make it possible to judge quantitatively the turbine state and detect its deterioration. However it would be rather difficult to reveal by these data some specific damages and shortcomings of the turbine steam path.

On the other hand, such a goal seems to be more reachable if traditional comparing the current and design values of the heat-rate and internal efficiencies and tracing their changes with time is supplemented for advanced analysis of redistribution of steam parameters and enthalpy drops lengthwise of the steam path. Application of this concept was partially materialized by Central Turbine-Boiler Institute (TsKTI) in computerized ASTD or DAS of the 800-MW supercritical-pressure units of the Zaporozhskaya and Surgutskaya-2 power plants of the former Soviet Union,[25,26] however the accuracy and definiteness of such diagnosis

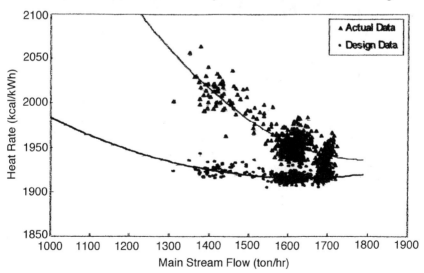

Figure 14-12. Dependences on the main steam flow of the actual and design heat-rate for a Korean 500-MW subcritical-pressure power unit. Source: By courtesy of Doosan Heavy Industries and Constructions.

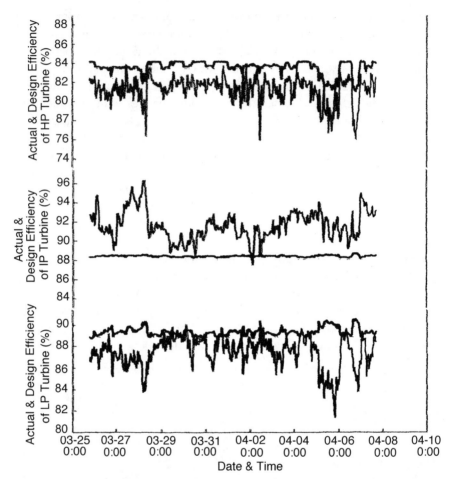

Figure 14-13. Changes with operating time of the actual (1) and design (2) efficiencies for the HP, IP, and LP sections of a Korean 550-MW subcritical-pressure power unit. *Source: By courtesy of Doosan Heavy Industries and Constructions.*

was insufficient because of low accuracy of the original measurements and affecting their readings by unavoidable variations and fluctuations of the operating conditions.

The changes in the turbine state to be sought reflect on the thermal process in the turbine and hence on steam parameters in the steam path. However, these changes, even caused by such serious damages as, for example, breakage of individual blades or even stages upon the whole, often turn out rather minor, quite comparable with the instrumental errors of measuring the monitored steam parameters. It is all the more true for such changes in the steam path condition as, for example, deposits on the blading surfaces or their erosion. So the changes in steam parameters within the turbine caused by these damages or shortcomings commonly remain unnoticed by the operational personnel against the background of the changes caused by variations in the turbine's operating conditions: main and reheat steam temperatures, load, and position of the control valves. Thus, for example, a failure of all the rotating blades of one stage in a large reaction-type steam turbine's HP section changes the measured steam temperature at the section outlet by not more than 2-3°C (4-5°F). Because these blades are not too long and bulky, vibrational diagnostics, if fulfilled, does not commonly allow revealing such failures, too. Therefore, to diagnose the steam turbine path condition, it is necessary to recognize and separate the changes of the measured values caused by variations in the operating conditions for the undamaged turbine.

This way, it is important to specify at least two independent goals. The first one is associated with defining the current normative values of the diagnostic parameters for the current operating conditions, and the second goal comes to minimizing the errors in measuring the variables employed for diagnostics and defining the mentioned normative values. It can be done by means of special statistical processing of the measured data. Such a methodology was developed as applied to 228-MW subcritical-pressure power units with their existent computerized DASs of the Israeli power plant Eshkol.[27] At this power plant, there are four reaction-type turbines of the considered type put into operation between 1974 and 1978. During the last decade, three of them had serious failures of several stages in the HP cylinder, and all these failures remained unnoticed by the operational personnel until the turbine was opened during the overhaul. Therewith, it is obvious that the failures progressed with time, and this progress could be prevented if the original failures had been revealed timely. In addition, it would prevent the turbine from operating with a substantially decreased efficiency.

The developed approach is based on constructing some empirical multidimensional dependences of the internal steam parameters on the outside conditions. The dependences are constructed via statistic processing of the data accumulated during the first operation period after the overhaul when the turbine steam path is considered to be in its undamaged state. As applied to the considered turbine's HP cylinder with its existent regular measurements, three internal steam parameters were chosen to diagnose on-line the cylinder's steam path condition. They are the steam temperature and pressure at the HP outlet and the steam pressure at the control stage. As the outside parameters characterizing the turbine's operating conditions, the following measured variables were taken: the steam pressure at the LP cylinder inlet as an index of the steam flow amount through the turbine, vacuum in the condenser, main steam pressure and temperature in the HP stop-valve steam-chests, and the hydraulic fluid pressure in the turbine's governing system as an index of the HP control valve position. The dependences sought for the normative values of the diagnostic parameters were constructed based on the common operating data recorded by the unit's DAS. The mathematical models for the considered diagnostic parameters were developed with the use of a regressive approach in the form of multidimensional quadric polynomials coupled with quotient variables.

The model structure was determined with the use of a two-stage procedure. The task of the first stage was to find the structure which would be the best in a certain sense for the taken model size, varying within certain limits: 2, 3, 4, and so on. This search was fulfilled with the use of an adaptive Monte-Carlo method, unsusceptible to the existing considerable correlation between the variables. At the second stage, the models of different sizes were compared with one another. In doing so, a sequential F-test of comparison was employed with a crucial addition: the variances were calculated with regard to the error of measuring the parameters. This made it possible to compensate the actual observance scheme's dissimilarity from the classical regression scheme implying that the initial variables are measured with the complete accuracy.

The regression equation's coefficients were assessed with the help of an enhanced square method, and the robustness of the gotten results was estimated with the help of an improved jackknife method. The mentioned procedures were materialized in the form of a software package that successfully passed through long-term tests. The models were gotten for operating conditions varying within a wide range corresponding to typical daily load changes for the considered power units. With these mathematical models, the currently measured values of the internal parameters can be on-line compared with its normative values for the current operating conditions. The resulted difference becomes the diagnostic symptom telling about the steam path state of the cylinder. Noteworthy is that these symptoms are released from the influence of the varying op-

Diagnostic Monitoring of Turbine Heat Rate and Flow Capacity   399

erating conditions, since these variations are already taken into account while calculating the normative values.

The above described procedures of testing the model robustness, as well as standard examinations of the model adequacy, are necessary but not sufficient to be sure in the model applicability. The most crucial examination is the model verification with the use of "fresh" data that have not been employed before in the identification process (so called cross-validation). So, every model was additionally verified on the basis of three months of turbine operation under both stationary and transient operating conditions—Figure 14-14. The resulted average standard deviations did not go beyond the threshold of 95%-confidence interval. This approves a quite sufficient adequacy and applicability of the models. As expected, the biggest deviations were observed at the transients, and for stationary and near-stationary operating conditions the models were more accurate.

As pointed out above, the differences between the actual measured values of the monitored internal steam parameters and their calculated normative values are considered to be the diagnostic symptoms. Being monitored on-line, these variables comprise random errors for both the calculated normative values and measured internal parameters. The robust smoothness methods make it possible to overcome the affect of probable excursions. Coupled with the subsequent averaging of the measured values at certain time intervals, these methods provide a considerable decrease of the final error. The resulted difference, as a diagnostic symptom, has a higher signal-to-noise ratio. Thus, for example, as applied to the steam temperature at the HP outlet, such a difference has the standard deviation of 0.8°C (1.44°F) which permits recognizing reliably the changes in the monitored symptom about 2-3°C (3.6-5.4°F). The diagram of Figure 14-15 showed emulation of a hypothetical failure of the turbine's HP stage that, according to steam expansion process calculations, would manifest itself by the increase of the diagnostic symptom by about 2.6°C (4.7°F) that looks quite recognizable. Different combinations of diagnostic symptoms variations (for different diagnostic parameters) allow recognizing different failures or shortcomings in the steam path. It seems advisable to attract the methodology of expert systems for distinguishing such events.

References

1. Bornstein B. and K.C. Cotton. "A Simplified ASME Acceptance Test Procedure for Steam Turbines," *Combustion*, March 1981: 40-47.

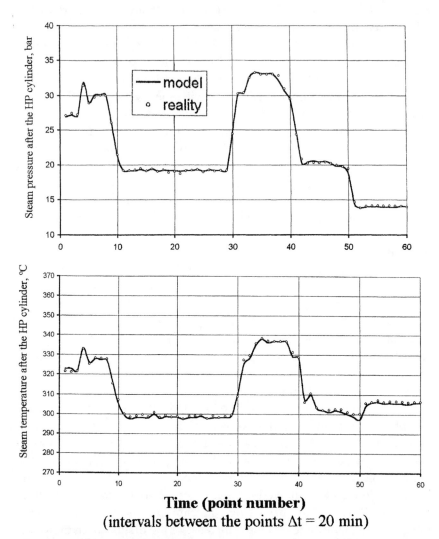

Figure 14-14. Verification of the mathematical models for the steam temperature and pressure at the HP cylinder's outlet of the 228-MW turbine for diagnostic purposes.

2. Sakharov A.M. *Heat-Rate Performance Tests of Steam Turbines* [in Russian], Moscow: Energoatomizdat, 1990.
3. Friedman J.R. "Certification and Accreditation for Power Plant and Equipment Performance Testing per ASME Performance Tests Codes," presented at the 2001 International Joint Power Generation Conference, New Orleans, 2001: JPGC2001/PWR-19016.

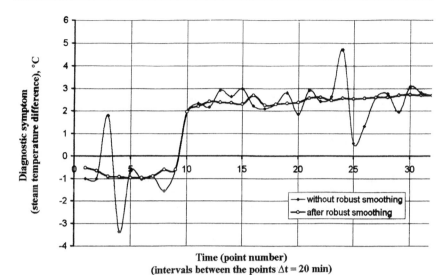

Figure 14-15. Emulating a hypothetical failure of an HP stage and resulted changes of the diagnostic symptom (the difference between the measured and normative values of the steam temperature at the HP outlet) for the 228-MW turbine.

4. Tortello E. and G. Bleakley. "Moving from planned to predictive maintenance," *Modern Power Systems* 18, no. 8 (1998): 55-57.
5. Swanekamp R. "Predictive maintenance technologies come together," *Power* 145, November/December 2001: 43-50.
6. Dundics D. "Incorporate RCM into Acceptance Testing to Detect Latent Defects," presented at the 2001 International Joint Power Generation Conference, New Orleans: 2001, JPGC2001/PWR-19032.
7. Roemer M.J. and T.H. McCloskey. "Advanced Steam Turbine-Generator Maintenance Planning Using Prognostic Models," presented at the 2001 International Joint Power Generation Conference, New Orleans: 2001, JPGC2001/PWR-19118.
8. Georg E.W. and F.A. Strum. "Effective Plant Management through Intelligent Diagnostics in the Power Plant," *VGB PowerTech* 83, no. 9 (2003): 84-88.
9. Woyke W. "Functions and Concept of the Process Information System for a Power Station," *VGB PowerTech* 81, no. 8 (2001): 67-71.
10. Mylonas J., L. Schreiber, and W. Woyke. "Process Efficiency Monitoring in EKW Power Stations" [in German], *VGB Kraftwerkstechnik* 83, no. 12 (2003): 64-67.
11. Smith D.J. "Heat-Rate Optimization Pays Dividends," *Power Engineering* 106, no. 1 (2002): 34-40.
12. Zörner W., H. Müller, K.-H. Andreae, and H. Emshoff. "Diagnostics System for Monitoring the Steam Turbine Generator Set" [in German], *VGB Kraftwerkstechnik* 71, no. 6 (1991): 487-498.
13. Sill U. and W. Zörner. *Steam Turbine Generators Process Control and Diagnostics*, Erlangen: Publicis MCD Verlag, 1996.

14. Young J.T. "Retrofitted color graphics system monitors Tennessee Valley Authority's coal-fired units," *Power Engineering* 89, no. 1 (1985): 54-55.
15. Kromhout J., E. Goudappel, and P. Pechtl. "Economic Optimization of the Cooling Water Flow of a 540-MWel Coal Fired Power Plant Using Thermodynamic Simulation," *VGB PowerTech* 81, no. 11 (2001): 43-46.
16. Czylwik M. and H. Fink. "On-Line Optimization of Cooling Water Operating Regime of a River Water-Cooled Cogeneration Station," *VGB PowerTech* 79, no. 7 (1999): 51-54.
17. Coit R.L. "Design trends use utility feed-water heaters and condensers," *Combustion* 46, no. 8 (1975): 14-27.
18. Akhtar S.Z. "Parallel Condensing Systems Getting Fresh Look," *Power Engineering* 104, no. 10 (2000): 63-68.
19. Eulinger R.D. "What to Look in a Performance Monitoring Systems," *Proc. of the American Power Conference* 50, Chicago, 1988: 366-369.
20. der Marel J. and I. Bins. "Benchmarking the Energy Efficiency of Power Plants Using Operational Efficiencies," *VGB PowerTech* 85, no. 11 (2005): 74-80.
21. Simon M. and E. Presutti. "On-Line Economic Control (OEC) in Fossil-Fired Thermal Power Plants—EDF's Approach," *VGB PowerTech* 80, no. 7 (2000): 45-47.
22. Anderson T.C. "Browser-based performance monitoring integrates thermal, financial information," *Power* 145, September/October 2001: 79-84.
23. Blachly S.R. and M.E. Foley. "Testing for Turbine Degradation and Improving Performance with Seal Optimization," presented at the POWER-GEN 2005 Conference, Las Vegas, 2005.
24. Chung Y.-K, J.-Y. Park, J.-J. Jung, and O. Park. "Development of On-Line Turbine Performance Diagnostic System," presented at the POWER-GEN Asia 2005 Conference, Singapore: 2005.
25. Perminov I.A. and V.G. Orlik. "Diagnosis of a Technical Condition of the HP and IP Steam Paths Based on Regular Measurements of Steam Temperatures and Pressures within the Turbine" [in Russian], *Elektricheskie Stantsii*, no. 6 (2003): 38-41.
26. Leyzerovich A.Sh., L.P. Safonov, A.M. Zhuravel', et al. "Automated Diagnostic Systems and Subsystems for Fossil Power Plant Units: the Experience of Designing and Mastering in the Former USSR," *Proc. of the American Power Conference* 55. Part 1. Chicago: 1993. 881-885.
27. Gordinsky A.A., E.R. Plotkin, E.I. Benenson, and A.S. Leyzerovich. "A New Approach to Statistic Processing of Steam Parameter Measurements in the Steam Turbine Path to Diagnose Its Condition," *Proc. of the 2000 International Joint Power Generation Conference*. New York: ASME Press, 2000,1-5.

Chapter 15

Diagnostic Monitoring of Turbine Temperature and Thermal-stress States

GENERAL FEATURES OF THIS DIAGNOSTIC FUNCTION

The considered function of monitoring the temperature and thermal-stress states of the turbine's major design elements can be eventually characterized as operating, functional, continual, and current, necessary for operating control.

On the other hand, results of this monitoring is used for calculating low-cycle thermal fatigue and high-temperature creep damages accumulated in turbine metal in the operation process with assessment of these elements' residual lifetime. Some developers believe reasonable to present this information for the operator along with results of operating diagnostic monitoring. The corresponding specialized computing means, so-called *resource counters*, have been developed and widely implemented since the mid-1970s.[1,2] By way of illustration, "Service Life Control" is a constituent of the monitoring/diagnostic system of Siemens for steam turbines and generators (see Figure 13-8) alongside the "Thermodynamics" and in turn includes both the "Turbine Stress Evaluator" (TSE) and a module for "Calculation of Life Expenditure."[3] Such confusion looks rather misconceptional from the standpoint of operating thought and engineering psychology, because the data referring to the lifetime expenditure are irrelevant for the operator and, like any excessive information, should be considered even harmful. These calculations are a typical example of maintenance, occasional, prognostic diagnostics with the forecast depth measured in years, and their results in the first place should be intended for maintenance staff and even utility authorities who have to make decisions about upgrading power equipment and/or replacement of its elements. In addition, it is worth noting that any estimation

403

of accumulated metal damages and residual lifetime bears in principle a probabilistic character and is usually perceptibly aggravated by the lack of some essential initial data. As a result, the obtained results cannot be used directly, but should be interpreted by specialists with regard to the mentioned circumstances. True, someone says that the estimation of specific metal damages at different transients serves for evaluation of the operation quality in reference to the observed equipment. However, first, it is not necessary to do this on-line, and second, there exist much simpler and more obvious methods of such evaluations (see below).

For all these reasons, the two mentioned diagnostic functions must be separated. The former (diagnostic monitoring of the temperature and thermal-stress state of the turbine elements) should be carried out on-line with operational presentation of its results to the operators and used to arrange advanced informative support for them at the transients (see Figure 13-9). In addition, under conditions of automated control, just these results are used for shaping the programs for analogous control and limitations for logical control.[2,4] At the same time, these results should be memorized for subsequent calculations of thermal fatigue and creep damages accumulated in equipment's metal in the process of exploitation. This second diagnostic function should be carried out occasionally off-line and probably in outside computers (see Figure 13-11).

All the above-mentioned circumstances condition a special character of the considered diagnostic function amongst others. This function does not indicate the faulty or damaged elements but only prepares the required data for such diagnosis. In addition, it reveals the operating conditions that in the greatest degree contribute to damages caused by thermal stresses.

THE SCOPE OF TEMPERATURE MEASUREMENTS

The scope of monitoring the temperature and thermal-stress states should embrace all the major turbine elements. In particular, the temperature measurements should give comprehension of main correlations between characteristic metal temperatures of the turbine cylinders, valve steam-chests, steam-lines, and crossover pipes, as well as between the steam and metal temperatures. Along with this, it is desirable to minimize the number of temperature measurements to facilitate their observance, simplify post-operating processing and analysis of the measured

Diagnostic Monitoring of Turbine Temperature/Thermal-stress States 405

and recorded variables, as well as facilitate service and maintenance of measurements. In reference to the thermal-stress state, it is sufficient to control and limit its indices for the most thermally stressed (critical) elements to be sure that thermal stresses in other elements do not exceed their admissible limits as well. Naturally, this makes sense only for those elements whose state is influenced by the same outside actions.

As noted before, for modern large steam turbines the high-temperature rotors are the most-stressed, critical elements, and accumulation of low-cycle (thermal) fatigue in their metal is one of the main quality criteria for the turbine transients. The unsteady-state thermal stresses in the HP and IP rotors are mainly determined by changes of the main and reheat steam temperatures, respectively, as well as by the heat transfer conditions from steam to the rotor surfaces. In turn, these heat transfer conditions are mainly determined by the steam flow amount through the turbine, that is, the turbine load or, until the generator is switched on to the grid, the turbine rotation speed. As to the HP rotor, its heating steam temperature is also influenced by the steam flow amount through the turbine and the mode of controlling this flow. A characteristic axial metal temperature distribution along the external and bore surfaces of a typical forged HP rotor (or the HP section of an HP-IP rotor) of an impulse-type turbine is presented in Figure 15-1a. The highest thermal stress takes place in the vicinity of stress concentrators caused by the first stage disc's fillet or thermal grooves at the inlet of the front or central gland seal. These places are simultaneously characterized by maximum radial temperature differences in the rotor. As the first approximation, this maximum thermal stress is taken in proportion to the "effective" radial temperature difference, $\overline{\Delta t}$, between the surface metal temperature and average integral temperature, $\overline{t}$, in the most stressed section. Both these values, Δt and $\overline{t}$, can be calculated based on the measured heating steam temperature t_{st} at the first HP stage. In other words, they are determined by means of on-line mathematical modeling of the heating up process for the rotor based on the measured heating steam temperature. Main notions of this approach are presented in the next subsection of this chapter.

For turbines with throttle steam admission control and with HP valve steam-chests settled directly beside the cylinder, the heating steam temperature measurement can be substituted by measurements of steam temperatures after the valves with the input of a calculated correction for the distinction of the measured steam temperatures from the heating steam temperature for the rotor in the monitored zone. The presence of

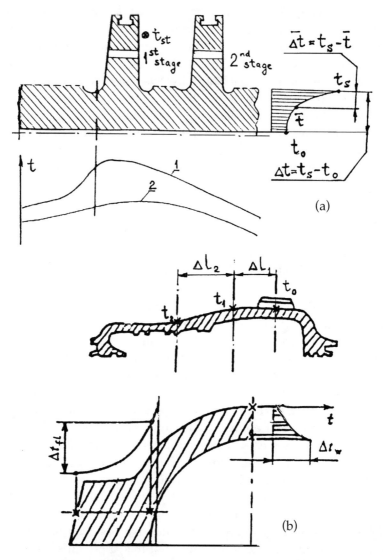

Figure 15-1. Characteristic axial metal temperature distributions along the external (1) and bore (2) surfaces of a forged HP rotor (a) and metal temperature differences in a turbine casing characterizing its thermally stress state (b).

Diagnostic Monitoring of Turbine Temperature/Thermal-stress States 407

crossover pipes between the HP valve steam-chests and cylinder can bring serious uncontrollable errors, and for turbines with nozzle-group steam admission control the difference between the steam temperatures in the valve steam-chests and within the cylinder practically cannot be calculated with due accuracy. That is why it is so important to measure the heating steam temperature in the right point, even though arranging this measurement often brings serious difficulties. In addition, the measuring shell should be sufficiently thin-wall and should come forward into the steam flow to be sufficiently low-inertial and trace the actual steam temperature variations causing thermal stress in the rotor. On the other hand, there exists a danger that such a thin-wall measuring shell can be damaged by the dense high-velocity steam flow. It seems that the use of a special thermometric cable with a thermocouple shielded by a thin stainless tube can be the best design solution for the steam temperature measurements providing their necessary accuracy and reliability.

For IP rotors (or IP sections of the HP-IP rotors), the steam temperature can be taken either after the intercept values or in the steam admission chamber with some calculated corrections for the static and dynamic differences between the measured and heating steam temperatures (see Figure 8-4). For rotors with artificial steam cooling of the heated outer surface around the first stages, the heating steam temperature should be taken with account of this effect or measured in the direct vicinity of the rotor surface in the most thermally stressed zone.

For large steam turbines with welded or forged LP rotors, great steam flow amounts, and, especially, increased steam temperatures downstream from the IP cylinder, it may occurs necessary to monitor additionally the thermally stressed state of the LP rotors, too. In this case, the most dangerous are the tensile stresses caused by superposition of centrifugal stress and thermal stress in the metal depth, that is, on the surface of the central bore or at the rotor axis in the steam admission section vicinity. This thermal stress can be taken in proportion to the "entire" metal temperature difference, Δt, between the metal temperatures on the heated and bore surfaces (or at the rotor axis for solid and welded rotors). The heating steam temperature for the temperature monitoring of LP rotors can be taken measured in the LP crossover pipes. Features of the mathematical modeling of the heating up process as applied to the LP rotors are also considered in the next subsection.

As distinct from the HP rotor, heating up of the HP valve steam-chests practically depends on only the main steam temperature, whereas the dependence on the steam flow amount is rather secondary. Because of this, the thermal stress indices for the HP valve steam-chests and HP rotor can be considered independent of one another, and at the early stages of start-ups the rate of raising the main steam temperature can be limited more by the thermal stress in the HP valve steam-chests than that in the HP rotor. That is the reason why, for example, in the Turbine Stress Evaluator (TSE) of Siemens there are three channels for monitoring the most thermally stressed turbine elements limiting the transients: for the HP valve steam-chests, HP rotor, and IP rotor—Figure 15-2.[3,5]

In some cases (for example, for turbines with artificial cooling of the rotor surfaces in the highest-temperature zones or with some other features of the turbine design), thermal stresses in the turbine casings can limit the rate of the transients as well as those in the turbine rotors do, and the thermal-stress state of the high-temperature (HP and IP, or HP-IP) casings need their own thermal-stress monitoring.

As a rule, for the turbine stator elements (cylinder casings, valve steam-chests, and so on), their maximum thermal stresses, σ_t, are determined by some temperature differences characterizing the complex temperature fields of these elements.[2,4,7] This influence can be approximately described by means of a linear combination of these temperature differences:

$$\sigma_t = \sum_{i=1}^{n} (a_i \times \Delta t_i), \tag{15.1}$$

where a_i ($i = 1, 2, \ldots n$) are the influential factors for the individual temperature field constituents. Since one of the acting temperature differences in (15.1) commonly turns out to be dominant in its influence on the thermal stress (for example, Δt_1), it is convenient to estimate the thermal-stress state of the considered casing by means of monitoring a definite "effective" temperature difference $\overline{\Delta t_1}$ to be in proportion to σ_t. This value is obtained as the monitored value Δt_1 with correction for the other temperature differences:

$$\overline{\Delta t_1} = \Delta t_1 + \sum_{i=2}^{n} \left(\frac{a_i}{a_1} \times \Delta t_i \right). \tag{15.2}$$

As applied to the cylinder casing walls—Figure 15-1b, it is the metal temperature difference across the wall thickness Δt_w that more often than

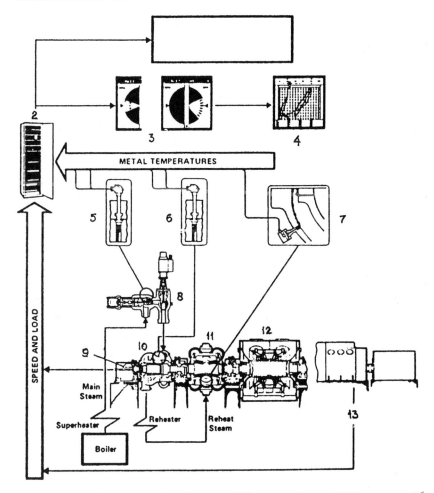

Figure 15-2. Turbine Stress Evaluator of Siemens. *Source: B. Hoerster et al.*[6]
1 – rotation speed and load limiters of the turbine's electro-hydraulical system, 2, 3, and 4 – TSE computing device, indicator, and multipoint recorder, 5, 6, and 7 – TSE detectors of temperature differences in the HP valve steam-chests, HP and IP rotors, 8 – HP combined unit of stop and control valves, 9 – rotation speed sensor, 10, 11, and 12 – HP, IP, and LP cylinders of the turbine, 13 – electric output sensor.

not can be considered the dominated temperature difference Δt_1, and the other temperature differences influencing the wall's maximum thermal stress are those across the flange width Δt_{fl} and between the wall and flange, as well as, in many cases, the "derived" temperature difference lengthwise the wall that determines the bending thermal stresses in the wall caused by the "breakage" of the axial metal temperature distribution

$$\Delta t_z = \frac{t_{w0} - t_{w1}}{\Delta l_1} - \frac{t_{w1} - t_{w2}}{\Delta l_2}$$

(15.3)

Of importance is that the true, complete temperature differences across the wall thickness and the flange width are considerably greater than the differences of the measured metal temperatures near the inner, heated surface and on the outer, insulated surface, if the "inner" metal temperature is measured at a certain distance (about 10-20 mm and even more) from the heated surface. That is why, these "measured" temperature differences should be corrected for the distance from the measured point to the heated surface. The temperature differences across the flange width and between the flange and bolts, together with the stress because of tightening the bolts, can also cause plastic deformations on the casing's flange split surface. Keeping this in mind, it can be also necessary to monitor and limit the mentioned temperature differences in the flange joint in the most heated zones. Along with this, the temperature difference between the flange and bolt commonly can be considered proportional to the temperature difference across the flange width, which makes it possible to abandon measuring the bolt temperatures.[4]

Metal temperature unevenness in the cylinder casings can also make them bend with resulted brushing and rubbing in the steam path because of a "humping" effect under action of temperature differences "top-bottom." As a rule, the casing cover has higher metal temperatures. This results in thermal bend of the casing upward, often named "cat's back." The maximum casing sag can be estimated with the use of equation (8.15) via the average top-bottom temperature difference $\overline{\Delta t}_{t-b}$. For practical purposes, it is commonly sufficient to monitor and limit the top-bottom temperature difference in one or two (mainly for integrated HP-IP cylinders) most characteristic section(s) of the casing, without measuring metal temperatures in several sections to find the genuine average value of the temperature difference.

This way, the number of metal temperature measurements at the turbine's high-temperature casings can significantly vary depending on the design features and whether it is necessary to monitor the thermally-stressed state of their individual components (for example, walls and flanges or LP rotors) or not.

Diagnostic Monitoring of Turbine Temperature/Thermal-stress States 411

Maximum thermal stresses in valve steam-chests can be taken in proportion to the metal temperature differences across the wall thickness with correction for the axial constituent of the temperature field according to (15.1) and (15.2).[2] Yet as applied to valve steam-chests, it is advisable to monitor the "major" temperature differences across the wall thickness by calculating them relying on the measured steam temperature in the considered steam-chest and metal temperature at the external, insulated surface, that is, by means of mathematical modeling of the heating up process. This became possible thanks to a possibility to calculate the heat transfer conditions from steam to the inner valve steam-chest surface with a sufficient accuracy with the use of dimensionless criterion equations and owing to a high level of these heat transfer conditions which make the results of heating up calculations less sensitive to the errors of setting the boundary conditions. Mathematical modeling of the heating up process not only heightens the monitoring accuracy (compared with finding the temperature differences across the wall thickness based on measuring the metal temperature near the heated surface at some distance from it) but also allows abandoning a special hole in the steam-chest wall. Such a method of temperature monitoring for the HP valve steam-chests was first used in the PC-based local subsystem of diagnostic monitoring for the 250/300-MW supercritical-pressure turbine at one of the Moscow cogeneration plants (see Figure 13-9),[8,9] and then this approach was repeated as applied to 228-MW turbines at the Israeli power plant Eshkol.[10] Metal temperature measurements on the outer, insulated surfaces are used for setting the boundary conditions on this surface, as well as to determine the axial temperature difference along the steam-chest height used for estimating the thermal stress in the steam-chest.

In the past, all the results of metal temperature measurements in the turbine were brought out on multipoint recorders, and the operator calculated in his head the needed differences of the recorded temperatures. In the best cases, as it has been practiced by Siemens, the turbine was furnished with differential thermocouples measuring directly the temperature differences (across the wall thickness or the flange width, between the flanges and bolts, and so on), and these measured temperature differences were directly shown for the operator and recorded. The recorded tape diagrams were also used for post-operating analysis of the turbine's transients. Presently, with computerized DASs and DACSs installed at power plants everywhere, these multipoint recorders are done away of the control board, and more often than not the only way to

412   *Steam Turbines for Modern Fossil-Fuel Power Plants*

see the measured metal temperatures is to request them on the display. With regard to modern rich possibilities of processing and presenting information data, on the one hand, and limited possibilities of stationing and perceiving these data on the display screens, the measured values should be pre-processed and presented in the more obvious and compact graphic forms. This general concept with its application to the data of monitoring the temperature and thermal-stress states of the turbine is considered below in a subsection especially devoted to this problem. The same monitored data for the turbine transients should be also selected and stored for subsequent analysis.

MATHEMATICAL MODELING OF THE HEATING UP PROCESS FOR TEMPERATURE MONITORING OF TURBINE ROTORS

The problem of monitoring the temperature and thermal-stress states of the turbine rotors first appeared in the 1960s to prevent the rotors of newly designed large power steam turbines from cracking caused by cyclic thermal stresses arising at the turbine transients, mainly start-ups.[11-17] Since direct thermometry of turbine rotors, like that for turbine casings, valve steam-chests, and steam-lines, is too complex and unreliable, this problem had to be solved by means of indirect methods of physical or mathematical modeling.

Physical modeling of the rotor temperature state implies settling a so-called "thermometric probe" within the cylinder. This probe to a degree imitates the shape of the rotor sector, and its head is swept by steam of approximately the same temperature as that for the rotor near its most-stressed section. Such probes were first developed and introduced by Brown Boveri (BBC), and then similar devices were employed by some other European turbine producers.[12,16,18] Physical models of the rotor temperature state have some inherent shortcomings. First, settling the probe within the turbine cylinder in the proper place involves serious design difficulties. Second, the heat transfer conditions from steam to the probe head and the rotor surface are significantly distinct from each other. Finally, there exist some hardly removed heat flows on the side probe surfaces which cause essential methodical errors of modeling. Mathematical models, materialized on the basis of analog or digital computing technique, do not have these vices. Only the absence of simple and reliable computing means acceptable in the power plant conditions until the 1970s explained addressing to physical models.

The developed mathematical models can use diverse calculation schemes. Most of them are based on approximate methods of finite differences (more often in its explicit mode) and approximate transfer functions. In the first case, the mathematical modeling results in a more or less detailed temperature field in the considered rotor section. In the second case, the model outputs are some characteristic temperatures in a few chosen points and temperature differences between them.

A rather simple analogous device for modeling the rotor temperature and thermal-stress states was developed as applied to the HP and IP rotors of LMZ turbines K-300-240, the most widespread in the former Soviet Union.[19-21] Some examples of these devices have been employed at power plants for more than 25 years, and some results of monitoring with the use of these devices and their later modifications at different power plants are presented in Figures 7-8, 9-4, 12-2, and 12-14. The model is constructed using the method of approximate transfer functions, taking into account variation of heat transfer conditions on the rotor surface with the steam flow through the turbine. The corresponding functional chart of such a model in its simplest form is presented in Figure 15-3, and the corresponding transfer function for the effective radial temperature difference in relation to the heating steam temperature looks like:

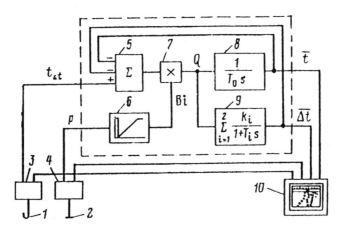

Figure 15-3. Functional chart of a simple analogous model for monitoring the temperature and thermal-stress states of a rotor.
1- heating steam temperature sensor, 2 – steam pressure sensor, 3 and 4 – normalizing converters, 5- inlet summer, 6 –non-linear converters, 7 – multiplier, 8 – integrator, 9 –non-periodic element, 10 – multipoint recorder

$$W(s)_{\overline{\Delta t}} = \frac{T_0 s \times \sum_{n=1}^{m} \dfrac{k_n}{1 + T_n s}}{1 + \dfrac{T_0 s}{Bi} \times \left(1 + Bi \times \sum_{n=1}^{m} \dfrac{k_n}{1 + T_n s}\right)}.$$

$$(15.4)$$

Here, s is the Laplace transfer parameter; m is the approximation order (mostly, m is taken equal to 2 or 3); Bi is the dimensionless heat transfer coefficient to the rotor surface (the Biot number) taken depending on the steam pressure in the cylinder; T_0 and T_n are the time constants, and k_n (where $n = 1, 2,...m$) are the influence factors. The heat transfer conditions for the rotor are set based on results of preliminary calculations with the use of dimensionless criterion equations for the considered rotor surface.

The pilot sample of such a device was installed at the turbine which was also furnished by LMZ with direct measurements of the HP rotor metal temperatures (see Figure 8-14). This makes it possible to check an accuracy of the modeling device as a whole and its calculation scheme, in particular, by comparing the recorded results of monitoring with the data of direct thermometry of the rotor at the same operating conditions, calculated temperature fields of the rotors, and results of calculations which emulated functioning of modeling devices with different calculation schemes. Results of these comparisons were admitted quite satisfactory by both the turbine producer and the plant operators.[20]

Appearance of microprocessor-based programmable controllers made it possible to extend this experience to other turbine types and raise an accuracy of modeling due to some complication of the applied models. In particular, the advanced models took into approximate consideration the dependence of the metal thermal conduction on temperature, influence of the turbine rotation speed on the heat transfer conditions, and the axial heat flux from the considered rotor section to the front seal and bearing. The last factor is especially important for rotors of single-flow cylinders. With all these improvements, the approximate models provide the error of monitoring the effective radial temperature difference in characteristic HP and IP rotors not exceeding 10-15% which seems quite acceptable for operational purposes.[4,20,21]

Rich opportunities taken by microprocessor-based controllers made it also possible to monitor the temperature state of the LP rotors with modeling both their heating up in the operation process and cooling down during outages, because this is the only way to set their initial, pre-start thermal state. For these rotors, the monitored values are the "entire" temperature difference across the rotor radius and the metal temperature at the rotor axis or on the bore surface in the most thermally stresses section near the first stages.[22]

With transition to digital computing technique, it seems more advisable to use mathematical models based on the method of finite differences rather than that of approximate transition functions. The former gives even more freedom in taking into consideration different factors influencing the model accuracy up to the use of two-dimensional calculation schemes if necessary. Similar mathematical models can also be applied to monitoring some turbine stator elements if they are poorly suitable for accurate temperature measurements. This predominantly refers to the HP valve steam-chests. In this case, beside the measured heating steam temperature inside the steam-chest, the applied model also uses the measured metal temperature on the outer insulated surface for setting the boundary heat conditions at this side. The heat transfer conditions from steam to the inner, heated surface are set based on the known empirical equations as a function of the steam flow amount through the valves and steam pressure at the valve entrance.[10,23]

A similar concept of mathematical modeling as applied to the heating up process based on measuring the heating steam temperature can also be used for monitoring the thermal-stress state of thick-wall components of boilers.[24-26]

Diverse models based on the method of finite differences for diverse types of turbines were developed for temperature monitoring of the rotors with the use of microprocessors of the electro-hydraulic governing system of the turbine, computerized DASs, DACSs, and ASTDs, and PC of the PC-based LSDM-T.[7-10,23,27-29]

In the last cases, temperature monitoring of the rotors is further used to trace the influence of the actual operating conditions on low-cycle, thermal, fatigue of the rotor metal. Just this step can be regarded as a transition from an advanced temperature monitoring to a diagnostic monitoring of the turbine temperature and thermal-stress states. An example is the algorithmic module "CONTUR" developed as common for various supercritical-pressure turbines with the single capacity of 250, 300, 500, and 800 MW and for diverse types of computing means and program languages (including FORTRAN and C++). The algorithm is adapted to the specific turbine type by means of setting the "key" input constants. It was also modified for some turbines of smaller output, substantially different in their design solutions. Besides calculating the temperature and thermal-stress state indices for the HP and IP rotors, the algorithm also models the heating up of the HP stop-valve steam-chests, computes the temperature state indices as applied to other stator elements on the basis of their measured metal temperatures, and checks the validity of the measured input values. The algorithm also selects and filtrates the extreme output values of the thermal stress-strain state indices in the rotors for the subsequent (post-operating) evaluation of the low-cycle thermal fatigue, identifies the transients and their stages, records the major operating parameters and turbine indices for post-operating analysis, and diagnoses thermal insulation of the turbine cylinders and steam-lines based on characteristics of their cooling down at the stopped turbine.

This algorithm is also applied as a basic module for other algorithmic modules—see Figure 13-9. One of them shapes in real-time some pieces of advice and messages for the unit's operator on handling the transients based on the current turbine temperature state (algorithm "ADVICE"). The other algorithms off-line calculate accumulation of the low-cycle thermal fatigue in metal of the turbine rotors (algorithm "RESOURCE"), statistically process the stored data of the turbine operating conditions and violations of the limits that are assigned for the turbine state indications (algorithm "SUM"), and analyze the transients' quality and reveal the operating errors making the monitored indices exceed their limits (algorithm "ANALYSIS").

The choice of the mathematic model scheme is to a large degree determined by diverse subjective and specific factors. The world practice demonstrates a great variety of the applied solutions. In particular, advanced mathematical models for monitoring the turbine rotors were developed by specialists of ABB.[30,31] A specialized computing device "Tensomax" was developed for retrofitted steam turbines without steam temperature measurements available in the most thermally stressed rotor zones to be monitored. In this case the heating steam temperature and heat transfer conditions for the rotor are calculated based on the available steam pressure and temperature measurements inside an adjacent cylinder chamber. Along with this, in the 1980s for turbines furnished with thermometric probes, the principle of physical modeling was partially replaced with mathematical modeling, even though not completely accurate. In the computing devices known commercially by their product

Diagnostic Monitoring of Turbine Temperature/Thermal-stress States 417

names "Tensomarg" and "Turbomax 6," the measured temperature at the probe head was taken as if it was equal to the rotor surface temperature, but the average integral temperature of the rotor and the effective radial temperature difference are calculated.

Unfortunately, the same concept has been accepted for more up-to-date computing devices of Turbine Stress Evaluator (TSE) and Turbine Stress Controller (TSC) of Siemens (Figures 15-2 and 15-4) as constituents of its monitoring and control systems: the measured temperature on the heated surface of the probe installed into the cylinder casing is considered to be representative for the rotor surface, and the temperature difference characterizing the rotor's thermal-stress state is calculated based on this temperature.[3] It is needless to say that the heat transfer conditions from steam to the inner surface of the cylinder casing considerably differ from those to the rotating surface of the rotor. It would be more reasonable to arrange measuring the heating steam temperature and calculate the rotor temperature on the heated surface with the use of known empirical dependences for the heat transfer conditions.

Thermal Stress Evaluator mainly intends for information support

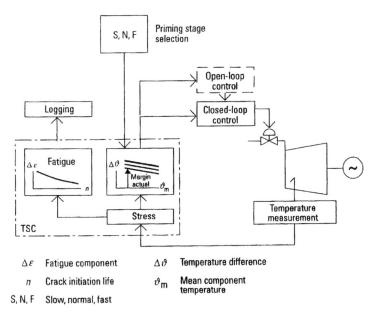

Figure 15-4. Configuration of Turbine Stress Controller of Siemens.
Source: U. Sill and W. Zörner[3]

of the operational personnel at transient operating conditions and additionally can be used to block automatically the process of rolling up or loading the turbine, if the thermal-stress state index for any of three monitored "critical" design element of the turbine goes beyond the permissible margin varying with the metal temperature. The same functions, even though in somewhat extended and advanced forms, are also fulfilled by Turbine Stress Controller (see below, the next section), but it is also used for calculating the accumulation of lifetime expenditures because of low-cycle, thermal, fatigue and creep fatigue. For rotors, additional calculations are made for accumulation of stress cycle fatigue (a possible crack's growth). By way of illustration, Figure 15-5 shows the contents of these assessments as applied to the HP and IP rotors.

It should be also noted that according to the general concept of Siemens, calculations of the life-time expenditures because of creep and thermal fatigue are supplemented with some monitoring functions of the Diagnostic Expert System for Turbomachinery (DIGEST) also related to the turbine lifetime. These functions are fulfilled by special modules of the DIGEST—for windage under influence of increased exhaust steam temperatures in the HP and LP sections and blade vibration for the LSBs. They can also be supplemented with an additional sub-module for calculating fatigue because of torsional oscillations of the turbine shaft caused by interaction between the generator and grid under certain specific conditions (short-circuits with subsequent successful or unsuccessful automatic high-speed reclosing, subsynchronous resonance in the grid, and some others).[3]

ON-LINE OPERATIVE SUPPORT FOR
TURBINE OPERATORS AT THE TRANSIENTS

Thermal Stress Evaluator of Siemens (see Figure 15-2) can be considered as one of the most well thought out and appropriate examples of devices for informative support for the turbine operators at transient operating conditions, developed in the last decades of the past century. This device continuously let the operational personnel clearly see how much they can vary the turbine's load (or rotation speed, if the turbine is rolled up during start-up), without exceeding by the monitored metal temperature differences in three critical elements of the turbine (HP valve steam-chest, HP rotor, and IP rotor) their admissible values, set

Diagnostic Monitoring of Turbine Temperature/Thermal-stress States

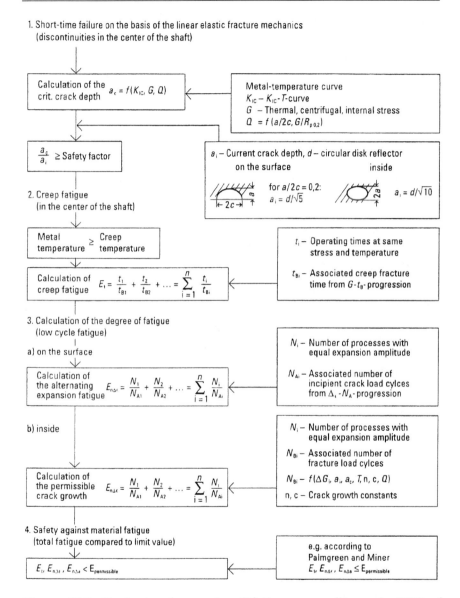

Figure 15-5. Contents of assessing lifetime expenditures in TSC of Siemens. *Source: U. Sill and W. Zörner*[3]

420 Steam Turbines for Modern Fossil-Fuel Power Plants

from the reasoning of low-cycle (thermal) fatigue of the turbine metal. Processed in a special computing device, these temperature differences give the upper and lower admissible values of the turbine's load, or rotation speed, continually presented to the operator by means of a special circular indicator in comparison with the actual current value of the controllable parameter. Input into the turbine's control system, these calculated values limit possible movements of the turbine valves. The open sector of the circular indicator just corresponds to the admissible range, and the current value of the controlled parameter should not go beyond this sector's borders. The monitored values are continually recorded at the multipoint recorder for possible subsequent analysis.

In succeeding years, the same function were passed to the Turbine Stress Controller (Figure 15-4) as a part of the new Monitoring, Control, and Protection System developed and employed by Siemens for its turbines—see Figures 13-6 and 13-8. The measured temperatures in the turbine's HP valve steam-chests and HP and IP cylinders are recalculated into the temperature differences that are considered as indices of their thermally stressed state. These temperature differences are compared with their upper and lower temperature-dependent admissible values, yielding the temperature margins for varying the turbine's operating conditions by increasing or reducing the turbine load or rotation speed. The smallest upper and lower margins of those for the three monitoring channels are transmitted to the set-point controller of the turbine's control system as the limits for loading and unloading. The admissible values of the temperature differences vary depending on the operation mode selected and set by the operator from three options: slow (S), normal (N), and fast (F). In practice, this means that, for example, relatively infrequent cold start-ups taking considerable time can be performed more rapidly by setting the "fast" rate and accepting in this way a somewhat higher accumulated fatigue rate. The resultant on-time increased life-time expenditure can be compensated in other cases with less fuel expenditures by selecting the slower operation mode. All the monitored temperature differences and margins are displayed in the form of bar-graphs on the TSC's monitor together with the selected operation mode (slow/normal/fast), and the current value of the controlled parameter is continuously indicated by a flashing cursor against the margins. The TSC also provides long-term storage of the monitored and calculated values.[3]

With all the merits of the considered devices and systems, it should be noted that they consider the turbine as if it worked with invariable

Diagnostic Monitoring of Turbine Temperature/Thermal-stress States 421

steam conditions, as it takes place for non-reheat turbines fed with steam from a common power plant header. In reality, especially during start-ups, changes of the main and reheat steam temperatures have significantly greater effects on the temperature and thermally stressed states of the turbine than variations of the turbine load or rotation speed, and in practical situations such devices as TSE and TSC can only help the turbine operator to compensate changes of the main and reheat steam temperatures which are considered as if uncontrollable. Such an approach, quite understandable from the standpoint of a turbine producer, is not quite acceptable for the power plant operator who cannot and should not consider the turbine as if isolated from the boiler.

Many modern power units are already operated by one operator per shift and, even if there are two operators at the control desk with some subdivision of functions between them, it is no good considering them as the one responsible for the turbine and another responsible for the boiler. Likewise the start-up instructions for the turbine and boiler should be united into one common document, informative support for the "turbine operator" should take into consideration all the operation actions influencing the turbine state, no matter if they refer to the boiler (changes of the main and reheat steam temperatures, for example) or to the turbine itself (the increase of the steam flow amount through the turbine, that is, its rotation speed or load).

As mentioned above, at modern fossil-fuel power plants transition to the use of advanced computerized automated systems acquiring, processing, and presenting information has led to the situation when practically all the current operating data, including those relating to the turbine state, occur to be available only via screens of several displays and survey panels—see Figures 13-2—13-5, and the operative access to any information becomes possible only by means of requesting relevant windows (slides, images, videograms) on one of these screens. Because different information items referring to different equipment components and different aspects of their technological processes are dispersed over numerous windows, they come into competition to one another. In many cases, the operators experience difficulty in choosing among numerous options and, as a result, often miss information needed for them. This is especially true for the transient operating conditions, when the operator has to act under conditions of time shortage, looking through and comprehending a great amount of diverse pieces of information substantially varying in time.

422 *Steam Turbines for Modern Fossil-Fuel Power Plants*

Under these conditions, of extreme importance is to arrange in a reasonable way the whole information bulk which can be needed for the operator and requested by them. In the first place, it means that all the excessive data should be abandoned, and the remainder, being potentially useful, should be preprocessed and presented for the operator in the most clear, obvious graphic forms. Then, it seems advisable to arrange the operator's access to different windows based on a hierarchic principle with the use of a problem-oriented menu displaying the needed data in the most appropriate forms (see Figure 13-12).

Such an approach was materialized and tested in the course of developing a few above-mentioned computerized systems—DASs, ASTDs, and PC-based LSDM-T for different types of turbines: from a group of 110-MW non-reheat turbines up to 300-MW and 800-MW supercritical-pressure power units.[7-9,23,28,29] This approach can be illustrated by the example of the LSDM-T with several functions of on-line and off-line informative support for the operator shown in Figure 13-9. The general menu allows the operators to request the main window of any on-line informative support functions intended for them. If the system serves a few objects operated from the common control desk, the general menu lets the operator request any function for each of the monitored objects—see Figure 13-10.

In the case illustrated by Figure 13-9, during the transients the operator is provided with informative support from two modules: "CONTUR" and "ADVICE." The first one is the basic module for the current and predictive operative monitoring of the temperature and thermally stressed states of the major high-temperature turbine elements: rotors and casings of the HP and IP cylinders, HP valve steam-chests and crossover pipes after them, as well as the main and reheat steam-lines. Characteristic temperatures and temperature differences for both the rotors and HP valve steam-chests are determined by means of mathematical modeling of their heating up based on the measured heating steam temperatures. Temperature monitoring of the rest turbine elements is fulfilled with the help of regular metal temperature measurements. Schematic of information brought onto the first window of this module is shown in Figure 15-6, presenting a hard copy of such a window for a 250-MW supercritical-pressure cogeneration turbine with captions translated into English. The lower toolbar lines name other windows of this module and functional keys to request them, as well as the functional keys for transition to the windows with the output results of another

Diagnostic Monitoring of Turbine Temperature/Thermal-stress States   423

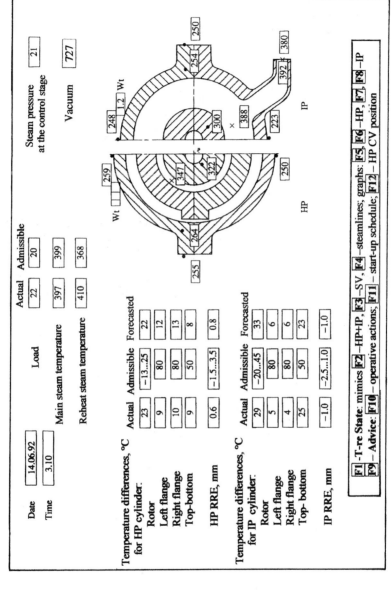

Figure 15-6. A sample window for monitoring the temperature state of the turbine's HP and IP cylinders.
Source: A.Sh. Leyzerovich et al.[8]

module. Information in all the diagrams is renewed each 30 seconds.

The considered window is a mimic diagram with the current numeric values of the monitored temperature and temperature differences for the HP (there is on the sketch its left-hand half) and IP (the right-hand half on the sketch) cylinders and the table with the current (actual), admissible, and forecasted values of the temperature differences and relative rotor expansions (RREs). The forecasted values are taken by extrapolating the past changes of the calculated values with the forecast depth equal to 3 min. In addition, the window shows the current values of the major controllable steam conditions: the turbine rotation speed or load, main and reheat steam temperatures averaged over the measured values, as well as their values which could be taken at this instant not making the monitored indices exceed their admissible values. In doing so, the admissible values of the effective temperature differences in the rotors are automatically set different for start-ups and turbine operation in the governed load range. If the actual or forecasted values of any of the monitored indices exceed its admissible values, the corresponding figure is marked by red. Simultaneously, the designation on the bar line of the window referring to the turbine elements where this happens is also marked by color to attract the operator's attention.

Along with the mimic diagram like shown in Figure 15-6, change in time of the effective temperature differences in the HP and IP rotors together with the controllable operating parameters influencing these indices (the turbine rotation speed and load, main or reheat steam temperatures), is presented by trend graphs. The change of the monitored temperature difference is plotted against the field of its admissible values. An example of such a graph is shown in Figure 15-7. Two additional trend graphs intended for the top-bottom temperature differences and RRE for the HP and IP cylinders. Hard copies of a few other windows are given in the author's previous book.[2]

A special submodule of "CONTUR" continually identifies the operating conditions, and if the current situation is recognized as a start-up, another module, "ADVICE," begins drawing the diagrams of rolling-up, loading, and raising the main and reheat steam temperatures according to the power plant instruction as applied to the specific initial temperature state of the turbine. The diagrams are displayed for the operator against the actual changes of the controlled parameters with time.

One of possible ways for identifying the operating conditions is shown in Table 7-2 presented in a matrix form. Labeling the turbine's

Diagnostic Monitoring of Turbine Temperature/Thermal-stress States 425

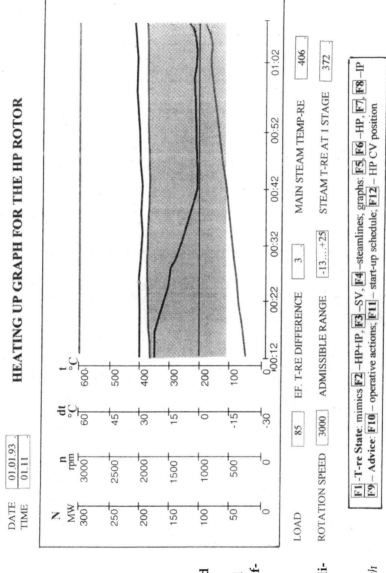

Figure 15-7. A sample window with trend graphs for the effective radial temperature difference in the HP rotor and controllable operating conditions influencing it. *Source: A.Sh. Leyzerovich et al.*[8]

426   *Steam Turbines for Modern Fossil-Fuel Power Plants*

current operating conditions as stationary under load triggers the process of on-line heat-rate monitoring (module "HEATRATE" in Figure 13-9). The mentioned function of identifying the current operating conditions is also extremely important to separate and store different bulks of information for its post-operative processing and analysis. For the considered case of LSDM-T in Figure 13-9, these functions are fulfilled by modules of "SUM," "ANALYSIS," and "FATIGUE."

In the start-up process, module "ADVICE" also continually shapes messages with pieces of advice (recommendations, suggestions), displayed at the operator's request. These messages draw the operator's attention to revealed abnormal deviations, give their probable explanations, and suggest the required or admissible technological operations corresponding to the current technological situation. These suggestions can be, for example: "Hot Reheat Steam-Lines Should Be Heated Further" or "Heating of Reheat Steam-Lines Can Be Finished," "Opening of HP Valves Is Wrong," "Main Steam Temperature Is Too Low," "Switch On Reheater's Steam Bypass," "Switch On Steam Heating of HP Flanges," "Close HP Pipes' Drainage," and so on. Some of these recommendations (for example, referring to switching on the reheater's bypass or steam heating systems of flanges) are preceded by warning notices: "Be Prepared to..." The mechanism of shaping these messages is close to algorithms of automated logical control.[2,4]

In addition, at every start-up instant, the upper admissible values of the main and reheat steam temperatures are calculated and exposed on the mimic diagram(s)—see Figure 15-6. If the actual average steam temperature differs from the admissible level too much (for example, by over 20°C, or 36°F), the corresponding figure is marked red. As the turbine load reaches the nominal lower boundary of the governed range (50% MCR), the operator is continually being informed about the upper admissible steam flow amount through the turbine. (In the specific considered case, because for cogeneration turbines there is not a single-valued correlation between the steam amount and load, the former is characterized by the steam pressure at the control stage, whose current value is also displayed.)

References

1. Leyzerovich A.Sh. "Operational Monitoring of Fatigue Accumulation for Steam Turbine Details" [in Russian], *Energokhozyajstvo za Rubezom*, no. 1 (1979): 6-10.
2. Leyzerovich A. *Large Power Steam Turbines: Design & Operation*, 2 vols., Tulsa (OK): PennWell Books, 1997.

Diagnostic Monitoring of Turbine Temperature/Thermal-stress States 427

3. Sill U. and W. Zörner. *Steam Turbine Generators Process Control and Diagnostics*, Erlangen: Publicis MCD Verlag, 1996.
4. Leyzerovich A.Sh. *Technological Fundamentals for Steam Turbine Start-up Automation* [in Russian], Moscow: Energoatomizdat, 1983.
5. Termuehlen, H. *100 Years of Power Plant Development. Focus on Steam and Gas Turbines as Prime Movers*, New York: ASME Press, 2001.
6. Hoerster B., W. Gorzegno, and H. Termuehlen. "Two-Shift Handley Station Units 4 and 5—Experience with 2000 Starts," *Proc. of the American Power Conference* 47, Chicago, 1985, 62-71.
7. Leyzerovich A., V. Berlyand, A. Pozhidaev, and S. Yatskevich. "Continuous Monitoring as a Tool for More Accurate Assessment of Remaining Lifetime for Rotors and Casings of Steam Turbines in Service," *Proc. of the American Power Conference* 60. Part 2, Chicago, 1998, 1021-1026.
8. Leyzerovich A.Sh., B.R. Beyzerman, N.F. Komarov, et al. "First Experience of Applying the PC-Based Local Subsystem of Diagnostic Monitoring for Steam Turbine" [in Russian], *Elektricheskie Stantsii*, no. 4 (1993): 18-22.
9. Leizerovich A.Sh. "Information Support for the Operating Personnel of a Thermal Power Station during Transients and Their Post-Operative Analysis," *Thermal Engineering* 43, no. 10 (1996): 817-823.
10. Plotkin E., Y. Berkovich, A. Leyzerovich, et al. "Improvement of Pre-Start Heating Technology for Steam-Turbine Units," presented at the Electric Power Conference 2005, Chicago, 2005.
11. Pahl G. "Safe and Fast Start-ups of Steam Turbines" [in German], *BBC-Nachrichten* 40 (1958): 340-347.
12. Pahl G., W. Reitze, and M. Salm. "Observance of Admissible Thermal Stresses in Steam Turbines" [in German], *BBC-Nachrichten* 46: no. 3 (1964): 139-147.
13. Berry W.R. and I. Johnson. "Prevention of Cyclic Thermal Stress Cracking in Steam Turbine Rotors," *Transactions of the ASME* series A, no. 3 (1964): 361-368.
14. Timo D.P. and G.W. Sarney. "The Operation of Large Steam Turbines to Limit Cyclic Thermal Cracking," *Paper ASME 67-WA/PWR-4.*
15. Leyzerovich A.Sh. "Limitations at Start-Ups of Large Steam Turbines Caused by Thermal Strength of Their Rotors" [in Russian], *Teploenergetika* 18, no 10 (1971): 88-91.
16. Šindelář R. "Control of Thermal Stresses in Metal of Steam Turbines at Start-ups and Load Changes" [in German], *Škoda-Revue*, no. 4 (1972): 19-30.
17. Spencer R.C. and D.P. Timo. "Starting and Loading of Large Steam Turbines," *Proc. of the American Power Conference* 36, Chicago, 1974: 511-521.
18. Leyzerovich A.Sh. "Experience in Application of Thermometric Probes to Monitoring the Heating of Steam Turbine Rotors" [in Russian], *Energokhozyajstvo za Rubezom*, no. 1 (1978): 10-12.
19. Leizerovich A.Sh., V.N. Kozlov, V.D. Mironov, et al. "Monitoring the Heating up Process for Rotors of Large Steam Turbines with the Use of an Analogous Model" [in Russian], *Teploenergetika* 24, no 8 (1977): 10-13.
20. Leizerovich A.Sh., B.D. Ivanov, Yu.N. Vezenitsyn, et al. "Commercial Tests of a Prototype for Operational Monitoring of the Warming of Rotors of High-Capacity Steam Turbines," *Thermal Engineering* 23, no. 11 (1978): 29-33.
21. Leizerovich A.Sh. "Mathematical Modeling of Large Steam Turbine Rotor Heating-Up for Operating Monitoring," *Proc. of the Academy of Sciences of the USSR. Power Engineering and Transport*, no. 2 (1981): 123-131.
22. Leizerovich A.Sh. "Monitoring the Thermal Stress State of LP Rotors of Steam Turbines" [in Russian], *Teploenergetika* 27, no. 8 (1980): 17-21.

23. Leyzerovich A.Sh., Ledaschev A.V., Rusanova N.A., et al. "Automated Monitoring Thermal State of the 250-MW Turbines at Transients" [in Russian], *Elektricheskie Stantsii*, no. 4 (1987): 14-18.
24. Lehne F. and R. Leithner. "Calculation of the Wall Temperature Difference of Thick-Walled Steam Generator Components from Measured Steam Temperature Values," *VGB PowerTech* 80, no. 1 (2000): 32-35.
25. Talet J., B. Weglowski, S. Gradziel, et al. "Monitoring of Thermal Stresses in Pressure Components of Large Steam Boilers," *VGB PowerTech* 82, no. 1 (2002): 73-78.
26. Lehne F. and R. Leithner. "Calculation of Temperature Differences in the Walls of Thick-Walled Pressure Parts for Calculating the Life-Time Consumption" [in German], *VGB PowerTech* 82, no. 10 (2002): 77-80.
27. Leizerovich A.Sh. "Algorithm for Monitoring Temperature State and Defining Admissible Load Changes for LMZ Turbines with the Use of Microprocessor-Based Electro-Hydraulical Governing System" [in Russian], *Teploenergetika* 34, no. 10 (1987): 29-32.
28. Leyzerovich A.Sh., A.V. Antonovich, V.I. Berlyand, et al. "Temperature and Thermal-Stress State of Steam Turbines: Complete Diagnostic Monitoring as a Part of Functions of an Automated Technical Diagnostics System" [in Russian], *Elektricheskie Stantsii*, no. 10 (1992): 32-38.
29. Leizerovich A.Sh., L.P. Safonov, A.V. Antonovich, et al. "Developing and Mastering the Automated Systems for Diagnostic Monitoring of Power-Generating Units at Thermal Power Stations," *Thermal Engineering* 42, no. 2 (1995): 154-160.
30. Sindelar R. and W. Toewe. "Thermal Stress Monitoring in Steam Turbines" [in German], *BWK* 51, no. 10 (1999): 46-49.
31. Sindelar R. and W. Toewe. "TENSOMAX—A Retrofit Thermal Stress Monitoring System for Steam Turbines," *VGB PowerTech* 80, no. 1 (2000): 41-43.
32. Leyzerovich A.Sh, D.I. Bukhny, Yu.A. Radin, et al. "PC-Based System of Information Support for Fossil-Fired Non-Unit Power Plant's Group Board Operator" [in Russian], *Elektricheskie Stantsii*, no. 7 (1994): 27-31.

Chapter 16

Post-operative Analysis Of the Turbine's Operating Conditions

COMPUTERIZED POST-OPERATIVE ANALYSIS OF TURBINE OPERATING CONDITIONS

The measured and calculated data of on-line diagnostic monitoring should be recorded and stored to be analyzed off-line by plant engineers and supervisors. The main goals of this analysis are evaluation of the equipments state, as well as the quality of its service—in particular, the quality of control at stationary and transient operating conditions. The resultant comprehension of the equipment state provides the basis for scheduling maintenance—inspections, current repairs, and overhauls. Together with results of maintenance diagnostics at the stopped unit, this analysis is used for planning replacement of individual equipment elements or more general upgrades. Evaluation of the operation quality is basic to teaching and training the operational personnel to prevent repetition of the revealed characteristic operating errors. All these measures are necessary to keep and improve the unit's efficiency, reliability, and flexibility. Effectiveness of this off-line analysis appreciably depends on an opportunity to access the stored results, their completeness and representativeness, and the possibility of processing them.

For power units furnished with computer-based DASs, this plant management process can be taken over to a great degree to the unit's or plant's computer systems. So, for example, as applied to the computerized system for fossil-fuel power plant units developed by Hitachi and shown in Figure 13-2, the plant management computer gathers all the data from the plant units' networks. These data are stored on magnetic discs and can finally be transferred to an optical disc of a large memory capacity for long term storage. The main purposes of processing the stored operating data are supposed to be the followings: statistical processing of the operation time periods, revelation of heat-rate performance variations in time, evaluation of start-up energy and fuel expenditures, analysis of start-ups and shut-downs and their contribution to accumulation of thermal fatigue of metal for the most thermally stressed design elements, assessment of the creep fatigue accumulation while the power unit is operating under load, analysis of vibration characteristics for the turbine, and so on. All the obtained data for the observed unit and its equipment are compared with similar data taken during the prior operating conditions and for other sister units, subjected to correlation analysis. The prepared reports and processed data are displayed and/or printed in the matrix and graph forms.

A similar approach, although with the use of much simpler hardware, was accepted for the PC-based Local Subsystem of Diagnostic Monitoring for the Turbine (LSDM-T)—Figure 13-9. All major measured and calculated operating data that were processed on-line by the above mentioned modules titled "HEATRATE," "CONTUR," and "ADVICE" were stored in the same operating PC. They can be processed in the same PC immediately at the end of every shift or day, but more often these data were brought into another PC and processed there with the use of modules "SUM," "FATIGUE," and "ANALYSIS."[1-3] For Automated Systems of Complex Technical Diagnostics (ASCTD),[4,5] both the functions of on-line and off-line processing are commonly executed within a single computer system, even if the stored on-line diagnostic monitoring data can also be transferred into another computer system (for example, in the utility's computing center) and processed there (see Fig. 13-11).

The algorithmic module "SUM" is intended for processing statistically the data of the turbine's operating conditions (their number and length-time), as well as deviations of the turbine's operating parameters and state indices from their admissible values. Correlation of this information with the heat-rate performance data varying in time allows the plant engineers to establish some associations between these factors and foresee the expected efficiency deterioration. Deviations of the steam temperature over their rated values are also substantial for assessing the creep fatigue accumulation for metal of the turbine's high-temperature design elements and steam-lines.

Due to module "FATIGUE," the user can trace and extrapolate accumulation of low-cycle, thermal, fatigue in metal of the turbine's most thermally stressed, critical elements. Together with assessment of the

Post-operative Analysis of the Turbine's Operating Conditions 431

creep fatigue accumulation, this makes it possible to estimate the residual resource of these elements taking into account its actual operating conditions and realistic estimations of the operation quality.[6,7] To a great degree, the methodology of these calculations is similar to that employed by Siemens in the TSC for assessing the turbine's lifetime expenditures and illustrated by Figure 15-5.[8] Presently, such calculations are done with a satisfactory validity only for the rotor bodies, even though with certain cautions these calculations could also be extended to cover the valve casings and rotor wheel discs with regard to the influence of high-temperature creep. Naturally, it is supposed that the initial metal state and pre-history of loading (before the diagnostic monitoring was implemented) are known or, at least, can be assessed relying on the monitoring data of present operation.

As distinct from "SUM" and "FATIGUE," module "ANALYSIS" by its nature is rather close to the logical algorithms of informative support, such as "ADVICE." It is intended to reveal the reasons of why the main state indices of the turbine have gone beyond their admissible range. This problem can be solved using different methodologies, in particular, with the use of expert systems based on event (fault) trees. However, successful solution of this problem even more decisively depends on what the body of the acquired data is available to discriminate between the numerous potential causes. For the inventory of the measured data accessible for the DASs employed for modern power units, it seems possible to reach the distinct solution only as applied to the leading indices which are influenced by changes of the major operating conditions, that is, variations of the steam flow amount through the turbine (its load or rotation speed), main and reheat steam temperatures, and steam admission control. In other cases (as, for example, as applied to the RRE or temperature differences in high-temperature cylinder casings), the monitored indications are additionally influenced by too many secondary factors without due representative measurements input into the unit's computer system.

For the leading indices (temperature differences in the high-temperature rotors and HP valve steam-chests), there are about 12-18 potential causes affecting each indication. In diverse operating conditions and at their different stages, these causes can be different or, at least, can manifest themselves in different manners. Appearance of the algorithm module "ANALYSIS" can be seen from a fragment of its schematic block-chart shown in Figure 16-1.

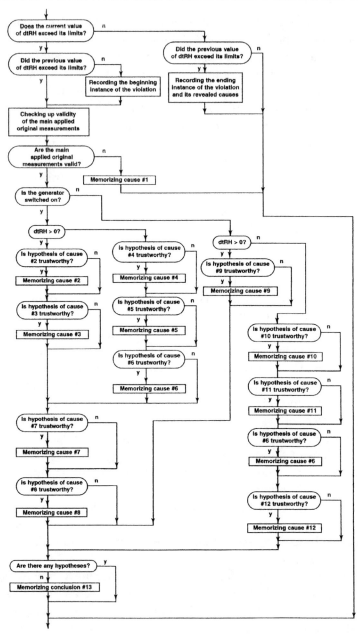

Figure 16-1. Fragment of a block-chart for the algorithmic module ANALYSIS referring to revealing potential causes why the monitored effective radial temperature difference dtRH in the HP rotor exceeded it admissible values.

EVALUATION OF THE CONTROL QUALITY AT THE TRANSIENTS

In many cases, it is desirable to have a possibility to evaluate the operation quality in a quantitative comparative manner for different shifts, units, and so on. This evaluation can be fulfilled in the process of post-operating analysis of the recorded operating conditions data. One of possible approaches to reach this goal as applied to controlling the turbine transients is described below.

It is a four-grade scale that is used for such evaluation based on criteria of the turbine's flexibility and reliability, that is, correlation between the actual process duration and its scheduled length-time (according to the power plant instruction or the dispatch chart) and between the actual extreme values of the monitored state indices and their admissible values.

In doing so, it is expedient to subdivide the monitored indices into those of explicit and implicit influence. Violations for the former ones manifest themselves immediately. Among such indices are, for example, the RRE and top-bottom temperature differences, whose abnormal values can result in brushing and rubbing in the steam path, increased vibration, and other immediate consequences. Some other temperature differences causing plastic strains and distortion could also be regarded as having explicit consequences. For these indices, only their extreme values are important if they went beyond the admissible range. As distinct from them, consequences of violations for implicit indices are postponed in time—they are accumulated in the process of operation and depend not only on the extreme values of the considered variables, but also on the pattern of their variations in time: sign alternation, amplitude and number of cycles.

So, for example, the considered transient operating conditions can be treated as having been performed with grade "A" provided that in their course all the monitored indices of the turbine's state did not go beyond their limits and duration of these transients did not exceed the scheduled length-time by more than 10%. The grade is "B" if, with observance of all the limitations for the turbine indices, the considered transients lasted more than their scheduled time within 10-50%. The process is also "B"-graded providing there were not violations for the explicit indices, but one of the turbine's thermal-stress (implicit) indices once went out beyond the range of its admissible values, and the depth of this violation did not exceed 30%. If either the number of these violations was more than one or the operating conditions lasted more than

434   *Steam Turbines for Modern Fossil-Fuel Power Plants*

1.5 times longer than it was scheduled, the grade is "C." Any explicit violation or implicit violation with the depth more than 30% decreases the grade to "D."

It is understandable that the mentioned criteria are not a dogma and might be corrected to fit the specific operating conditions. Moreover, the criteria of such evaluations can be supplemented with the values of the calculated heat losses and specific low-cycle thermal fatigue accumulation in metal at the considered transients.

Similar approaches can be applied to evaluations of the control quality at stationary operating conditions from the standpoint of the turbine efficiency deviations to worse from its possible targeted values and for the equipment state from the standpoint of its reliability, that is, for example, for the depth and total duration of deviations of the steam parameters from their rated values.

POST-OPERATIVE ANALYSIS OF THE TRANSIENTS FOR TURBINES WITHOUT COMPUTERIZED DAS

Many older power units in service have not been furnished with computerized Data Acquisition Systems, and their turbines do not have any computing subsystems (like the above mentioned PC-based LSDM-T) or devices to monitor on-line their temperature and thermal-stress states at the transients with sufficient representation. This commonly causes an appearance of significant temperature differences and thermal stresses in their critical elements (the high-temperature rotors, in the first place) that usually remains unnoticeable by both the operational personnel and plant engineers and supervisors. As a temporary solution for such turbines (until arrangement of the representative operating monitoring), it is expedient and possible to arrange a post-operative calculation analysis of their transients on the basis of common measurement readings having been registered at the transients in tape diagrams of regular multi-point recorders. These data in a schematized, piecewise linear form can be used as initial information for a specialized computer program calculating the main thermal-stress indices for the turbine's critical elements and evaluating the obtained results. The critical prerequisite for implementation of such a program is the presence of the heating steam temperature measurements for the HP and IP rotors (or HP and IP sections of the integrated HP-IP rotor) in their most thermally

Post-operative Analysis of the Turbine's Operating Conditions   435

stressed sections.

The scope of initial data for this program does not have to be too large, and the program itself should be simple enough to be daily applicable by both the plant engineers and operational personnel. This conceptually distinguishes such a program from more sophisticated software developments intended for sufficiently accurate thermal-stress calculations for the same critical elements.

A series of such programs, named TSAR (Thermal Stress Analysis of Rotors), had been developed by us for some most widespread steam turbines of the former Soviet Union, being quite adaptable for other turbine types, and gained fairly wide acceptance at power and cogeneration plants of Russia and Ukraine.[10] Calculation of the rotor temperature and thermal-stress states in these programs is based on the use of approximate mathematical models which require a minimum number of initial data, but take into account the main factors influencing the heating process in the rotors. Each program also includes static dependences between the measured operating parameters and boundary heat transfer conditions for the rotors. A set of auxiliary programs for calculating these dependences was also developed.

The program requires inputting the following values varying in time: turbine rotation speed and load (or steam pressure at the first stage) and steam temperatures at the HP first stage and in the IP steam-admission chamber (or in the IP inlet sleeves). Sometimes, if necessary, the last variable can be replaced with the measured reheat steam temperatures after the boiler's attemperator. In this case, the mentioned list is to be supplemented with variation in time of the hot reheat steam-line metal temperatures. For turbines with throttle steam admission control, the heating steam temperature for the HP rotor can be calculated based on the steam temperature measurements at the HP valves with account to the metal temperatures of the crossover pipes (if necessary). All these data are set in the piecewise linearized, tabulated form. Usually, the number of the specific points in these dependences varies from 2-5 (commonly for the rotation speed) up to 20-25 (mainly for the steam temperatures). In addition, the user should also input the HP and IP cylinder temperature values before the considered process begins, and they are used as the initial temperature conditions. Some program modifications also included calculations for the HP stop-valve steam-chests, and the mentioned input data are to be supplemented with the initial metal temperature of the steam-chests and variation in time of the main (throttle) steam temperature and pressure.

The output results are displayed on the screen in the graphic form (Figure 16-2) and printed up (Table 16-1) on the user's request. The extreme values of the effective temperature differences in the critical elements exceeding their admissible values are marked, separated, filtered, and recorded. They can be employed for subsequent estimations of low-cycle thermal fatigue of the metal. The number and depth of these violations, together with the process duration compared with the scheduled length-time, are used for evaluating the control quality according to the above described approaches.

It is essential that on the basis of the gained results the user has a possibility to "correct" the input data for repeated calculations until gaining satisfactory results. Thus, the program can be used for teaching and training the operational personnel as the simplest simulator.

TEACHING AND TRAINING THE OPERATIONAL
PERSONNEL WITH THE USE OF SIMULATORS.

According to numerous investigations, the share of accidents, forced outages, and failures at fossil-fuel power plants caused by human

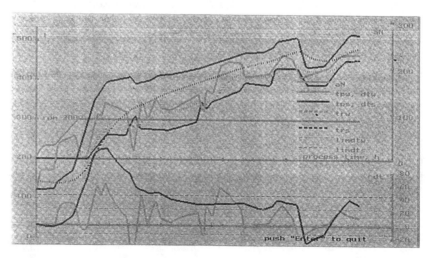

Figure 16-2. Hard copy of graphic presentation for post-operative processing the recorded data for a 300-MW supercritical-pressure turbine of Turboatom at warm start-up.

Post-operative Analysis of the Turbine's Operating Conditions 437

Table 16-1. Fragments of post-operative processing of the recorded data for a 300-MW supercritical-pressure turbine of Turboatom at a warm start-up.

Time, min	n, rpm	aN, MW	TstHP, °C	TrHP, °C	ΔTrHP, °C	TstIP, °C	TrIP, °C	ΔTrIP, °C
0	0	-	100.0	178.0	0.0	122.0	142.0	0.0
5	625	-	160.8	176.5	-3.6	122.5	141.8	-1.2
10	1000	-	172.7	176.1	-1.6	123.0	141.1	-2.9
15	1400	-	172.4	175.6	-1.5	123.5	140.3	-3.4
20	1800	-	172.0	175.2	-1.4	124.0	139.6	-3.5
25	2200	-	241.1	183.0	25.9 >	145.5	139.6	-0.0
30	2600	-	246.7	196.2	32.6 >	154.4	140.6	4.0
35	3000	-	252.2	207.8	33.3 >	163.2	142.8	8.9
40	3000	-	257.8	218.0	32.2 >	172.0	146.0	14.2
45	3000	-	233.0	223.7	11.4	212.2	153.0	34.4
50	3000	-	205.7	216.6	-7.8	252.5	165.0	53.7 >
55	3000	-	214.9	213.3	-0.4	292.7	180.2	71.4 >
60	3000	16	264.5	224.5	33.7 >	332.9	200.2	99.8 >
65	3000	29	314.0	246.1	61.0 >	359.8	223.1	112.4 >
70	3000	41	310.0	262.5	45.0 >	377.9	243.8	115.4 >
75	3000	54	341.9	281.7	57.3 >	396.0	263.1	117.2 >
80	3000	54	336.7	293.2	41.8 >	398.6	279.7	106.7 >
85	3000	53	331.5	300.1	30.2 >	401.2	293.0	97.5 >
90	3000	53	326.3	304.5	21.1	403.8	304.7	89.6 >
95	3000	52	321.0	307.0	13.7	406.4	315.0	82.6 >
100	3000	74	259.4	297.8	-34.7 <	409.0	324.6	77.7 >
105	3000	96	244.7	275.6	-30.8 <	392.0	331.1	58.4 >
110	3000	67	321.3	287.0	30.5 >	394.7	336.0	54.4 >
...	...	...	...	...	...	...	...	...
270	3000	200	456.0	442.4	13.5	508.6	472.8	34.2
275	3000	200	427.2	438.5	-10.5	511.0	476.4	33.0
280	3000	182	411.6	430.0	-17.9 <	468.8	474.8	-3.1
285	3000	164	415.6	426.7	-11.0	435.3	466.6	-27.1 <
290	3000	164	419.6	425,7	-6.0	436.6	460.7	-22.1
....	...	...	...	...	...	...	...	...
330	3000	217	467.0	450.4	16.4	519.0	475.7	40.4
335	3000	217	467.5	453.2	14.1	518.5	481.1	35.6
340	3000	218	468.0	455.6	12.2	518.0	484.9	31.7

Extremes of the effective temperature differences in the HP and IP rotors exceeding their admissible values (25...–13°C for the HP rotor and 45...–25°C for the IP rotor)

at the HP rotor (kH = 6), °C:	65.2	-63.7	50.6	-26.1	59.5	-22.1
time of the extremes, min:	73	103	112	177	195	278
at the IP rotor (kI = 2), °C:	117.2	-28.4				
time of the extremes, min:	75	284				

Evaluation of the process quality – **D**

errors in the last decades have steadily risen, whereas the portion of failures caused by purely technical factors has continually decreased.[11,12] Unskilled actions of the operational personnel also frequently result in deterioration of the unit efficiency and reliability, as well as power plant operation with some casual energy losses at the stationary operating conditions and undue fuel expenditures at the transients. Although these occurrences are not as dramatic as equipment failures or even forced outages are and they do not attract such a close attention of plant managers, these factors also substantially affect the power plant operation performances. According to a special study of EPRI fulfilled in the mid-1980s, the impact of the operational personnel's errors on the power plant availability entails the annual loss of approximately 122,000 MWh per unit which translates into an average annual cost of over $3,000,000 per unit.[13] New power units with their elevated steam conditions, advanced design features, and state-of-the-art DACSs need even more thoughtful and comprehensive operation, whereas any operating errors and mistaken actions entail more serious or, at least, more expensive consequences. Against this background, of importance is a risen shortage of skilled and experienced operational personnel at power plants. This situation aggravates with aging of the very highly qualified operators with valuable experience, knowledge, and skills. According to another EPRI survey, in the mid-2000s the major and continually growing segment of workforces at the U.S. power plants consists of workers in the age between 55 and 65, while the number of operators aged 25 to 44 remarkably declines.[14]

All these circumstances lead to a problem of preparing new workforces for power plants, teaching and training their operational personnel. For many years, the main form of such education has been an apprenticeship at actual plants and their control boards, as well as the training in classes with instructors. In order to prepare an operator for activities in unexpected abnormal situations, the instructor "presented" these situations to the trainee in the form of distinct scenarios of varying the observed values. Even in the mid-2000s, according to the American Society of Training and Development (ASTD), about 74% of all the corporate trainings in the United States is administered in a classroom or in the on-the-job-training environment.[14] More often than not, this practice provides only hands-on training of how, but not why, things are done. It is also noted in many cases there is little or no control over how or what the trainees are taught, or whether or not the trainer is teaching the correct approaches.

Along with teaching the beginners, even more difficult task is the retraining of experienced operators referring to new equipment, new technologies, new controlling and measuring means, and new operation modes. It is well known that it is much more difficult to overcome former dynamic behavior stereotypes than even create them. In particular, this human factor is to a large measure responsible for slow introduction and low effectiveness of many novelties. The deep-rooted stereotypes, stored in the human subconscious, come into effect especially intense when the operator has to act under conditions of time shortage.

The most effective way to overcome these difficulties is the use of computer-based *simulators* which model the actual equipment and methods of controlling it. A trainee, working with a simulator, receives a sensual comprehension of the object, gets used to it, gains skills in controlling it, and works up new dynamic stereotypes adequate to the object. Experience of "operation" with a simulator allows the trainee to work up automatic reactions and comprehend dynamics of the controlled processes at the sensual level.

Early simulators were mainly intended to provide a simulated hands-on experience in extraordinary, emergency situations that can rarely be encountered in the routine operation practice. With these simulators, the operators were trained in the way they should properly respond to failures at the power plant and recover from them. Further, power plant simulators were used to solve the tasks of a wider profile; in particular, they were offered as a tool to learn how a power plant and its equipment work, to feel its responses on operation actions under most characteristic operating conditions.[15,16] A well-thought-out, high-fidelity simulator is commonly thought to provide a verisimilar replica of the power unit's major parameters at the steady-state operating conditions and periods of dynamic response to control actions and malfunctions, different changes in the equipment states, to teach trainees to comprehend the power unit's control by giving them a thorough understanding how their actions impact plant operations and overall performances.

The simulator should function in a real-time scale of the modeled processes. If the models are too complex, this requirement can cause some problems because of insufficient internal performance of the applied computing means and often forces to simplify the models. At the same time, any experienced power plant operator well knows that no two power units behave alike in operation, and each one is unique in its responds on the control actions. In addition, even sister power units of

440 *Steam Turbines for Modern Fossil-Fuel Power Plants*

the same type are frequently differ from one another in their control tools and contours. So, at all times it has been a serious problem how to adapt the simulator to the specific object, its distinguishing static and dynamic features, methods and tools of control. Fast progress in power plant technology, appearance and replacement of new automatic technique generations, variety of equipment and means of controlling it, variations in an operating mode—all these factors make solution of this problem extremely important and, at the same time, difficult.

It is always meant that the processes of teaching and training include an interaction between the trainees and instructor. As applied to teaching and training the plant's operational personnel with the simulators, the instructor poses the problem for the trainee, set the initial conditions of the task, and evaluates its solution, analyzing the current and recorded results of the process. Frequently, the instructor can additionally interfere in the modeled controlled process inflicting some disturbances, as well as varying some static and dynamic characteristics of the model to near it to some specific objects. Along with this, analysis of the trainee's results, revelation of the made errors, and explanations of the correct behavior in the modeled situations might be the most important part of the instructor's activity. All these purposes impose special demands on the instructors who, however, often suffer from the same dynamic stereotypes and misconceptions as the trainees. It is quite characteristic that many of today's training programs are seeking to obviate the need for the instructor and are designed to be self-directed and self-taught. So, many modern training centers use so-called *intelligent tutoring systems*, and the trainees can receive directed training even if instructors are unavailable.

Prior to the 1980s, most simulators used to be based on the use of analogous technique, but all the modern simulators are computer-based. There are three main directions in creating the power unit simulators.[9,13] The first is based on creation of a *full-scope* (*full-scale*, or *full-replica*) computer model of the unit as a whole in combination with the natural model of the unit's control board with all its consoles, switches, keys, indicators, CRT screens, and all other human-machine interface means. Moreover, the extent of the simulator is done as a one-to-one *replica* of the actual control room even down to desks, chairs, and lights. Each simulator of this type is unique and oriented to the single, specific unit type, commonly the largest and the most complicated in the power system or, by contrast, the most typical there. As a rule, these simulators are

Post-operative Analysis of the Turbine's Operating Conditions   441

constructed on the basis of special regional training centers, and their number can be counted on fingers even in the most industrially developed countries.

The main disadvantage of this direction is an impossibility to remake and adjust the developed simulator for other units. In addition, such a simulator is a multi-purpose by duty, but this great advantage has its back side, namely, some of the imitated processes are forcedly modeled too approximately and do not give a good comprehension of their dynamics. For this reason, the full-scaled simulators are commonly supplemented by some *local simulators* modeling individual technological processes or control of individual functional groups of equipment. The local simulators usually run in more simple computing means and are furnished with natural models of local control desks or panels. Sometimes, these simulators also include some natural circuits of automated (or automatic) governors.

Another direction does not pretend to train the motion activity of the personnel and is intended to teach the operator not to handle the unit and its equipment but rather to optimize its performance. For this reason, a simulator makes no attempts to duplicate the natural control room. In this case, all the human-machine interfaces usually go through the CRT screens. With this exception, the simulators of this (*non-replica*) type do not differ from the first-type ones and, that is the most important, they remain full-scope, that is, embrace the unit as a whole with all the advantages and disadvantages of this approach. At the same time, these simulators are somewhat more adaptable and can follow the object's alterations more easily. Commonly, such simulators are materialized in microcomputers linked into a network. By the late 1980s, a PC- and CRT-based full-scale non-replica simulator could be purchased for under $200,000 as compared with a $5-million price tag for a full-replica one.[13]

Finally, the third direction is presented by simulators of a special duty, intended to solve not the general problems, but only individual tasks of teaching and training the plant personnel. Such simulators, received the name of *compact simulators*, are the set of the program modules which can run in individual PCs or their local networks and can be adapted to the specific distinguishing features of the modeled object much more easily than the full-scope simulators. This reduces their cost by a factor of five to ten compared with the full-scope ones, and, what is even more important, such simulators can be easily installed at any power plant and adjusted to its specific needs.[17,18]

Compact simulators offer more flexibility than conventional ones in application, modeling capability, and the training timing. On the same hardware platform, the compact simulators can model various types of the control systems by loading the appropriate models and replacing the custom keyboard templates. As applied to each specific problem, compact simulators can imitate the modeled processes and equipment with high accuracy and validity. In addition, it can be said that emulation of control systems and operator interfaces in compact simulators is becoming easier and less expensive, too.

The programs of training and the set of compact simulators can be substantially different for different plants. So, for example, according to EPRI, the operators of base-load fossil-fired and nuclear power plants do not have to go through many scheduled start-ups and shut-downs and should pay more attention to optimizing the heat-rate performance and actions in extreme unexpected situations.[18] On the other hand, the operators of exactly these plants do not have sufficient practice in start-ups and other transients and are not prepared to act correctly in these operating conditions at any deviations from the scheduled patterns. So they need special training.

The compact simulators intended for different training tasks substantially differ in their models, monitored output variables, imitated input actions, and scenarios of training. As a sample, a simulator for heat-rate training should be capable of emulating the following exercises: sliding versus constant throttle pressure, circulating-water pumps out of service or operating with varied speed, cooling towers at reduced capacity, removal of feedwater heaters from service, operating with decreased main and/or reheat steam temperatures, operating with worsened vacuum in the condenser and increased terminal temperature differences in FWH, and so on. The training scenarios should predetermine unit operation at different levels of output with different equipment state and various disturbances.

In the last decade, many electric power utilities, along with other methodologies of teaching their operational personnel, have increasingly more used so called web-based training (WBT).[14,19] According to the ASTD's survey, in 2001 it was 10.5% of the power plant workforces involved in WBT, and by 2004 this portion increased to 25%. Presently, the web-based formats of teaching the operational personnel mainly consist of common educational courses and courses for qualification improve-

Post-operative Analysis of the Turbine's Operating Conditions   443

ment. Yet it is quite conceivable that in the nearest future some of these courses will include lessons with the use of virtual simulators based on the methodology employed in compact simulators.

Along with teaching and training the operational personnel, simulators can also be additionally used for some engineering purposes, including, for example, configuring and debugging the new control systems before they are put into service.[20,21] So, for example, a new full-scope simulator was developed by Intermountain Power Service Corporation and installed at its training center to replace the existent full-replica hard-panel simulator developed in the early 1980s to prepare the operational personnel for the newly constructed coal-fired power plant with two units rated at 750 MW each.[22] In the early 2000s, it was decided to upgrade these units with the use of advanced DCS and CRT-based data presentation on the control board. The new simulator was primarily considered to be a tool to aid in engineering studies on new DACS, plant equipment and operation modifications, and training the plant's personnel for operation under new conditions, with the use of new control system is only the second objective. The new simulator was installed in January 2005, whereas the process of changing the control system began in April 2006, and preliminary works at the new simulator, its use as a proactive tool allowed the power plant to minimize the time required for mastering the new system. The new simulator, as distinct from the original one, to a great degree based on the principle of "virtual simulation" with modeling the actual DCS circuits by several PCs. This greatly reduced the simulator's hardware cost, and the new simulator was installed at the cost of approximately $1 million instead of $6 million of the original one.

Power plant simulators can be also widely used for examining the personnel's professional skills and knowledge. For this purpose, for example, Russian electric power utilities conduct special competitions of the operational personnel of different power and cogeneration plants with power units of the same types: from non-reheat boilers and turbines with a common steam header up to 300-MW and 800-MW supercritical-pressure power units.[23,24] Commonly such competitions include the following tests: 1) conduct a hot start-up, 2) operate for two hours under load following complicated variable dispatch demands, and 3) operate in emergency situations created by a failure in main or auxiliary equipment. The actions of the competing teams are evaluated automatically by a special computer program built in the simulator's software, as well as a special commission of power plant experts. Although the competitors commonly displayed a quite high level of understanding the essence of the processes they have to control, more often than not all of them commit some errors, rather typical and common for different teams. It is characteristic that most of these errors fall in the category of "implicit" whose consequences are accumulated with time and manifest themselves in power equipment damages only in some time interval. Such competitions are very useful to determine the main directions for further training and teaching the operational personnel to raise the quality of operation at power plants.

Regardless of the simulator type, the simulated object is typically divided into a number of components and elements. Then, the independent models are created for each of them, and later these models are combined into the full model of the unit as a whole or its modeled part.

The models of individual components are commonly based on the conservation equations for mass and energy (or heat).[25-28] In addition, these models also include numerous empirical and semi-empirical static and dynamic dependences between the modeled parameters. In particular, this concerns equations for flow amounts, heat transfer conditions, and others. In many models, non-linear differential equations are substituted for linear ones without regard to variation of the influencing factors with the modeled variables. However, sometimes such a simplification can cause serious errors. Often, the separated elements, distributed in the co-ordinates, are substituted for concentrated elements, and therefore the partial differential equations for them transform into ordinary differential equations (only with time derivative and without derivatives by a spatial co-ordinates).

Mathematical models applied in different simulators considerably differ not only in numerous assumptions that are forcedly accepted while the model is being constructed and in the manner of modeling, but also in calculation methods to solve the simultaneous differential equations. These methods varies from the simplest methods of Euler and Runge-Kutt to different iterative integration ones.

It is understandable that different simulation goals need different models and pose different requirements to their accuracy. This is one of the most important advantages of compact simulators when the same object can be modeled in absolutely different modes, depending on the set specific task of training. So, for example, it is obvious that for training in heat-rate optimization at the steady-state operating conditions the

Post-operative Analysis of the Turbine's Operating Conditions

processes of heating the turbine elements can be ultimately simplified. By contrast, these processes have to be the focus of attention and require the most trustworthy modeling for training the operator in controlling the transients. For many tasks, modeling the logical connections between the valve and auxiliary state and the fluid parameters at their inlets and outlets are the most significant.

There is a common opinion that, to be effective, the model must have a high degree of trust so that even experienced operators and engineers cannot recognize the difference between the response of the simulator and that of the unit it is based on.[20] In reality, this requirement sounds too indefinite because even experienced operators and engineers have rather qualitative notions of the controlled processes and can sooner judge whether one or another action or variation of an operating parameter influence the monitored values in the correct direction. To judge quantitatively the adequacy and fidelity of the model, results of their functioning should be compared with results of calculations with the use of testing, "standard" programs or with results of field tests at the modeled unit. The permissible difference in dynamic and static characteristics between the modeled and "true" values usually set at the level of 5 to 10 %. At the same time, this limitation can also change depending on the purposes of modeling.

SPECIALIZED PROGRAM-SIMULATOR FOR TEACHING AND TRAINING OPERATORS IN RUNNING THE TURBINE TRANSIENTS

By way of illustration, it seems reasonable to consider a specialized compact simulator intended to teach and train the operational personnel in handling steam-turbine-based power units at their transients. Such a simulator was first conceived in the very beginning of the 1990s and then was developed as applied to 300-MW supercritical-pressure steam turbines operated of power units with one-bypass start-up systems.[29-31] In the development process, the simulator was made maximally adaptable to be applied to other types of turbines. The simulator should give the trainee a sensual comprehension of dynamic variations of the major temperature and thermal-stress state indices of the turbine at general types of the scheduled transients. The operator should comprehend and sense the relationships between the variable controlled operating parameters and the turbine's indices and practice in handling the transients with the maximum fast-acting without violating the limitations for the monitored indices. The trainee should learn to handle the processes in the cases of arbitrary variations in the initial temperature state of the turbine and steamlines, as well as after any forced deviations from the schedules. It is also important to train the operator in interaction with the automated control and informative support systems. Later, a specialized simulator of a similar destination—for optimizing power unit start-ups—was developed by Siemens PG.[32] It was generated on the basis of the MATLAB/SIMULINK program system as applied to a fast processor with the frequency more than 500 MHz. The simulator's layout is shown in Figure 16-3.

As distinct from the simulator of Siemens, the considered one mainly covers the ties between the power unit's operating parameters and indices of the turbine's temperature and thermally stressed states. The simulator was originally thought out on the basis of freely-programmable controllers, but eventually it was developed as special software for standard IBM-compatible PCs with the use of a special graphic editor and control system as a superstructure over the operating system MS DOS. The simulator includes a logical-dynamic model of the turbine

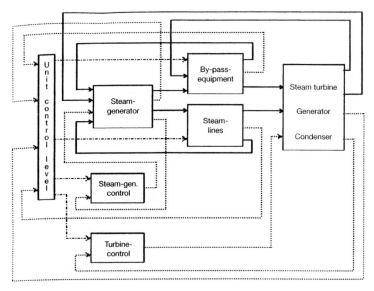

Figure 16-3. Layout of a start-up simulator of Siemens PG. *Source: O. Zaviska and H. Reichel*[32]

Figure 16-4. General block-chart of a logic-dynamic model of a turbine at transients.
I – set initial data; II – analog and discrete control actions of a trainee: 1 – blockade of opening the IP valves, 2 – switching on/off the generator, 3 and 4 – switching on/off steam heating systems of the HP and IP flanges; III – modules of simulating the changes of the controlled operating parameters: 5 – position of the bypass valves, 6 – steam flow capacity from the boiler, 7 – position of the turbine's control valves, 8 – back pressure in the condenser, 9 and 10 – main and reheat steam temperatures; IV – modules of simulating the monitored operating parameters and state indices of the turbine: 11 – steam flow amount through the turbine and throttle steam pressure, 12 – turbine rotation speed, 13 – turbine load, 14 – steam temperature in the stop-valve steam-chests and mean metal temperature of the main steam-lines, 15 – wall metal temperature and effective temperature difference across the wall thickness for the HP stop-valve steam-chests, 16 – mean metal temperature of the HP cross-over pipes,

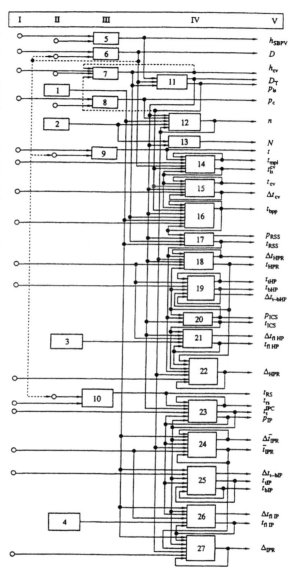

17 – steam pressure and temperature at the control stage, 18 and 24 – average integral metal temperatures and effective radial temperature differences for the HP and IP rotors, respectively, 19 and 25 – metal temperatures and temperature differences for the top and bottom walls of the HP and IP casings, 20 – steam pressure and temperature in the HP intercasing space, 21 and 26 – metal temperatures on the external surface and temperature differences across the width of the HP and IP casing flanges, 22 and 27 – RRE for the HP and IP cylinder, 23 – steam pressure and temperature at the IP cylinder inlet and mean metal temperature of the hot reheat steam line

448 *Steam Turbines for Modern Fossil-Fuel Power Plants*

and human-machine interface. The former is shown in Figure 16-4. It is a complex of individual modules that are connected to one another through a single data base. Each module comprises certain logic rules and simultaneous algebraic and differential equations. The algebraic equations mostly describe the static characteristic of the object: correlation between its operating parameters, metal properties, dependences of the boundary heat transfer conditions on the operating parameters, and so on. Dynamics of the boundary heat conditions and indications characterizing the temperature and thermal-stress states of the turbine elements is modeled with the use of non-linear rational transfer functions with regard to the dependences of these dynamic characteristics from the operating parameters.

The transients are simulated in real time or with an accelerated rate to pass through individual portions of the process for minimal time. The simulator can be tuned to function in one of three following modes: with "passive" monitoring of the turbine state, with informative support of the trainee, and with "automatic" change of the controlled parameters. At any time, the program may be stopped for altering the mode of operating and time scale. For each type of the transients, the initial temperature state of the turbine components (metal temperatures of the high-temperature cylinders, valve steam-chests, steam-lines, and cross-over pipes, temperature differences in these elements, and relative rotor expansions), as well as the initial values of the controlled parameters are given individually or from the list of "standard" initial conditions (as an example, for the most characteristic start-up types).

If the initial conditions are set individually, the program checks their correspondence to each other and to the type of the simulated process. The program can also include inflicting some disturbances in change of the controlled parameters. All the initial data, mode of operating, and variations of the controlled operating parameters and monitored state indices during the process are recorded and printed out for the following analysis and evaluating the control quality. This analysis and evaluation can be done with the use of a special program like the above-mentioned module "ANALYSIS."

Current output information is displayed for the "operator" on a screen in the form of an "instrument panel," mimic diagrams, and trend graphs similar those that are used in reality. If the simulator functions in the mode of informative support, the trainee can see the current admissible boundaries of step-changes for the main operating conditions and

Post-operative Analysis of the Turbine's Operating Conditions 449

textual suggestions about necessary or required technological operations.

The following three types of the transients are mainly simulated: start-ups from arbitrary initial temperature state (beginning from the pre-start warming of the main steam-lines and HP valve steam-chests), load changes inside the governed range, and shut-downs with conservation of the turbine temperature state or cooling the turbine with decreasing the steam parameters. During these operating conditions, the trainee changes the following main operating parameters: steam flow amount from the boiler, main and reheat steam temperatures after the boiler's attemperators, backpressure in the condenser, state (position) of the turbine's and bypass' control valves. Besides these analogous actions, the trainee also carries out some discrete operating actions, namely: switches on (at start-ups) and off (at shut-downs) the generator, blocks up or de-blocks opening the IP control valves (at start-ups), and switches on and off the steam systems of heating the flanges of the HP and IP cylinders. In addition, the operator can set the "automatic" or "manual" mode of controlling the steam flow rate, main and reheat steam temperature, turbine and bypass control valves. Symbols of the corresponding control bodies and their position are displayed at the screen.

The simulator's logic-dynamic model (see Figure 16-4) embraces main and hot-reheat steam-lines and HP crossover pipes with their influence on the outlet steam temperature because of heat transfer from steam to metal and heat accumulation in their metal. As applied to the HP cylinder with its loop-flow design scheme (see Figures 7-4a and 12-15), the model takes into consideration the steam temperature variations from the state in the stop-valve steam-chests to the control-stage chamber and intercasing space depending on the current steam flow amount through the turbine. The heat transfer conditions from steam to the rotor and outer casing also vary with steam flow amount and rotation speed of the turbine. All these factors influence the variations of the monitored effective radial temperature difference in the rotor, temperature differences between the top and bottom and across the flange width of the outer casing, and relative rotor expansion. The RRE and temperature difference in the flange also depends on the steam heating of the flange. Similar models are applied to the IP cylinder. For the HP stop-valve steam-chests, the monitored values are the metal temperature and temperature difference across the wall thickness. Dynamic connections are also included, for example, in the module simulating variations of the turbine rotation speed at running-up.

Most of other dependences between the operating conditions are regarded as static in distinction from other simulator models. This seems possible because the time constants of the considered processes of heating the turbine elements are much greater than those for the processes of changing the load and steam pressures along the steam path. However, for other compact simulators, intended for imitating the fast-acting processes like, for example, load discharges, such an assumption would be mistaken.

The logic-dynamic model comes to the system of differential and algebraic equations which are solved by numeric integration of finite differences with the discrete (invariable) time step (time marching). For the considered model, with its time constants counted in minutes, this time step can be set sufficiently great about 10-30 seconds. It seems reasonable to use the same "universal" models for different elements. So, such general models were developed for heating the turbine elements, change of the outlet steam temperature and mean metal temperature of both the high-temperature steam-lines and cross-over pipes, and RRE of the cylinders.

Of importance is that not only can such simulators be used for training the operational personnel and post-operating analysis of actual operating conditions of the past, but also be helpful for teaching the students thinking of the carrier in power plant operation.

References

1. Leyzerovich A.Sh., B.R. Beyzerman, N.F. Komarov, et al. "First Experience of Applying the PC-Based Local Subsystem of Diagnostic Monitoring for Steam Turbine" [in Russian], *Elektricheskie Stantsii*, no. 4 (1993): 18-22.
2. Leyzerovich A. "PC-Based Operative Information Support at Transients and Their Postoperative Analysis in Reference to Steam Turbine," *Proc. of the American Power Conference* 56, Chicago, 1994, Part 1: 36-40.
3. Leizerovich A.Sh. "Information Support for the Operating Personnel of a Thermal Power Station during Transients and Their Post-Operative Analysis," *Thermal Engineering* 43, no. 10 (1996): 817-823.
4. Leizerovich A.Sh., L.P. Safonov, A.V. Antonovich, et al. "Developing and Mastering the Automated Systems for Diagnostic Monitoring of Power-Generating Units at Thermal Power Stations," *Thermal Engineering* 42, no. 2 (1995): 154-160.
5. Leyzerovich A.Sh., L.P. Safonov, A.M. Zhuravel, et al. "Automated Diagnostic Systems and Subsystems for Fossil Power Plant Units: the Experience of Designing and Operational Mastering in the Former USSR," *Proc. of the American Power Conference* 55, Chicago, 1993, Part 1: 881-885.
6. Leyzerovich A., V. Berlyand, A. Pozhidaev, and S. Yatskevich. "Continuous Monitoring as a Tool for More Accurate Assessment if Remaining Lifetime for Rotors and Casings of Steam Turbines in Service," *Proc. of the American Power Conference* 60. Chicago: 1998, Part 2, 1021-1026.

Post-operative Analysis of the Turbine's Operating Conditions 451

7. Berlyand V., A. Pozhidaev, E. Plotkin, et al. "Some Problems of Steam Turbine Lifetime Assessment and Extension," *Proc. of the American Power Conference* 61. Chicago, 1999: 795-800.

8. Sill U. and W. Zörner. *Steam Turbine Generators Process Control and Diagnostics,* Erlangen: Publicis MCD Verlag, 1996.

9. Leyzerovich A. *Large Power Steam Turbines: Design & Operation,* 2 vols., Tulsa (OK): PennWell Books, 1997.

10. Leyzerovich A.Sh., A.I. Strel'tsov, and N.A. Rusanova. "Post-Operating Analysis of the Operation Quality for Power Plant Unit Transients on the Criteria of Thermal-Stress State of the Steam Turbine's HP and IP Rotors" [in Russian], *Energetik,* no. 7 (1991): 13-15.

11. Makansi J. "Powerplant training. Ensure your team's survival in the trenches," *Power* 139, January 1995: 23-27.

12. Wood W.A. "The Brave New World in Power Plant Operations," presented at the 2000 International Joint Power Generation Conference, Miami Beach, 2000.

13. Coffee M.B. "Computerized simulators offer quality training at modest cost," *Power* 131, January 1987: 55-58.

14. Garrity R. "Web-Trained Plant Workers," *Power Engineering* 109, no. 2 (2005): 48-52.

15. Kaji A., T. Aramaki, and K. Tsushima. "New Operator Training Simulators," *Hitachi Review* 41, no. 3 (1992):153-160.

16. Blankship S. "Simulator Enhances Startup/Operator Training for PacificCorp," *Power Engineering* 109, no. 7 (2005): 7-16.

17. Herzog von R., D. LeFebve, H. Olia, et al. "Compact Simulators—the Efficient Way towards Deeper Understanding of Processes" [in German], *VGB Kraftwerkstechik* 74, no. 6 (1994): 527-531.

18. Fray R. and S.M. Divakurini. "Compact Simulators Can Improve Fossil Plant Operation," *Power Engineering* 99, no. 1 (1995): 30-32.

19. Boschee P. "Operator training goes the distance with Internet program," *Electric Light & Power,* no. 3 (2001): 1, 6, 8.

20. Elliot T.C. "Use of powerplant simulators goes beyond training," *Power* 138, March 1994: 32-44.

21. "Simulators do more than train," *Power* 144, no. 1 (2000): 4-5.

22. Burr J. "Out with the Old, In with the New," *Power Engineering* 110, no. 5 (2006): 48-54.

23. "Simulators win in Russian competition," *Electric Power International,* Winter 1998: 8.

24. Remezov A.N., B.V. Lomakin, V.K. Krajnov, et al. "Competition of Mosenergo's Boiler-Turbine Operators" [in Russian], *Elektricheskie Stantsii,* no. 5 (2001): 32-33.

25. Rubashkin A.S. and M. Khesin. "Dynamic Models for Fossil Power Plant Training Simulators," *Proc. of the American Power Conference* 55, Chicago, 1993, Part 1: 871-876.

26. Rubashkin A.S. "Computer-Based Training Simulators for Operators at Thermal Power Stations," *Thermal Engineering* 42, no. 10 (1995): 824-831.

27. Rubashkin A.S., V.L. Verbitskii, and V.A. Rubashkin. "Methods for Modeling the Technological Processes in Power Equipment," *Thermal Engineering* 50, no. 8 (2003): 659-663.

28. Rubashkin A.S. and V.A. Rubashkin. "Developing Technology for the Simulation of Dynamic Processes at Thermal Power Stations," *Thermal Engineering* 51, no. 10 (2004): 812-816.

29. Leyzerovich A.Sh "Development of Specialized Simulators as Applied to

Controlling Power Plant Unit Steam Turbine at Transients" [in Russian], *Elektricheskie Stantsii*, no. 2 (1991): 41-44.

30. Leizerovich A.Sh. and Y.A. Elizarov. "A Specialized Program for Simulators Intended for Training Operators to Control Steam-Turbine Unit Transients," *Thermal Engineering* 42, no. 9 (1995): 754-759.

31. Leyzerovich A. "Program-Simulator for Teaching and Training Powerplant's Operational Personnel in Handling Unit Steam Turbines at Their Transients," *Proc. of the American Power Conference* 59, Chicago, 1997, Part 2: 1130-1135.

32. Zaviska O. and H. Reichel. "Siemens Optimizes Start-up Sequences in a Power Plant Unit—New Unit Start-up Simulator Reduces Costs," *VGB PowerTech* 81, no. 7 (2001): 36-38.

33. Schiegel G. and K.A. Theis. "Highly Qualified Personnel—A Key Factor for Successful Power Plant Operation," *VGB PowerTech* 83, no. 3 (2003): 47-49.

Part IV

Lifetime Extension for Aging Steam Turbines and Their Refurbishment

Chapter 17

Assessment and Extension of Steam Turbine Lifetime

POSSIBLE SOLUTIONS FOR A PROBLEM WITH
MASS STEAM TURBINE AGING

Since the late-1980s, increasingly more new power steam turbines throughout the world have been designed keeping in mind a possibility, or possible necessity, of their operation for 200,000 hours, that is, about 25-40 years. The same lifetime duration of 200,0000 hours is laid into projects of many new power plants upon the whole, even though the lifetime of individual components (for example, tubes of the boiler's high-temperature heated surfaces) can be set at a lower level of about 100,000 hours. Along with this, as to contemporary fossil-fuel power plants in service, their main power equipment had mostly been designed for the service time, or lifetime, equal to only 100,000 hours, that is, about 15-20 years. It means that, with the rated stresses in major power equipment design components, the strength properties of metals, having been set according to traditional power industry standards, insure reliable and safe operation of these components for at least 100,000 hours. Therewith, the actual strength properties of metals, as a rule, allow much longer operation. So, according to some statistic studies of the long-term strength of castings for steam turbine casings, 99% of these casings designed following the industry's strength standards could survive 200,000 hours of operation, and 90% could exceed even 700,000 hours.[1] In addition, all the strength calculations according to the standard industry methodologies are conducted with certain conservative assumptions and safety margins, and the actual stresses and metal damages caused by these stresses are supposed to be less than the calculated ones. On the other hand, deviations of the actual operating conditions from the rated ones can cause substantially greater actual stresses and damages and make some equipment components failure well before their designed lifetime expiration.

455

Statistics shows that, as the power plant reaches its rated lifetime, the number of its forced outages begins to grow substantially (Figure 17-1), and its reliability and availability fall.[2] Simultaneously, in the operation process the turbine's efficiency lowers more and more, even though it is partially restored at overhauls—Figure 17-2. Some factors of this aging deterioration as applied to steam turbines and its auxiliaries are conventionally presented in Figure 17-3. Nevertheless, nobody thinks the turbine or other power equipment elements should be immediately taken out of service as soon as their lifetime expires.

At all times, individual steam turbines and other power equipment elements have been left in operation well beyond their lifetime limits. At many power plants of diverse countries, it is possible to find steam turbines successfully operated for 40 and 50 years, and even more. However, in the recent years the problem of power equipment lifetime extension has substantially gained in its scope and acquired much more serious significance. So, for example, according to a report of the EPRI of 2003, in two years (that is, in 2005) over 50% of U.S. coal-fired power generating capacities would be 30 and more years old, and it was supposed that between 2001 and 2010 about 45,000 MW of oil- and gas-fired power generating capacities and 15,000 MW of coal-fired capacities should be retired.[5]

In the last decade, many industrially developed and developing countries, even in different scales, have come across a similar situation

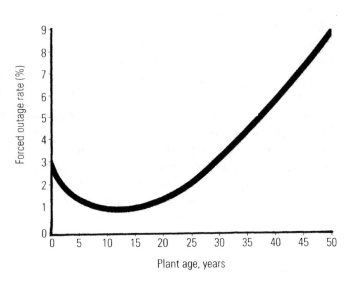

Figure 17-1. A typical change in a power plant's forced outage rate over operation time.

Assessment and Extension of Steam Turbine Lifetime

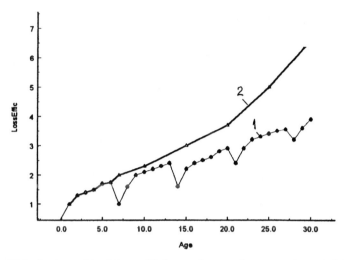

Figure 17-2. A rise of turbine efficiency losses for an aging turbine because of its deterioration with (1) and without (2) regard to its partial restoration, according Ansaldo Energia. *Source: L. Lazzeri*[3]

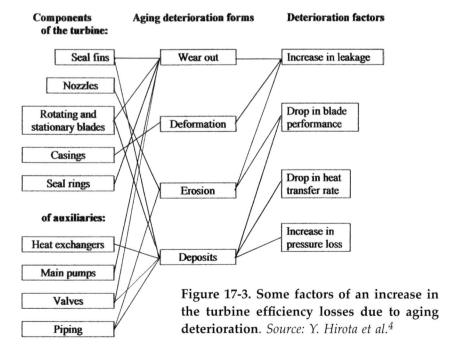

Figure 17-3. Some factors of an increase in the turbine efficiency losses due to aging deterioration. *Source: Y. Hirota et al.*[4]

because a significant portion of steam-turbine units in these countries was put into operation in the 1960s, 1970s, and 1980s. In many countries, just these power units made up the power industry backbone. With time, they have been materially exceeding their designed lifetime, but there exist serious technical and financial difficulties replacing them on a mass scale in the nearest years.

Therewith, noteworthy is that the aging steam-turbine power units in service are, as a rule, substantially inferior to newly designed ones in their efficiency and environment friendliness. Under these circumferences, with more rigid environmental requirements, skyrocketing rise of prices for oil and natural gas, serious legislative reforms in many countries, and more heated competence at the power generation market, some aging power units should be undoubtedly taken out of service and replaced by new, more up-to-date and efficient capacities. In many cases, it can be done with the use of the infrastructure, buildings, foundations, and even some auxiliaries of the existent power plants. Along with this, there can be quite convincing both technical and economical reasons to leave in service a significant part of the existent aged power plants with essential refurbishment of their equipment by means of replacing their most important and most vulnerable equipment elements with new ones of the better efficiency and new lifetime. (Such a possibility in its technical aspects is to be considered below in the next chapter.) And the third, most palliative solution, with the least capital expenditures, most acceptable on numerous occasions, although not the most cost-effective, is merely to prolong temporarily the lifetime of the existent equipment elements to the new figures on the basis of inspections of the current state of these elements, non-destructive tests of their metal, more scrupulous assessments of their residual lifetime, and with the wide use of diagnostic means to reveal in time possible dangerous deteriorations of the equipment state and prevent imminent failures.

In doing so, noteworthy is that many of aging steam-turbine units, including the largest ones with elevated steam parameters, have been operating in load-following and cycling manners, that is, with a great number of transients, and thermal fatigue has greatly contributed to their wear out. Of significance is also that under present power plant O&M conditions the periods between overhauls and regular current repairs have become substantially longer and longer, resulting in a more significant role of diagnosing and heightened demands to the accuracy and certainty of technical diagnostics for power equipment. All these factors are to be taken into consideration in the case of making decision to leave power units in operation with the extended lifetime.[6-17] For example, German experts suppose that normally, that is, with due operation quality, thorough inspections, and timely maintenance and repair measures, steam turbines of the vintages of the 1960s-1980s can be kept in operation up to forty years, which is three times as much the originally assessed lifetime.[11,12]

The last solution of the problem with mass steam turbine aging has gained sufficiently wide acceptance in different countries. Noteworthy is that, even if the turbine is upgraded and retrofitted by replacing, for example, its blading and rotors, some important turbine elements are left without replacement. More often than not, this concern the turbine casings (especially, outer ones—see examples below), and it is important to be sure that they will not cause turbine failures and forced outages during the new lifetime span provided by the replacing elements. So the importance of assessing the actual residual lifetime of the major turbine design elements and extending their allowable service time beyond the initially set limits without losses in reliability cannot be underestimated. The developed techniques of residual lifetime assessments are also necessary for setting up the inspection intervals and optimizing the entire O&M schedules for aging steam turbines.

STEAM TURBINE LIFETIME LIMITATIONS

The main factor determining the lifetime of steam turbines at modern fossil-fuel power plants is the strength and reliability of their high-temperature design components. They can fail due to excessive deformations or bursts in result of long-term influence of pressure or centrifugal force, or cracking under pure creep or creep-fatigue conditions. Failure by cracking may further be viewed as consisting of three stages: crack initiation, crack propagation, and final failure of the component once the crack reaches its critical size. Life assessment technique in this case is aimed at quantifying characteristics of these processes: the incipient period prior to crack initiation, the rate at which the cracks grow, and the critical crack size that will lead to the final failure. Eventual failure of the component can occur either by leakage (for casings, steam-lines, or other stator elements) or rupture at the operating temperature or by rapid brittle fracture at lower temperatures during start-up/shut-down transients. In the latter case, the critical size of the crack is defined by the material's fracture toughness, which, in addition, can lower in the service-exposed conditions because of embrittlement phenomena.

Figures 17-4 and 17-5 present softening trend curves of Cr-Mo-V forged steel mainly employed for high-temperature rotors of steam turbines with the main and reheat steam temperatures not exceeding 566°C (1,050°F) and 2-1/4Cr-Mo cast steel used for some high-temperature inner casings. These curves for the turbine materials are obtained by laboratory, bench tests. In Figures 17-4a and 17-5, the metal hardness ratio (related to that before service) is plotted against a temperature-time parameter G, taking additionally into account the influence of stress. This parameter is derivative from the Larson-Miller parameter:

$$P_{LM} = T \times (C + \log t), \tag{17.1}$$

where T is the absolute temperature in °K, and t is the time in hours. Constant C depends on the material but more often than not is taken invariable and equal to 20. With time and temperature the steel hardness lowers, representing changes in the metal's creep rupture strength. Figure 17-4b shows the softening of the rotor steel resulting from high-temperature low-cycle fatigue, as if caused by cyclic exposure of unsteady-state

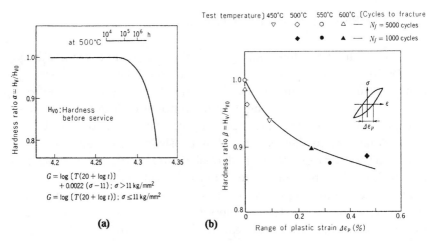

Figure 17-4. Softening curves of Cr-Mo-V forged rotor steel by time-temperature influence (a) and high-temperature low-cycle fatigue (b). Source: Y. Hirota et al.[4]

Figure 17-5. Softening curves of 2-1/4Cr-Mo cast casing steel. *Source: Y. Hirota et al.[4]*

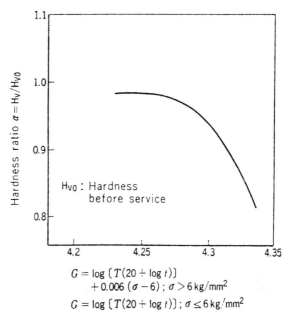

$G = \log [T(20 + \log t)] + 0.006 (\sigma - 6)$; $\sigma > 6\,\text{kg/mm}^2$

$G = \log [T(20 + \log t)]$; $\sigma \leq 6\,\text{kg/mm}^2$

thermal stresses.

Of significance is that different types of metal damages for different equipment components potentially pose treats of various danger extents. So, cracks in steam-lines and valve steam-chests can entail such severe emergencies as their rupture with possible casualties. Along with this, cracks in turbine casings, even through, do not make the casings rupture; in the worst cases, appearance of a plume because of steam leaking through the crack leads to the forced shutdown and unscheduled outage of the turbine for repairing the casing by welding up the crack. Similarly, for low-cycle (thermal) fatigue cracks on the surface of high-temperature rotors, the rate of their growth is not so great to make the rotor suddenly rupture. However, hidden cracks in the rotor depth, descended from unrevealed initial flaws in the metal, can be a more serious hazard if they remain unnoticed in service.

Not only high-temperature rotors are prone to this threat but the LP rotors, too, especially if the total tensile stresses in their depth, caused by both centrifugal and thermal stresses, reach their maximum values at relatively low metal temperatures, lower than the fracture appearance transition temperature (FATT). The events of such brittle fractures of turbine rotors are fortunately few in numbers. (One of the most well-known failures of this type happening in 1987 at the German power plant Irshing with the 330-MW four-cylinder turbine of Siemens/KWU[18,19] was noted in Chapter 12.) Brittle fracture of the turbine rotors can also come from accumulation in the operation process of fatigue damages caused by transient torsional oscillations of the turbine-generator shaft due to interaction between the generator and grid. (An example of such destruction of the 300-MW supercritical-pressure turbine was considered above—see Figure 12-15.[20])

Major critical zones for high-temperature rotors (as applied to impulse-type turbines with forged rotors with a central bore) from the standpoint of their long-term strength are presented in Figure 17-6. They are: the surface of the central axial bore under wheel discs, fillets of wheel discs and gland seals, thermal grooves of gland seals, disc rims, and the edge of pressure balance (equalizing) holes in the wheel discs.[21] As a rule, these critical zones mainly refer to the first HP and IP stages swept by steam of the highest temperatures. Along with this, such critical zones in long-term strength terms can also appear at the last HP stages of power units with two-bypass start-up systems, if these stages are overheated because of windage. For rotors with reaction-type blading, main stress concentrators on the rotor surface are blade root grooves and blade groove shoulders. For welded or solid rotors (without a central bore), a critical zone appears near the rotation axis, although the maximum tensile stresses in this case occur significantly less than those

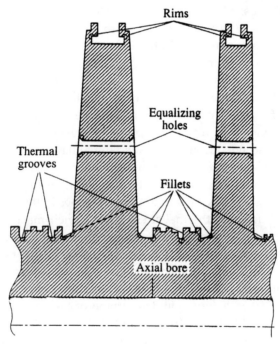

Figure 17-6. Critical (in terms of long-term strength) zones for forged high-temperature rotors (as applied to impulse-type turbines).

Assessment and Extension of Steam Turbine Lifetime

on the axial bore surface, all other things being the same.

For high-temperature turbine casings (casings of the HP and IP cylinders and valve steam-chests), cracks initiation is frequently found in the course of their early periodic inspections. So, Figure 17-7 shows

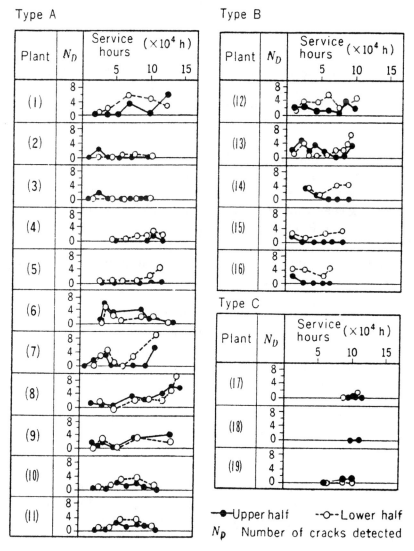

Figure 17-7. Number of cracks in the inner HP casings of MHI's steam turbines found at their overhauls in the process of long-term service. Source: Y. Hirota et al.[4]

number of cracks in the inner HP casings of MHI's steam turbines revealed in the process of their overhauls in long-term service. Cracks appearing at early operation stages are judged to be due to the remaining casting flaws (cavities, pores, non-metallic inclusions, etc.) and neither to the damage accumulation nor to material deterioration. These cracks are usually removed by grinding, and the largest ones are welded up. With years, there can appear new cracks just caused by steady stresses because of the inner steam pressure and unsteady-state thermal stresses arising at the turbine transients. Zones of revealing these cracks commonly coincide with zones of the highest temperatures and stresses. On the internal surfaces, the appearance of these cracks is most probable in the vicinity of stress concentrators, such as, for examples, slots for aligning the casing rings, diaphragms, or stationary blading.

On the external surfaces, the crack appearance is contributed by sharp changes of the casing shape and thickness and such stress concentrators as the female threads for the stud bolts. It is considered especially typical for valve steam-chests influenced by higher steam temperatures than the HP and IP cylinder casings. Therefore, their metal can be found more deteriorated in tensile strength and hardness. The female threads become susceptible to shear fracture as a result of oxidation, thinning, softening by long-term influence of high temperatures, and accumulated creep damage.

The service life of the LP rotors, along with the hazard of brittle fracture, depends on corrosion fatigue and stress-corrosion cracking (SCC) of their metal. Susceptible locations of SCC as applied to the combined LP rotors with shrunk-on discs are shown in Figure 12-18.[22] They include the keyway and shrunk-on surfaces, wheel-disc face, the entry blading slots, pressure balance holes, and the disc rim blade attachments (steeples). Corrosion-assisted fatigue was responsible for cracking and failures of many rotors and wheel discs. As said before, the SCC appears and propagates as a result of combined actions of three factors: sufficiently high tensile stress, susceptible steel, and a corrosive environment. Experience shows that SCC incidence is much higher for power units with once-trough boilers, including those of supercritical pressure, compared with power units equipped by drum-type boilers. So, recently the EPRI conducted a survey of 757 fossil-fuel and 109 nuclear U.S. power units in service to determine the extent of SCC for their LP rotors and wheel discs.[23] Rim attachment cracking was found in 41 nuclear units (38%), 29 of 110 (26%) supercritical-pressure, and 20 of 647 (3%) sub-

Assessment and Extension of Steam Turbine Lifetime

critical-pressure fossil-fuel units. Under these conditions, there appears a necessity of a life assessment tool not only for high-temperature turbine elements, but also for the LP rotors in the context of their SCC, especially in the rim attachment zone.

ASSESSMENT OF THE TURBINE'S RESIDUAL LIFETIME

As the initially assessed turbine lifetime nears to the end, a possibility of its further extension is to be determined individually based on inspecting the state of the turbine's major design elements, estimating the actual wear of their metal, and assessing their residual lifetime by means of non-destructive examination (NDE) and new calculations, taking into consideration the actual history of the turbine's operating conditions, that is, actual duration of operating at the diverse levels of steam temperatures and load, as well as the number and characteristics of start-ups, load discharges, and other types of the transients. This information is supposed to be found in the power plant's archives in the form of stored computer data and recorder tapes. The more detailed these data are, the more definite and trustworthy the assessments could be. If these records occur to be insufficiently detailed or partially lost, the absent data have to be conjectured, and the resultant calculation certainty unavoidably lowers.

The residual lifetime assessment is supported by NDE, including tests on the metal electric resistivity, hardness, and metallographic replica. The electric resistivity and hardness methods (see, in particular, Figures 17-4 and 17-5) are considered most effective for detecting the creep and fatigue damages in the turbine components. The metallographic replica method shows the changes in the metal microstructure and, according to Hitachi, can be effective for detecting the metal fatigue damage. The NDE scope recommended by Hitachi for forged HP rotors is presented in Figure 17-8.

Among other NDE methods for rotors, noteworthy are ultrasound flaw detection, magnet-powder, and eddy-current methods. In particular, all of them are used for examining the central axial bore surface. Special automated devices for ultrasound inspection (USI) of the axial bore were developed and have gained wide acceptance, particularly, in the United States, Russia, Japan, and some European countries. Special attention is also paid to measuring the bore diameter changes in high-temperature

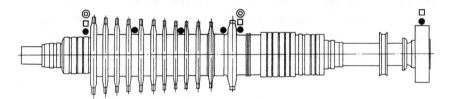

● : Hardness measurement
□ : Electric resistivity measurement
◎ : Replica measurement

Figure 17-8. The scope of non-destructive examination (NDE) for HP rotors, according to Hitachi. *Source: S. Mori et al.*[6]

zones due to creep deformation. If flaws, cracks, or a remarkable diameter increase are revealed, the bore is drilled out, that is, overbored, with removal of the surface metal layer damaged by creep. Some of USI devices allow examinations at an angle to the outer rotor surfaces; so they can be applied to USI of the solid rotors (without a central bore), and there have appeared USI devices for examining the fit surfaces of the shrunk-on wheel discs without their disassembling.

As distinct from the central bore surface, as well as wheel disc rims and edges of the pressure-balance holes, the body surfaces of high-temperature rotors (see Figure 17-8) mainly suffer from thermal fatigue, especially in the vicinity of such stress concentrators as the thermal grooves and fillets of the wheel discs and seals. For rotors with reaction-type blading, main stress concentrators on the rotor surface are blade root grooves. Appearance and propagation of cracks in these areas can be effectively prevented for long time by removing the surface metal layers in these zones with accumulated thermal fatigue damages. This skin peeling operation is also commonly used to reshape the stress concentrators with increasing their radii of curvature and reducing in this way the stress concentration factors—Figure 17-9.

Similar repair and maintenance remedies (skin peeling, overboring, and others) are applied to other turbine components—Table 17-1. All these components are subjected to regular, periodical inspections with NDT of metal. The terms when the remedy actions become necessary are assessed by calculation and based on results of inspections.

In particular, this refers to cast casings and valve steam-chests. Assessment of their lifetime is based on and, simultaneously, substantially

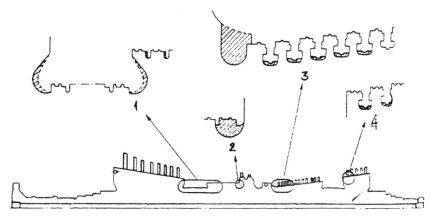

Figure 17-9. Skin peeling of an HP-IP rotor with reaction-type blading, according to Mitsubishi Heavy Industries. *Source: V. Hirota et al.*[17]

complicated by the fact that crack-like flaws may already exist in the original castings. That is why their lifetime is determined by the time needed for possible flaws to grow up to the maximum allowable size. This process of crack propagation and growth can be interrupted and stopped by grinding out the potentially damaged metal near the cracks and welding them up. If there appears creep damage of female threads for stud bolts, the thread holes are oversized, with the use of new stud bolts. The service life for turbine casings can be considered exhausted if the pattern and sizes of cracks and crack-like flaws are such that the defective casing is regarded as unsuitable for repairing, the steel properties degraded during long-time operation to such extent that the residual lifetime becomes less than the interval between repairs, and/or the casing overly distorts, preventing its proper assemblage. By applying NDE and well worked up weld-up repair techniques, power plants can save a lot of money and time by restoring the cracked turbine casings or valve steam-chests instead of their complete replacement.[24-27]

It is quite understandable that the rate of accumulating thermal fatigue damages for both the turbine rotor and stator components substantially depends on the number of transients and, even in a greater extent, on the quality of running these transients. This influence as applied to the HP and IP rotors of 300-MW supercritical-pressure steam turbines at their most characteristic cycles of transients was shown in Table 12-1. One "bad" start-up can contribute to thermal fatigue of metal more than tens of "good" start-ups. As this takes place, the con-

468 *Steam Turbines for Modern Fossil-Fuel Power Plants*

Table 17-1. Maintenance methods for increasing lifetime of individual turbine components, according to Mitsubishi Heavy Industries

Turbine Component	Position	Cause of Life Expiration	Remedy Action	Definition of Life Limit
HP and IP (HP-IP) rotors	Outer grooves and fillets	Fatigue	Skin peeling	Skin peeling is no longer practicable
	Center bore	Fatigue, creep crack propagation, and brittle fracture	Overboring	Overboring is no no longer practicable
	Blade groove shoulders	Creep	Detailed inspections & investigations	Crack initiation is confirmed
LP rotors	Center bore/ rotation axis	Fatigue, crack propagation, and brittle fracture	Detailed inspections & investigations	Crack propagation is confirmed
HP inner casing	Internal surface	Fatigue and creep	Weld-up repairing	Repair is considered no longer realistic
	Female thread	Creep	Oversizing	Oversizing is no longer practicable
Nozzle box	Root of vane	Fatigue and creep	Detailed inspections & investigations	Cracks are no longer repairable and/or deformation is significant
HP-IP blades	Tenons and roots	Creep	Detailed inspection & investigations	Cracks have initiated and/or any abnormality appears
Main valve steam-chests	Body	Fatigue and	Weld-up repairing	Weld-up repairing is no longer practicable and/or material has significant deterioration
	Female thread	Creep	Oversizing	Oversizing is no longer practicable
High-temperature bolts/stud-bolts	Thread	Fatigue and creep	Replacement	Crack initiation is anticipated

Source: Y. Hirota et al.[4]

Assessment and Extension of Steam Turbine Lifetime

tribution of one "bad" start-up can be accounted in tenths of percent, that is, fatigue cracks can appear after few hundreds of such start-ups, even without contribution of any other transients. At the same time, load variations within the governed range and beyond it, as well as steam temperature excursions during both the stationary and transient operating conditions, also substantially contributes to thermal fatigue damages of the turbine components. Even though the resulted thermal stress variation amplitudes are not so big, the numbers of such variations is counted in thousands and tens thousand. The contribution of these operating conditions also depends on the quality of running them. Of significance is also the quality of automatic governing systems controlling the steam temperatures entering the turbine. Temporary steam temperature excursions over the rated values could also substantially increase the metal damage because of creep.

Lifetime assessments for high-temperature turbine components are mainly based on a hypothesis of linear superposition (adding) of specific damages caused by fatigue and creep. This process, as it is laid into assessment of service lifetime expenditures for the Turbine Stress Controller (TSC) of Siemens, is illustrated by Figure 15-5. In principle, the metal resource is theoretically considered expired if the total of specific (relative) damages reaches 1.0. At the same time, because the taken hypothesis is not quite accurate, this limit is commonly set less than 1.0. On the other hand, in most cases, expiration of the metal resource does not mean any rupture—it is rather a sign of crack appearance, and the metal have a lot of additional time for subsequent growth and propagation of this crack. Therewith, all the employed calculation methodologies include numerous assumptions and their results should be considered rather conditional and having probabilistic nature.

With numerous diverse methodologies and computer software for residual lifetime assessments for individual design components of steam turbines and other power plant equipment, perhaps the most general and authoritative document guiding the order of such calculations is a special Code of ASME.[28] Some special standards and recommendations are also developed and issued in different countries, as, for example, some Russian methodical guidelines and industrial instructions.[29,30]

All these documents mainly refer to high-temperature steam turbine components subjected to creep and thermal fatigue. Along with this, they can be applied to the LP rotors subjected to fatigue, crack

470   *Steam Turbines for Modern Fossil-Fuel Power Plants*

propagation, and brittle fracture. As to the above mentioned problem of turbine parts subjected to SCC (or, to be more specific, the rims of the LP wheel discs), a special PC-based software called LPRimLife was developed to the EPRI's order by Structural Integrity Associates as a user-friendly tool for rapid assessment of the remaining lifetime for these rim attachments with axial entry.[23] This software integrates stress analysis and fracture mechanics algorithms with material degradation data. Either deterministic or probabilistic assessments can be conducted. Deterministic assessments based on the worst-case assumptions, whereas probabilistic calculations proceed from less pessimistic approaches. The same can be said about all the other lifetime assessment methodologies and techniques.

LIFETIME EXTENSION STRATEGY
FOR SETS OF THE SAME TYPE POWER EQUIPMENT

Noteworthy is the strategy of extending power equipment lifetime developed in the former Soviet Union and presently widely implemented in Russia.[7,12-14,16,31-33] Of significance is that by 2005 almost all power equipment of Russian fossil-fuel power plants has reached or exceeded its initial design lifetime limit of 100,000 hours, and their significant portion with the total capacity of 65,000 MW will have lived out their newly assessed so called fleet service life, which on the average is twice as much as the original design lifetime. In particular, this concerns 76 subcritical-pressure steam turbines with the single capacity of 200-210 MW, 52 supercritical-pressure 300-MW turbines, and 14 supercritical-pressure 800-MW turbines. The scheme of measures for arranging such a work is shown in Figure 17-10, and Figure 17-11 demonstrate the scope of the problem with the aging power equipment to be solved in the nearest years.

The developed strategy considers three stages of lifetime extension beyond the initially assessed value:

- until the fleet resource (or service life),
- until the individual resource, and
- beyond the individual resource.

By the fleet resource, it is meant the operating time of equipment

Assessment and Extension of Steam Turbine Lifetime

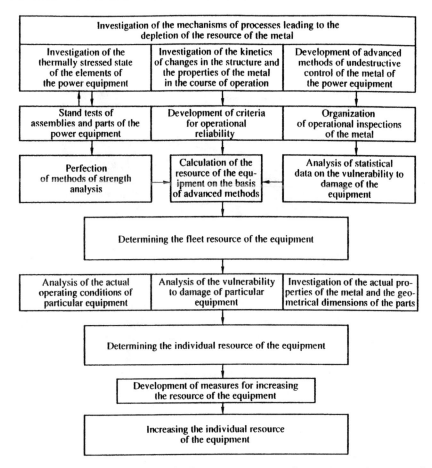

Figure 17-10. Arrangement of lifetime extension for main equipment of fossil-fuel power plants. *Source: A.G. Tumanovskii and V.F. Rezinskikh*[33]

units which are similar in their design, materials, and operating conditions. So, the fleet resource is set for each great group, or set, of standard equipment operated in a similar manner. Their failure-free functioning within this period is ensured provided that the standard O&M requirements are met. Permission to use any equipment beyond its fleet resource requires individual examinations of its metal and calculations as applied to its individual history and mode of operation. In certain cases, even the individual resource is not the end. The use of special methods of maintenance and repairs allows prolonging the physical life of the metal for critical design components even longer. Among such

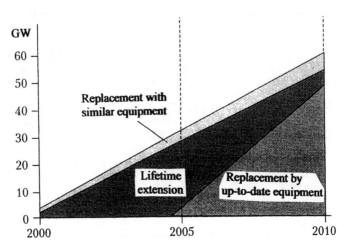

Figure 17-11. The scope of the problem with aging power equipment in Russia and ways of its solution. *Source: A.N. Remezov*[32]

methods that gain widespread acceptance are, in particular, removal of surface metal layers damaged by creep, fatigue, erosion-corrosion processes, etc., restoration thermal treatment, restoration welding, and so on.

As applied to standard main power equipment installed at fossil-fuel power plants of Russia, it was subdivided into three main groups. The first group "A" consists of supercritical-pressure steam-turbine units with the single capacity of 500, 800, and 1,200 MW. The second group "B" includes steam-turbine units with the single capacity of up to 300 MW, main steam pressure over 12.7 MPa and up to 23.5 MPa(1,840-3,400 psi), and main and reheat steam temperatures lying in the range of 540-560°C (1,000-1,040°F). And, finally, the third big group "C" comprises steam turbines with the output of up to 110 MW, main steam pressure of 9 MPa (1,305 psi) and less, and main steam temperature of 510°C (950°F) and less.

As to equipment of the first group "A", despite its less age compared with that of the second group, it grows old, wears out more rapidly due to higher stresses. It was concluded that after about 150,000 operating hours plus-minus about 30,000 hours it will be necessary to replace significant design components of these power units (in particular, the main and hot-reheat steam-lines, HP steam-admission valve units, and high-temperature HP and IP rotors). In particular, some recent researches showed that for 800-MW turbines after 100,000 hours of

Assessment and Extension of Steam Turbine Lifetime

operation there exists a failure probability for the first stage discs' rims of the IP rotors equal to approximately 30%.[34] Most likely, the service life expiration for individual components will not happen simultaneously. However, these terms will not differ more than a few years. Replacement of any one of the above-mentioned components requires great expenditures of time and labor. That is why these steam-turbine units are recommended to be generally reconstructed upon the whole. The reconstruction should not come to a mechanical replacement of the worn out components—they are to be refurbished to heighten the units' technical and economic performances. For the gas- and oil-fired power units, a possibility of topping them by gas turbines is considered with an increase of their capacity, for example, as applied to 800-MW units, up to 1,050-1,100 MW and their efficiency to 49-50%. The same approach is also applied to the gas- and oil-fired power units of the second group.

For the group "B", the fleet resource of these steam-turbine units was initially set at the level of 170-250 thousand operating hours for different turbine types. By means of replacement of the most vulnerable design components, it is supposed to prolong these terms individually up to 300,000 hours, on average. Beyond this term, complete reconstruction and refurbishment of these units with replacement of their main design elements will become necessary.

The turbine units of the third group "C" have become obsolete long time ago. It is not advisable to develop and implement special technical decisions aimed at substantial extension of their lifetime. They can be kept in operation without substantial additional expenditures up to 400,000 hours (or 270,000 hours with 900 start-ups). After these terms, they should be put out of commercial operation and replaced with newly designed equipment.

Of special interest is an experience with long-term operation of six 800-MW supercritical-pressure steam-turbine units at the Surgut-2 power plant beyond their rated resource.[35] According to the latest investigations, the fleet resource for these units was assessed as equal to the initially rated lifetime of 100,000 operating hours. Nevertheless by 2005 these power units, put in operation in 1985-88, worked a number of operating hours equal to 120,000-130,000. Such extension of their service life beyond the fleet resource was allowed owing to special NDE and calculations of individual remaining lifetime for all the major design components. As a result, after 100-110 thousand

474   *Steam Turbines for Modern Fossil-Fuel Power Plants*

hours, it became necessary to replace bends of the main steam-lines, whereas their straight parts were allowed to keep in operation up to 150-175 thousand hours. The same terms are planned for replacement of the HP and IP rotors of the turbines, whereas the casings and valve steam-chests can be left in operation for more extended time periods. A serious problem for these turbines is water-drop erosion (WDE) of their LP LSBs. To solve this problem, it is proposed to restore and protect their inlet and outlet edges with surface electric-spark alloying. The LSBs restored in such a way allow steam turbine operation between the overhauls without considerable damages. After 150-160 thousand operating hours, individual resource for many equipment components finally expires, and by this time it is planned to fulfill large-scale replacement of these components with new, more up-to-date ones, that is, upgrading the units.

References

1. "Estimating Steam Turbine Life Expectancy," *Turbomachinery International* 39, no. 5 (1998): 44.
2. Peltier R. "Steam turbine upgrading: Low-hanging fruit," *Power* 150, April 2006: 32-36.
3. Lazzeri L. "Refurbishing and Improving Old Plants," *Proc. of the American Power Conference* 61, Chicago, 1999: 789-794.
4. Hirota Y., Y. Kadoya, T. Coto, M. Wake, and H. Fuiji. "Changes of Material Properties and Life Management of Steam Turbine Components under Long Term Service," *Technical Review—Mitsubishi Heavy Industries* 19, no. 5 (1982): 202-212.
5. Horton W. and R. Peltier. "Realities restraining North American capacity expansion," *Power* 147, June 2003: 38-44.
6. Mori S., K. Tamura, and T. Tan. "Life Evaluation of Thermal Power Plant in-Service Exposure," *Hitachi Revue* 36, no. 6 (1987): 339-346.
7. Chizhik A.A., editor. *Problems of Long-Term Strength of Power Equipment* [in Russian], *Proc. of the Central Turbine-Boiler Institute* 246, Leningrad: 1988.
8. Parker J.D., A. McMinn, R.J. Bell, et al. *Condition Assessment Guidelines for Fossil Power Plant Components.* Topical Report GS-6724, Palo Alto (CA), EPRI, 1990.
9. Viswanathan R. *Damage and Life Assessment of High-Temperature Components*, New York: ASME International, 1994.
10. Viswanathan R. and R.W. Porter (editors). *Life Assessment Technology for Fossil Power Plants, Proc. of American Power Conference* 57-III, Chicago: 1995.
11. Mühle E.-E., E. Neuhaus, A. Sadek, and M. Siegel. "Assessment of Residual Lifetime of Steam Turbines after Long-Term Operation" [in German], *VGB PowerTech* 74, no. 1 (1994): 39-46 and no. 2 (1994): 128-135.
12. *Life Extension of Fossil Power Plants. Proceedings of the European Conference on Assessing the Residual Metal Resource and Lifetime Extension for Power Units of Fossil-Fuel Power Plants* [in Russian], Moscow: All-Russia Thermal Engineering Research Institute, 1994, Parts 1-3
13. Berlyand V., A. Pozhidaev, A. Glyadya, E. Plotkin, G. Avrutsky, and A.

Leyzerovich. "Some Problems of Steam Turbine Lifetime Assessment and Extension," *Proc. of the American Power Conference* 61, Chicago, 1999: 795-800.
14. Rezinskikh V.F., E.A. Grin, and V.F. Zlepko. "Reliability and Durability Enhancement for Fossil Power Units' Main Equipment Metal," *Proc. of the American Power Conference* 61, Chicago, 1999: 801-805.
15. Laire Ch. and M. Eyckmans. "Evaluating the Condition and Remaining Life of Older Power Plants," *VGB PowerTech* 81, no. 10 (2001): 98-102.
16. Rezinskikh V.F. and V.I. Gladshtein. "The Service Life and Reliability of the Metal of Steam Turbines at Thermal Power Stations," *Thermal Engineering* 51, no. 4 (2004): 257-261.
17. Kauer R. and H.-C. Schröder. "Concepts and Examples for Life Time Extension of Systems and Components in Power Plants," *VGB PowerTech* 86, no. 8 (2006): 61-66.
18. Merz A. and R. Reinfenhäuser. "Failure of the Turbine at Power Plant Irshing" [in German], *VGB Kraftwerkstechnik* 69, no. 3 (1989): 255-259.
19. Leyzerovich A. *Large Power Steam Turbines: Design & Operation*, 2 vols., Tulsa (OK): PennWell Books, 1997.
20. Zagretdinov I.Sh., A.G. Kostyuk, A.D. Trukhnii, and P.R. Dolzhanskii. "Destruction of the 300-MW Turbine-Generator Unit at the Kashira District Power Station: Causes, Consequences, and Conclusions," *Thermal Engineering* 51, no. 5 (2004): 345-355.
21. Kostyuk A.G. and A.D. Trukhnii. "The Long-Term Static Loading Strength of Solid-Forged Rotors of the 200-, 300-, and 800-MW Turbines Manufactured by the Leningrad Metal Works," *Thermal Engineering* 51, no. 10 (2004): 817-825.
22. McCloskey T.H., R.B.Dooley, and W.P. McNaughton *Turbine Steam Path Damage: Theory and Practice*, 2 vols., Palo Alto (CA): EPRI, 1999.
23. Viswanathan R., D. Gandy, and D. Rosario. "Rim Attachment Cracking Prompts Development of Life Assessment Tools," *Power Engineering* 104, no. 7 (2000): 50-51.
24. Knarr N. and H. Thielsch. "Casing repair adds years of service to vintage turbine," *Power* 135, January 1991: 55-56.
25. Herbst C.A. "Consider repair before replacing cracked steam-turbine shells," *Power* 137, March 1993: 69-70.
26. Gladshtein V.I. "Service of Turbines with the Main Steam Pressure over 9 MPa Having Cracks and Notches in Their Cast Casing Components" [in Russian], *Elektricheskie Stantsii*, no. 4 (2004): 30-33.
27. Gladshtein V.I. "The Effect of Prolonged Operation on the High-Temperature Strength of the Metal of the High-Pressure Cast Bodies of Valves and Turbine Casings," *Thermal Engineering* 48, no. 1 (2001): 285-288.
28. *ASME Boiler and Pressure Vessel Code, Case N47-29, Components in Elevated Temperature Service*, Section III. New York: ASME International, 1991.
29. *Methodical Guidelines on the Order of Carrying out Assessments of Lifetime for Individual Steam Turbines and Its Extension beyond the Fleet Resource* [in Russian], RD 153-34.1-17.440-2003, Moscow: VTI, 2003.
30. *Standard Instruction for Inspections and Lifetime Extension of Major Components for Boilers, Turbines, and Steam-Lines of Thermal Power Stations* [in Russian], RD 153-34.1-17.421-2003, Moscow: VTI-ORGRES, 2003.
31. D'yakov A.F., V.R. Nechaev, and G.G. Olkhovsky. "Repowering Existing Thermal Power Stations," *Proc. of the American Power Conference* 60, Chicago, 1998: 1033-1037.
32. Remezov A.N. "Problems of Technical Re-equipment and Service Time Extension

of Power Plants" [in Russian], *Elektricheskie Stantsii*, no. 9 (1999): 77-79.

33. Tumanovskii A.G. and V.F. Rezinskikh. "The Strategy of Prolonging the Service Life and Technical Reequipment of Thermal Power Stations," *Thermal Engineering* 48, no. 6 (2001): 431-439.

34. Kostyuk A.G. "Long-Term Strength of Steam-Turbine Rotors in the Stress Concentration Zone," *Thermal Engineering* 53, no. 2 (2006): 88-94.

35. Rezinskikh V.F., E.A. Grin', and Y.A. Bukin. "Operating Reliability and Service Time Extension Prospects for Thermal-Mechanical Equipment of the Surgut-2 Power Station" [in Russian], *Elektricheskie Stantsii*, no. 3 (2005): 11-15.

Chapter 18

Steam Turbine Upgrade

AN INCREASE IN THE TURBINE EFFICIENCY
AND OTHER BENEFITS OF TURBINE UPGRADES

Even if the lifetime of individual turbine elements expires, any possibilities to restore their operation capacity by means of palliative measures exhaust, and further lifetime extension for these elements becomes hazardous, many other turbine elements retain their serviceability and can be used further for many years. For example, it was mentioned before that according to some researches about 90% of steam turbine casings could survive 700,000 hours of operation,[1] and the lifetime limit of 100,000 or 200,000 operating hours concerns only few number of most vulnerable turbine components. Under these conditions, it is advantageous to preserve the turbine in service, replacing those elements whose lifetime expired with new, up-to-date ones. Simultaneously, it is advisable to renew the steam path of the refurbished cylinders to raise their efficiency to the level reachable with the use of modern approaches and technologies, no matter whether the cylinder's blading could still be left in operation for years or not. In this way, two problems are solved at once: 1) the turbine is retained in operation with minimum investments and 2) its efficiency is raised to the merited level making aged power equipment competitive again at the modern power industry market.

It is supposed that upgrading and retrofitting a turbine in the 500-MW to 700-MW capacity range of approximately 20-year age can reduce its heat-rate by a total of about 4% and increase its output by approximately 5% with the same main steam mass-flow rate.[2] Such an improvement is combined from the efficiency gains for the HP, IP, and LP sections (cylinders) by approximately 6%, 1%, and 2%, respectively. Their contributions to the turbine output increase account to about 2%, 1%, and 2%, respectively. An additional 0.5% increase in the turbine output can be achieved by modifying the condensers and replacing the main-steam and reheat valves with new ones of less pressure drops. In other cases, the expected improvements in the internal efficiency of the HP, IP, and LP sections are estimated as lying at the levels of 5%, 4%, and 2.5%, respectively.[3,4] Upgrading the aged steam turbines is considered to be the most cost-effective solution, with the best investment return. It is counted that a decrease in the heat-rate for a base-loaded 500-MW coal-fired power unit by 100 Btu/kWh (105.5 kJ/kWh), or an increase in its efficiency by 2.9%, will save as much as $10 million annually in the fuel costs alone.[5]

In this connection, upgrading the turbine or its individual components to improve its efficiency and increase the output can be reasonable even if the turbine state allows its further operation without replacement. It is understandable that in this case upgrading should be first applied to those elements whose efficiency worsens the operation process to a greater degree or can be most remarkably heightened by upgrading and, on the other hand, whose replacement needs minimum investments. In this connection, very instructive is an experience with long-term operation, revisions, and refurbishment of steam turbines at the 2,100-MW coal-fired German power plant Neurath consisting of three 300-MW power units ("A," "B," and "C") and two 600-MW ones ("D" and "E").[6] All the units are equipped with subcritical-pressure steam turbines of ABB (Figure 18-1) and started their commercial operation in the years between 1972 and 1975. The units have been operated in a base-load manner and envisaged to run over at least 45 years. After first ten years of operation, the heat consumption of the 300-MW turbines had increased only by about 0.69% on the average; the internal efficiency of the HP cylinders lowered by 1.26%-points on the average, and the internal efficiency of the IP cylinders had remained almost invariable (Figures 18-2 and 18-3). As to the 600-MW turbines, their heat-rate had increased somewhat more, but, as well as for the 300-MW turbines, mainly due to worsening the internal efficiency of the HP cylinders (Figures 18-2 and 18-4).

After careful assessments of the turbine state, diagnosing, and weighing possible risks, the first comprehensive turbine revision was carried out at the "B" power unit in 1987 after over 120,000 hours of operation. Up to that time, the unplanned unavailability of the turbine had amounted to only 0.39%. None of the revision results testified to any operational hazard, and based on the gotten results the strategy of further revisions was outlined. The calendar of carried out and planned revisions for all the power plant's five units is given in Table 18-1. Start-

Steam Turbine Upgrade

Figure 18-1. Longitudinal sections of 300-MW (a) and 600-MW (b) subcritical-pressure steam turbines of ABB. *Source: J. Schweimler[6]*

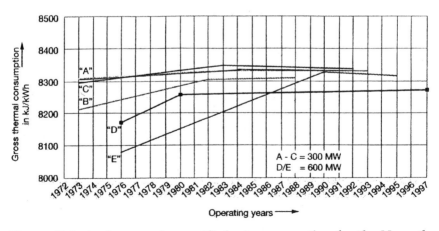

Figure 18-2. An increase in specific heat consumption for the Neurath power plant's units in the first 20 years of operation. *Source: J. Schweimler[6]*

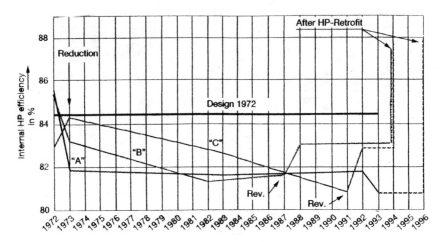

Figure 18-3. Changes of the HP cylinders' internal efficiency for 300-MW turbines of the Neurath power plant in the process of O&M. *Source: J. Schweimler[6]*

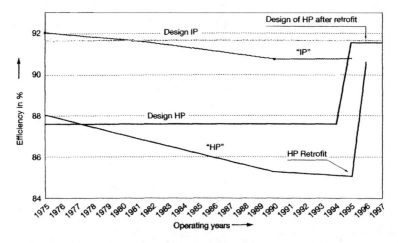

Figure 18-4. Changes of the HP and IP cylinders' internal efficiency for 600-MW turbine "E" of the Neurath power plant in the process of O&M. *Source: J. Schweimler[6]*

ing 1994, all the turbines were submitted to so-called "efficiency improving measures," that is, upgrading. In this connection, the HP and LP steam paths of all the five turbines were replaced, even though the IP steam paths were left untouched. Compared to the rated (designed) level

Steam Turbine Upgrade   481

of 1972, the internal efficiency of the HP cylinders for 300-MW turbines was improved by approximately 3.1% up to 87.7% (see Figure 18-3). For the HP cylinders of 600-MW turbines, the rated efficiency increased by about 4% up to 91.5% (see Figure 18-4). The further retrofits should be focused at the IP cylinders.

**Table 4-2. Revisions of the power plant Neurath's
300-MW and 600-MW turbines**

Turbine Unit	1st Revision	Years in Operation	Hours /Startups	2nd Revision	Years in Operation	Hours /Startups	Retrofit HP and LP	Years in Operation	Hours /Startups	Next Envisaged Revision
A	June 1989	17	134,889 167	March 1993	21	162,252 193	August 1996	24	191,595 220	2012
B	Sept. 1987	15 171	121,798	—			August 1994	22	176,266 222	2006
C	June 1988	15	121,232 176	March 1991	18	141,387 208	August 1994	21	169,784 244	2010
D	March 1988	13	106,652 117	July 1993	18	141,571 150	August 1997	22	173,900 195	2013
E	July 1991	16	125,051 112	Sept 1995	20	159,013 145	Sept. 1995	20	151,013 145	2015

Source: J. Schweimler[6]

Applying to a study carried out for 300-MW and 600-MW turbines, like those installed at the Neurath and other German power plants, ABB considered redesign of the LP cylinders to be the most profitable investment.[7,8] Their higher efficiency is reached from optimizing the steam flow path, including the outlet diffusers, applying modern blade profiles, and reduction in heat losses with the exit velocity by means of using the longer LSBs with increasing the annular exhaust area by about 25%. A cost-benefit analysis showed it makes sense to retain the outer cylinder casing and modify the inner one and diffusers, with replacing the blading and blade carriers, as well as the rotor if the new LSBs are applied. Such an approach is shared by other turbine producers as well. A possible refurbishment of an LP cylinder proposed by ABB is presented in Figure 18-5a (the original design shown below the turbine axis plane and the improved version above). The resulted efficiency increase, according to ABB bench tests can be seen in Figure 18-5b. Application of later LP cylinder versions (see, for example, Figure 4-12) can provide even higher efficiency gain.

Steam turbine upgrades additionally result in some secondary benefits. The efficiency gain translates into reduced environmental impact: less fuel for the same output (or the same fuel consumption for larger power generation), fewer emissions into the atmosphere, and less heat rejection to cooling water. The increased output reduces the need for constructing new power plant or enables the power utility to retire other, older, less efficient, and less environmentally friendly power plants. The

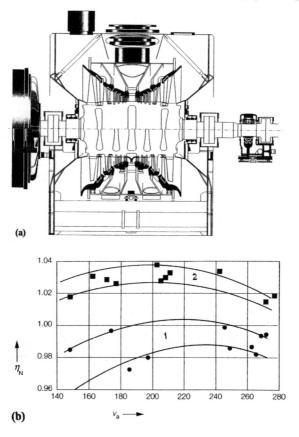

Figure 18-5. Proposal of ABB for LP cylinder refurbishment (a) and resulted gain in its efficiency (b). 1 – old turbine design, 2 - new turbine design. *Source: E. Krämer et al.*[7]

Steam Turbine Upgrade 483

increase in the power plant output gained by means of upgrading the aged turbines is typically achieved at much lower capital costs than introduction of "greenfield" power generation and, as a rule, at lower operation costs.[9]

Of significance is that refurbishment of steam turbines, with replacement of their most vulnerable design components, provides considerable extensions to the overhaul inspection intervals—from the present every four-to-six years to 10-12 years. In particular, this was especially emphasized by ABB for the refurbished LP cylinders.[7]

Steam turbine refurbishment strategies along with upgrading the turbine itself commonly include renewal of the turbine's control and protection systems. In particular, until the beginning of the 1980s most steam turbines were equipped with mechanical-hydraulic control and protection systems which are substantially inferior to modern electronic systems in their functionality, reliability, maintainability, and interaction with both the operator and power unit's other control systems. Numerous disadvantages peculiar to old systems become more obvious and intolerable if the aged power units are shifted to operation with more frequent and rapid transients, as well as under conditions of changes in the structure of power plant staff qualification, including the loss of senior, experienced personnel. That is why many turbine producers (original equipment manufacturers—OEM and non-OEM), specialized companies from the automation industry, and diverse institutions of the power industry have developed and proposed numerous projects of refurbishing and replacing the existent out-of-date control and protection systems for aging steam turbines in service based on modern state-of-the-art computerized I&C technologies.[10-15] For the customers, the challenge remains to select the most appropriate proposal. In turn, developers must focus on analyzing the options and requirements in close co-operation with the customer. For example, in due time ABB (then ABB Alstom Power, then ALSTOM Power) developed a standard complex project for refurbishing the control, protection, and monitoring systems of their aged turbines, including numerous ones of subcritical and supercritical steam pressure with the single capacity of up to 1,300 MW installed at diverse U.S. power plants in the 1960s-1980s.[11-13] For the specific process functions, the standard hardware and software packages may be used for all the steam turbine types. If required, new packages with specific functions can be developed, or standard functions can be modified. Along with this, it might be well to note that the standard project did not pay proper attention to upgrading the set of temperature measurements for the turbine; as a result, even after refurbishing these turbines have not had sufficiently representative temperature monitoring.

Upgrading the control and protection system by non-OEM can be presented by the example of an 175-MW steam turbine of Westinghouse operated at the E.W. Brown power plant Unit 2.[15] The turbine, put in operation in 1963, was originally equipped with a mechanical-hydraulic governing (control) and protection system. After evaluating different options, the power plant awarded a contract to TurboCare. The replacement itself required a five-week turbine outage in the fall of 2002. With preserving the turbine's original steam valves, the old mechanical-hydraulic system was completely replaced with digital electric-hydraulic contours. New digital servo actuators for the throttle, intercept, and governing valves were also installed. A redundant high-pressure hydraulic power unit has replaced the old hydraulic system. An important constituent of the new system is a MicroNet programmable digital controller. Commonly, in most of upgraded systems, its functions are executed by a distributed control system (DCS), but in this case the cost was prohibitive. The big difference from the operator's standpoint is the new human-machine interface: one screen gives a graphic overview of the entire system with its peripheral components, and current operating information, including that of the throttle steam pressure, status of each valve, lube oil and hydraulic oil pressure, and key operating data of the turbine's rotation speed and load—Figure 18-6.

If refurbishment of the turbine and its control and protection system is combined with upgrading the power unit's DACS upon the whole, as it frequently takes place, this offers a considerable scope for improving the quality of operation and increasing in this way additionally the turbine's future service time. Another important factor is a possibility of decreasing the operational personnel quantity. According to some considerations, in theory, twelve operators are required to run start-ups of a modern power unit without automation, and two or three operators are needed during normal steady-state operation. With automated control systems in place, embracing start-up and shut-down operations, the entire power unit can be reliably operated by just one person.[11]

Perhaps the most impressive case of the recent years with a complex steam turbine upgrade has been demonstrated by Siemens Power Generation at the 25-year-old 750-MW German coal-fired power unit Mehrum.[16,17] The unit was designed for operation in a load following manner, with regular shut-downs for weekends and nights. By 2003, it

Steam Turbine Upgrade 485

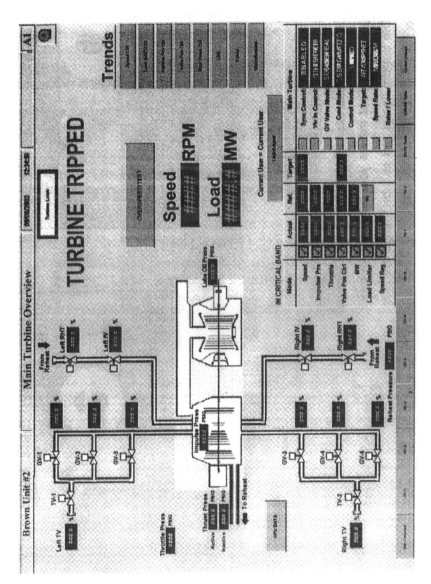

Figure 18-6. Hardcopy of the screen image with main data from the turbine control system for the 175-MW turbine of the Brown unit 2 after upgrading. *Source: J. Zwers*[15]

had undergone nearly 4,000 start-ups and has averaged no more than 4,100 full-load operating hours per year. The unit's turbine consists of separate single-flow HP and double-flow IP cylinders and two double-flow LP ones of Siemens' traditional design (see as example Figure 2-10). In the summer outage of 2003, the turbine was upgraded with replacing the blading, inner casings, and rotors of all the cylinders and optimizing the entire turbine steam paths. The new blades, both stationary and rotating, are twisted and bowed of Siemens' 3DS type, with a variable reaction degree. The LP annular exhaust area was increased from 6.2 m^2 to 8 m^2 per flow through replacing the original LSBs with new ones of 915-mm (36-in) length. The last three LP stages have free-standing blades made of 12% Cr steel, with the side entry root, while all the other blade rows are of an interlocked integrally shrouded design. The last stage's stationary blades (vanes) are made hollow with suction slots to avoid formation of large water drops, thus reducing WDE of the LSBs. The diagram of Figure 18-7 shows the efficiency gains achieved for each turbine cylinder, according to the heat-rate test's data before and after the upgrade. The gotten efficiency figure for the HP cylinder of 93.6% is believed to be the world's highest for retrofitted turbines.

As a result, the turbine efficiency was increased by 3.8 percent-

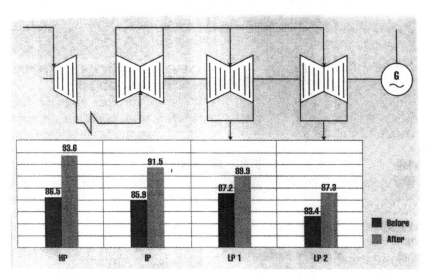

Figure 18-7. Efficiency gains for individual cylinders of 750-MW turbines of Siemens at the power plant Mehrum after upgrading. *Source: J. Varley[17]*

age points, and the overall plant output was increased by about 2% providing additional production of 38 MW without an increase in the steam flow amount and fuel consumption. In addition, modernizing the cooling tower and some retrofit of the boiler with regenerative flue gas reheating has raised the thermal efficiency of the 25-year-old power unit to a very meritorious level of 40.5%. The diagram of Figure 18-8 demonstrates the dependencies of the power unit's efficiency on its load before and after upgrading. The efficiency gain is equivalent to saving in carbon dioxide emissions of 193,000 tons annually. It is supposed that the upgrade will reduce the power unit's annual need for coal by 76,500 tons. The upgrade project also included replacement of the turbine I&C and stress-monitoring system, making the unit more dispatchable and thus more profitable, contributing to its more qualitative operation and greater operating flexibility, that is, faster start-ups, higher output ramp rates, lower minimum load, and improved partial load efficiency. Since 1998, the number of employees at the plant has decreased from 230 to 130, so the plant's productivity in terms of MW per employed people has nearly doubled. The production cost in 2003 accounted to around 2.8 euro cents per kW.

The whole turbine upgrade project was carried into effect during the scheduled overhaul of 2003, for less than three weeks (since May 28

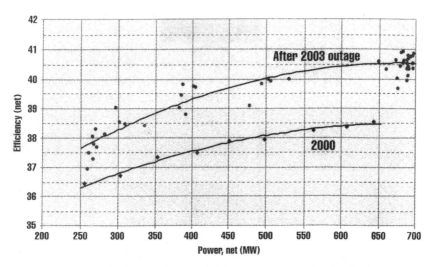

Figure 18-8. Variations of net efficiency with output of the power plant Mehrum before and after upgrading the steam turbine. *Source: J. Varley*[17]

488   *Steam Turbines for Modern Fossil-Fuel Power Plants*

to June 16). The upgrade cost amounted to 6,599 thousand euros, and, according to Siemens, all the new investments made up 12,676 thousand euros. In such a way, the total expenditures for the turbine upgrade were equal to 19,276 thousand euros.[17]

A similar refurbishment, even if in a less scale, was carried out by Siemens in October 2004 at another 35-year-old 350-MW German coal-fired power unit Farge.[18] The ultimate goal of the $30 million modernization project was to increase the unit's output with no increase in fuel consumption. Upgrading the turbine's IP cylinder, two LP cylinders, and condenser increases the unit's capacity by 18 MW. Besides making the Farge more efficient, competitive, and profitable, another goal of the project was to lower the unit's gas emissions. After the project was completed, the CO_2 emission was expected to drop by as much as about 100,000 tons per year.

Noteworthy is also another turbine upgrade project materialized by Siemens prior to the Mehrum one. It concerned the Danish power plant Enstedtværket's Unit 3 with the nominal gross capacity of 630 MW and subcritical-pressure steam conditions.[19] As distinct from the Mehrum's turbine, this one consists of single-flow HP and IP cylinders and two double-flow LP ones. The turbine upgrade took 44 days, instead of 65 days requested initially, and after 46 days the unit was recommissioned. The turbine refurbishment included replacing the HP, IP, and LP steam paths (based on the use of Siemens' 3DS blading), inner casings, and rotors. It was the first time when Siemens applied their 45-in (1,146-mm) long steel LSBs with the exhaust annular area equal to 12.5 m^2 per flow to replace the originally employed 43-in (1,093-mm) LSBs with the exhaust annular area of 10 m^2 per flow. The turbine upgrade resulted in the increase of the power unit's output of 30 MW, that is, up to 660 MW at the same fuel consumption.

In 2004, Siemens Westinghouse Power Corporation, as a part of Siemens Power Generation, completed upgrading the turbine of the U.S. coal-fired power plant J.P. Madgett Unit 1 entered into commercial service in 1979.[5,20-22] The subcritical-pressure steam turbine of Westinghouse has a nominal output of 365 MW and consists of the integrated HP-IP cylinder, common for many Westinghouse turbines—see Figure 3-13, and double-flow LP cylinder. Retiring the unit was not considered as an option because the power utility's resource planners insisted they need this capacity currently and for another 30 years, so upgrading the turbine was the only feasible solution. Besides retaining the unit in op-

Steam Turbine Upgrade 489

eration, the upgrade project pursued the goals of improving the unit's efficiency, increasing its output, and lengthening the turbine's maintenance period between overhauls to ten years. A 50-day outage was scheduled for refurbishing the turbine with replacement of the blading, inner casings, and rotors for both the HP-IP and LP cylinders in parallel with overhauling the boiler and upgrading the unit's control system to a modern DCS, as well as replacing the main transformer to handle the expected higher power output.

The BB44FA HP-IP cylinder refurbishment design targets the existent building block BB44 having been employed at numerous Westinghouse turbines with steam conditions of 16.6 MPa, 538/538°C (2,400 psi, 1,000/1,000°F) ranging in their size between 350 MW and 680 MW. With retaining the original outer casing, the new HP-IP cylinder, as compared to the existent design—Figure 18-9, provides full-arc HP steam admission and eliminates the separate nozzle boxes, impulse-type control stage, and 180-degree steam turnaround to the HP steam path. As before, the cylinder rotor in its IP steam admission part is cooled by steam taken from the HP exhaust chamber. The new fully integral inner casing has all the stationary components removed and replaced together, which reduces time taken for assembling and disassembling the cylinder. Along with this, it might be well to note that a complicate geometric form of the new inner casing, with numerous significant changes of its outline diameter, somewhat contradicts to Siemens' customary striving for using simpler, more "streamline" forms for all the turbine casings to make their unsteady temperature fields more even and avoid excessive thermal stresses at the transients. As applied to the new inner casing, significant lengthwise variations in its longitudinal rigidity can result in plastic deformations of the horizontal split surface caused by start-up thermal stresses.

The original five-stage LP steam path was replaced by a new seven-stage design BB73-8.7 with free-standing 958-mm (37.7-in) long LSBs and the annular exhaust area equal to 8.7 m^2 per flow. The LP last stage's vanes are made hollow with heating by steam (see Figure 5-27) taken from an upstream LP extraction to decrease steam wetness before the LSBs. The leading edges of the LSBs are laser hardened to additionally protect them against WDE.

The steam paths for both the HP-IP and LP cylinders are composed entirely of 3DS blading with variable reaction degree. All the rotating blades of the HP-IP stages and LP stages, excepting the LSBs, are made

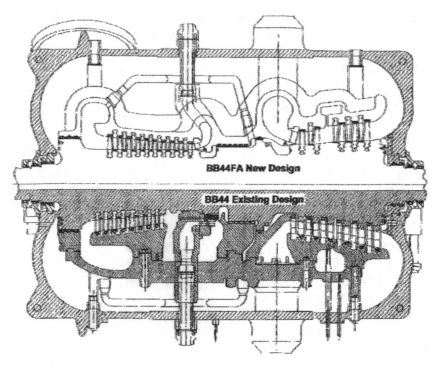

Figure 18-9. Comparison of existing (BB44) and refurbished (BB44FA) integrated HP-IP cylinders of Siemens Westinghouse Power Corporation. *Source: J.R. Cheski et al.*[21]

integrally shrouded. In addition to standard non-contacting seals such as labyrinth gland seals, advanced sealing technologies, such as abradable, brush, and retractable seals, are used in the turbine—see Figure 4-23—providing significant performance benefits. Apart from upgrading aging steam turbines, the advanced designs BB44FA and BB73-8.7, which were approved at the J.P. Madgett Unit 1, are currently proposed for newly constructed steam-turbine and CC power units with the main steam pressure of up to 17.6 MPa (2,550 psi) and main and reheat steam temperatures of 565°C (1,050°F) with a potential raise of them up to 600°C (1,112°F).[20]

The upgraded turbine was successfully recommissioned after the 50-day outage, and the heat-rate tests showed the increase in the HP efficiency of 8% to 10% over the whole load range, while that of the IP section rose by 2% to 4%. The total turbine output went up by 20 to 27 MW without an increase in fuel consumption and steam flow amount.

The LP cylinder's share in this increase accounted to 7-to-8 MW.

Prior to the J.P. Madgett Unit 1, in 2000 a similar refurbishment project was carried into effect by Siemens Westinghouse Power Corporation at the 420-MW Bonanza Unit 1; the gain in the output brought by the complete upgrade of the turbine was assessed as equal to about 32 MW.[23] In 2002, two HP-IP cylinders BB44 of Westinghouse were replaced at the AmerenUE's Labadie power plant (Units 1 and 2, with the nominal output of 571 MW each, put in operation in the early 1970s) with the resulted gain in the HP efficiency of 7%.[5] Three other HP-IP cylinders BB44 of Westinghouse steam turbines were also upgraded in 2004 with similar results, and two more upgrade projects were planned for implementation in 2005.[20]

Steam turbines of the same class of GE are proposed to be upgraded with replacement of the existing blading with so-called "dense pack" steam paths distinguishing with higher reaction, lower pitch diameters, longer blades, an increased number of stages without increasing the overall cylinder length, and often an increased exhaust annular area.[24] According to empirical studies of General Physics Corporation as applied to a turbine with a nominal output of 360 MW,[4] if the original internal efficiency of the HP section was equal to 84%, after upgrading this figure closes to 90%, and the IP section, originally designed to be 89% efficient, can reach 97%; the LP cylinder is also refurbished with installing the new rotor and replacing the existing 660-mm (26-in) LSBs with new buckets of the same length but with more efficient airfoils. The diagram of Figure 18-10 demonstrates the turbine cycle gross efficiency plotted against the main steam flow amount through the turbine. The upgraded turbine uses less steam per MW of power production. The unit's boiler can produce roughly 2,600 kpph (kilo pounds per hour), that is, 1,180 kg/h, of throttle steam, which enabled the turbine to produce about 360 MW; the upgraded turbine with the same steam flow amount produces 371 MW. With the assumed sale price of $45/MWh, this results in an additional $495 per hour of revenue. With the supposed load factor of 90% and 85% availability, the upgrade results in a potential $3.3 million increase in annual revenue. With a capital investment of $5 million, this upgrade should pay itself in about a year and a half.

In 2003, AmerenUE upgraded two 571-MW of GE at its power plant Labadie (Units 3 and 4) with the goal to extend the turbine lifetime, improve the unit efficiency and plant output, as well as to lengthen the turbine overhaul intervals.[5] The HP-IP cylinders were refurbished with-

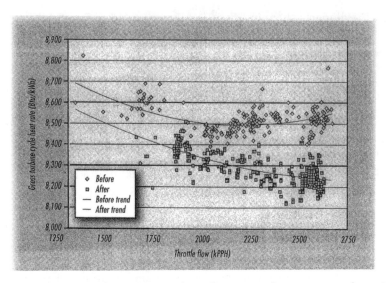

Figure 18-10. Gross heat rate against the steam flow amount for a 360-MW steam turbine of GE before and after upgrading. *Source: K. Potter and D. Olear[4]*

out replacement of the outer casings. The double-flow LP cylinders were to be retrofitted because of intense stress corrosion cracking of the next-to-last stage's wheel discs in the dovetail area. The LP refurbishment included replacement of the 30-in (762-mm) LSBs with more up-to-date 34-in (864-mm) buckets. Among others, it may be well also to note upgrading the turbine of GE of the U.S. power plant Crosby Unit 1, currently rated at 177 MW, which had accumulated over 361,000 operating hours.[25]

Thermodynamic effects of refurbishing the turbine's HP and IP sections set developers thinking of altering the thermal conditions of the power unit as a whole.[26] So, the increase in the HP steam path's efficiency makes the HP exhaust (cold reheat) steam temperature fall. This reduction is obviously dependant on the HP efficiency levels before and after refurbishing, but most often it amounts to about 10°C (18°F). Since the hot reheat steam temperature is assumed to be maintained at the former, design level, reduction of the cold reheat steam temperature requires extra heat to be added into the reheater. For turbines with integrated HP-IP cylinders, reduction of internal steam leakages through the central gland seal between the HP and IP sections owing to the use of advanced sealing technologies leads to an increase in the reheat steam flow, thus increasing the need in the extra reheater heat. In some cases, the steam leakage between the HP and IP sections, having accounted to 7-10% of the main steam flow, can be decreased by 3-4%.[27,28] If the reheat steam temperature is maintained by spray attemperators, the increase in the steam flow amount through the reheater and the decrease in the cold reheat steam temperature reduce the required amount of sprayed water and additionally improve the unit's heat rate, even though somewhat reduce its output. It might be well also to note that an increase in the reheat steam flow makes the hot reheat steam pressure fall because of the increase in the pressure drop. This results in some decrease of the steam wetness at the turbine exhaust which additionally improves the turbine efficiency. In turn, refurbishing the HP steam path allows for the main steam flow to be increased at the same throttle steam pressure, increasing the steam pressure levels upstream of the IP and LP sections. On the other hand, an increase in the flow passing capacity of the IP section reduces the HP exhaust pressure. In the case of the final feedwater heater fed from the cold reheat steam-lines, the final feedwater temperature can therefore be adjusted, and changes in this temperature significantly influence the heat input into the boiler and the power unit's entire heat rate and output. So, upgrading the turbine should not be considered in isolation from analysis of thermal conditions from the boiler, and optimal refurbishment solutions should be looked for the power unit upon the whole.[26]

In some cases, such a complex refurbishment can be even more drastic with complete reconstruction of power units. So, for example, according to Mitsui Babcock Energy Ltd., their innovative technology of vertical tube surfaces (as employed at the Yaomeng power plant in China[29]) makes it possible and realistic to turn the existent once-through subcritical-pressure boilers into supercritical-pressure ones. Together with modifying steam turbines and replacing main steam-lines, this would allow converting radically a significant portion of existent fossil-fuel power plants into advanced modern supercritical-pressure ones with preserving most of the existing power plant. This initiative is suggested in connection with the Carbon Abatement Technology strategy published by the UK government in 2005. In particular, according to the Large Combustion Plant Directive (LCPD), such a project is considered as applied to ten coal-fired power plants of the UK with the total capacity of about 12.5 GW.[30] The total capital costs for their conversion is estimated as equal to £2.5 billion with the resultant CO_2 saving of 10 million tons per year versus existent fossil-fuel power plants with the average efficiency of 36%. The capital costs for installing 5 GW of wind turbines with an equal resultant CO_2 saving would make up £3.6-5 billion. The power production cost for the converted fossil-fuel power plants is assessed as equal to 1.8 p/kWh (with the average annual capacity factor of 60% and 15 years of writing off the capital costs) compared to 5.5-7.2 p/kWh for wind turbines (with the average annual capacity factor of 25% and 20-year lifetime). It is estimated that conversion of a standard 600-MW power unit into an advanced supercritical-pressure unit would take a twelve-month outage.

Another popular solution for complex refurbishment of steam-turbine power plants is their conversion into CC ones by means of topping them by gas turbines with different schemes of recovering the heat of their exhaust gas.[31-34] For our subject, of importance is that such a conversion can be carried out without serious redesign and reconstruction of the existent steam turbines in service. As an example is the reconstruction of the Scottish Hydro-Electric's Peterhead power plant with two 660-MW gas-fired steam-turbine power units commissioned in 1980 and two simple-cycle gas turbines of 115-MW capacity each. In 1998 one of the steam-turbine units was selected for conversion to a CC system by topping it by three V94.3A gas turbines of Siemens with associated triple-pressure heat recovery steam generators (HRSGs).[35] All the three new gas turbines can be operated independently in an open-cycle mode with the total output of about 795 MW. The existent boiler, being in good condition, is retained in the new system and can still be used in conjunction with the original steam turbine to produce the full designed output in the original mode, with the gross efficiency of 39.7%. The maximum efficiency of 55% is gained when all the three gas turbines with their HRSGs are connected to the steam turbine, and the boiler is shut down. This CC scheme gives some 1,150 MW of the output, of which only 350 to 360 MW comes from the steam turbine. And, finally, the system can work in a hybrid mode, when the CC output is augmented by means feeding the steam turbine additionally by steam from the original boiler, which add another 300 MW, increasing the total output to 1,420 to 1,450 MW, restricted by the steam turbine's last stage blades. With account to the existent gas turbines and second steam turbine unit, the power plant's total maximum output could be as high as over 2,300 MW. The capital cost for such a reconstruction made up about 60% of that for a new equivalent CC unit.

STEAM TURBINE REFURBISHMENT BY NON-OEMS
AND WITH THE USE OF ALIEN DESIGN SOLUTIONS

Until recently, power steam turbines had been traditionally retrofitted and refurbished by their original equipment manufacturers (OEMs). The situation sharply changed literally in the last decade, and refurbishment of steam turbines in service at power plants became as important a field of activity and competition on the world's market for turbine manufacturers as is steam turbine production. Mergers of many former turbine manufacturers and redistribution of their spheres of influence in the world power market open wide possibilities for the use of new design approaches that can be even alien to the OEM. If an early experience in retrofitting steam turbines by non-OEMs mainly focused on separate turbine elements causing long standing reliability concerns, like replacement of combined (shrink-fit) LP rotors with welded or solid ones, later this experience has acquired a much wider character. The deregulation of electricity generation markets provided financial incentives for the power plant owners and operators to maximize the economic competitiveness of their steam turbines in service and raise their availability and efficiency using new approaches, which had not necessarily been offered by the OEMs.[36-39]

In contrast with the newly designed steam turbines and power units upon the whole, where a strong tendency towards standardization and modularization dominates, upgrading of existent power equipment is considered to be a much more individual task. It is generally characterized by an intensive preliminary phase with detailed survey of the object *in situ*, analysis of its current state and peculiarities of operating conditions, development and comparative consideration of different options before the actual retrofit works begins. With regard to these features, in 1996, Siemens formed a special service department to handle matters concerning reconstruction of steam turbines and other power equipment manufactured by both Siemens/KWU and other turbine producers.[40,41] According to Siemens, in the case of approximately 20-(and more)-year-old steam turbines in the 600-MW to 700-MW capacity range the use of advanced contemporary blading is capable of increasing the turbine efficiency by a total of about 4% and the electric output by about 5% with the same main steam flow amount. An additional 0.5% increase in the output can be achieved by modifying the condensers and by replacing the main and reheat steam valves with those of larger cross section, less

496   *Steam Turbines for Modern Fossil-Fuel Power Plants*

steam velocities, and less pressure drops and energy losses.

In some cases, a refurbishment occurs to be obligatory because the individual service time of the turbine or, to be more specific, its major design elements completely expired and further operation becomes risky. In other cases, the turbine refurbishment is considered to be rather arbitrary and dictated by purely economic reasons. An example of a forced, necessary refurbishment for prolonging the turbine lifetime may be a 100-MW turbine set of the Negvelli Unit 8 in India. In collaboration with BHEL, Siemens managed to carry out a combined retrofit of the turbine, generator, and condenser, which results in an increase of the availability factor from 68% to 90%. Another project of Siemens for prolonging the turbine's lifetime alongside with an increase of its efficiency and availability referred to 200-MW units at the Zmijev power plant in Ukraine with steam turbines of LMZ.[40,41]

MHI's Experience in Upgrading Steam Turbines of Other OEMs

An example of upgrading an old steam turbine by a non-OEM is an experience at the US power plant Cromby operated by Exelon Power.[25,42] While the Unit 1's turbine of GE was retrofitted by its OEM, the Unit 2's turbine was upgraded by Mitsubishi Heavy Industries (MHI), although the OEM was Westinghouse. This 200-MW turbine, commissioned in 1954, was originally designed for base-load operation but was mainly operated in a cycling mode and by 2004 had accumulated over 263,000 hours of operation. Steam turbine performance had degraded with time due to such factors as higher seal clearances, increased surface roughness in the steam path, WDE, and so on. But just replacing old components with their spares may not be enough to prolong its operation for at least another ten years. MHI proposed to renew the turbine's steam path with the use of modern, three-dimensionally calculated and manufactured blading which would make the turbine quite competitive as far as possible with regard to its moderate steam conditions of 12.5 MPa, 538/538°C (1,810 psi, 1,000/1,000°F). Besides improving efficiency and extending the lifetime, the turbine upgrade was aimed at making the turbine more resistant to WDE and SCC. In particular, it is a general approach of MMI to replace combined LP rotors with solid ones to mitigate SCC. To reduce unsteady-state thermal stresses at the transients and make the turbine more flexible, the HP forged rotor with a central bore was replaced with a solid one and the cylinder's inner casing was renewed for one of a simpler outline. The HP cylinder was

Steam Turbine Upgrade

radically renewed upon the whole with the fixed position of main steam inlets, extraction and exhaust piping, and retaining the existent steam admission valve system. In this case, the challenge was met by replacing the existing Curtis control stage with a modern impulse-type stage and improving the subsequent reaction stages based on state-of-the-art 3D technologies. Figure 18-11 compares appearances of the HP cylinder before and after upgrading.

As it frequently happens with non-OEMs, MHI had no detailed engineering drawings of the OEM to work with. So, during a scheduled two-week outage of 2002 the company engineers working with local technicians took all the necessary 3D measurements on the disassembled

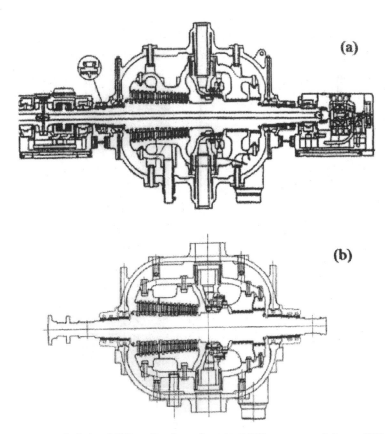

Figure 18-11. Original HP cylinder of a 200-MW steam turbine of Westinghouse (a) and its renewed version after upgrading by Mitsubishi Heavy Industries (b). *Source: By courtesy of Mitsubishi Heavy Industries*

turbine, inputting them directly into the CAD system for verification. An example of these measured dimensions is shown in Figure 18-12. The new HP cylinder installation was completed during a seven-week outage period in late 2003. Regular measurements and approximate calculations made on their basis showed that after refurbishing the cylinder, the turbine output increased by over 4%, the feedwater flow amount being the same.

In other cases, the scope of upgrading fulfilled by MHI as a non-OEM was significantly more. Thus, MHI successfully completed the upgrade of a 300-MW-class turbine for the Korean Yeosu thermal power plant's Unit 2 with the steam conditions of 16.3 MPa, 538/538°C (2,364 psi, 1,000/1,000°F).[42] The existent four-cylinder turbine, with its single-flow HP and IP cylinders and two double-flow LP ones (Figure 18-13a), was reconstructed into a three-cylinder design with an integrated HP-IP cylinder (see Figure 18-13b). Current engineering technology of MHI

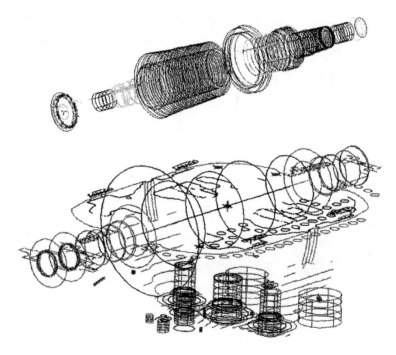

Figure 18-12. Measured dimensions for the turbine's HP rotor as input into the CAD system of MHI. *Source: By courtesy of Mitsubishi Heavy Industries*

Steam Turbine Upgrade

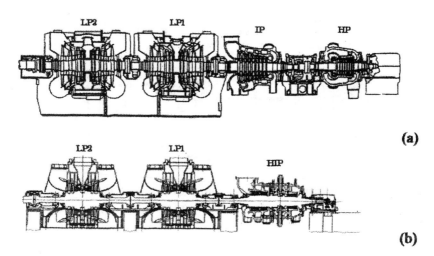

Figure 18-13. 300-MW-class steam turbine of the Yeosu Unit 2 before (a) and after (b) its retrofit by MHI as a non-OEM. *Source: By courtesy of Mitsubishi Heavy Industries*

dictates one LP cylinder for a 300-MW class turbine; however, considering the interface with the condenser, MHI designed the steam turbine to consist of two LP cylinders with the LSBs of the same length, but integrally shrouded. The turbine refurbishment included replacement of all the major turbine elements, including the rotors, inner and outer casings, and blading for all the cylinders, main steam valves, and HP feedwater heaters. The control oil system was separated from the lube oil system. To replace the existent turbine with a turbine having a less number of cylinders, the configuration and layout of the foundation had to be changed, too. Accordingly, the turbine foundation was modified where the turbine would sit—Figure 18-14. Commonly, in similar instances the original foundation is completely removed before the new one is installed to maintain the integrity and unity of the existing foundation with the new replacement members. In the case of the Yeosu project, only the top deck was removed and replaced to match the tight work schedule. The project was carried into effect according to the schedule, and the final acceptance tests verified that higher-than-design efficiency indices were achieved at all the guaranteed points. With the initial rated output equal to 311 MW, after upgrading it rose to 329 MW.

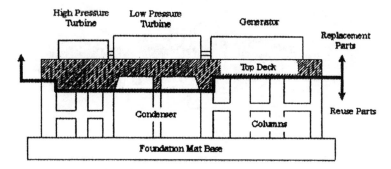

Figure 18-14. Modifying the base foundation of a 300-MW-class steam turbine of the Yeosu Unit 2 upgraded by MHI as a non-OEM. *Source: By courtesy of Mitsubishi Heavy Industries*

ALSTOM's Experience in Upgrading Steam Turbines of Other OEMs

Not to mention that ALSTOM Power has upgraded many aged steam turbines produced by manufacturers which later merged into this concern: Alsthom, Rateau, and CEM of France, GEC with its predecessors of AEI and English Electric of the UK, MAN Energie of Germany, and BBC (later ABB) of Switzerland and Germany, it also has accumulated a great experience in upgrading steam turbines of other OEMs. Upgrading these turbines installed at power plants of diverse countries, ALSTOM Power has mainly preserved the major traditional design peculiarities of the OEMs. Along with this, some design elements have been decisively replaced with versions inherent in the practice of GEC Alsthom or ABB whose effectiveness and reliability were well proved in the operation practice. Thus, in all the retrofit projects of ALSTOM Power the combined disc-type LP rotors have been replaced with welded rotors to avoid SCC of shrunk-on wheel discs. Simultaneously, the LP steam path has been refurbished with the use of up-to-date 3D technologies, longer, more efficient LSBs, and new, more spacious and efficient exhaust diffusers and hoods have been employed to reduce additionally the exit and hood losses in conjunction with a larger exhaust area—Figure 18-15.[38] For further refurbishment of LP cylinders, ALSTOM has also proposed its "optiflow" configuration, with several single-flow stages inserted in the central part of the cylinder and preceding its major double-flow steam path—see Figure 3-11. In particular, this concept based on earlier investigations of Alsthom,[43] was materialized in the process of reconstructing the 1,127-MW turbines of the San Onofre Nuclear Generation Station (SONGS).[44,45]

Steam Turbine Upgrade

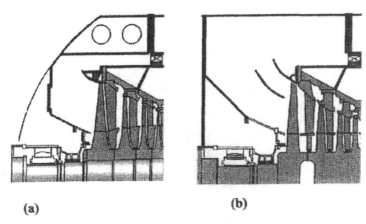

Figure 18-15. Comparison of original (a) and retrofit (b) design for LP exit sections of steam turbines upgraded by ALSTOM Power. *Source: By courtesy of ALSTOM Power*

A similar approach is also characteristic for other upgrade projects of ALSTOM Power as applied to steam turbines of OEMs alien to AL-STOM. This can be demonstrated with an example of a retrofit project for the integrated HP-IP cylinder BB-44 of Westinghouse turbines with the steam conditions of 16.6 MPa, 538/538°C (2,400 psi, 1,000/1,000°F) and the output ranging between 350 MW and 680 MW—see Figs. 3-13. The alternative design is made in an "ABB-style" (see Figs. 3-5 and 18-1) with reaction blading and welded rotor and eliminating the steam flow turnaround after the control stage but retaining the outer casing and all the conjunction sizes—Figure 18-16. Along with this, as distinct from Siemens-Westinghouse's project BB44FA (see Figure 18-9), in this case, the turbine retains its nozzle boxes and control stage, and the renewed inner casing preserves its simple, even outline.

ALSTOM Power has also actively contributed to upgrading U.S. fossil-fuel power units with steam turbines of GE, including supercritical-pressure units of TVA: the 500-MW Widows Creek Unit 7, 900-MW Bull Run Unit 1, and 1,100-MW Paradise Unit 3. By 1999, ALSTOM had had contracts with six U.S. electric power utilities to upgrade over 30 turbine cylinders, including five HP and four LP cylinders for TVA power plants.[46,47] The original design of the HP cylinder for a 700-MW subcritical-pressure steam turbine of GE is shown in Figure 18-17a, and its upgraded version of ALSTOM employed for the Paradise Units 1 and

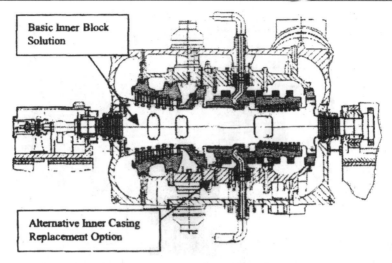

Figure 18-16. Alternative project of ALSTOM Power for retrofitting the HP-IP cylinder BB44 of Westinghouse turbines. *Source: By courtesy of ALSTOM Power*

2 is in Figure 18-17b. Earlier, ALSTOM upgraded in this way the HP cylinders of four similar turbines in Alabama. Along with the turbine's lifetime extension and efficiency improvement, upgrading the turbine has allowed the extension of overhaul interval periods up to 12 years. The original HP cylinders of GE had nozzle-group control with half-arc steam admission and double-flow nozzle-box configuration. The change to throttle control, with a single, forward-flow first stage and full-arc steam admission, using fewer, wider nozzles, together with the use of integrally shrouded rotating blades, significantly reduces the SPE problem, having been very topical for these turbines. The ALSTOM retrofit solution also has an optimized steam path, with four more stages compared to the original configuration, using airfoils of advanced 3D design. As well as in the above mentioned case with upgrading the GE turbine by MHI, prior to the retrofit, the existing cylinder was thoroughly surveyed, which results in detailed CAD models of the existing casings and rotor and those to be replaced with.[47]

A similar refurbishment was fulfilled by ALSTOM for the integrated HP-IP cylinder of the 550-MW-class GE turbine of the J.K. Spruce Unit 1 with the steam conditions of 17.4 MPa, 538/538°C (2400 psi, 1,000/1,000°F).[48] By that time, the unit was in service for only approximately ten years. The original cylinder (Figure 18-18a) featured nozzle

Steam Turbine Upgrade

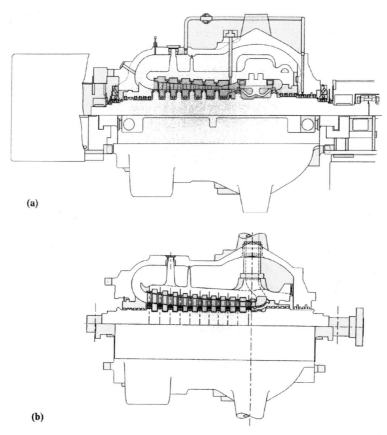

Figure 18-17. Original HP cylinder of a 700-MW steam turbine of GE (a) and its refurbished version of ALSTOM (b). *Source: J. McCoach*[46]

control, with partial-arc steam admission, nozzle-boxes, and a control stage with high aerodynamic losses. From the expected load profile for the unit's future operation, it was set advisable to pass to throttle control with full-arc steam admission and without nozzle-boxes and a separate control stage. Among other benefits, it also allowed accommodating additional stages in the HP steam path to increase its efficiency. The cylinder was provided with a new bladed rotor, HP inner casing, IP diaphragm carrier, and new diaphragms throughout the HP and IP steam paths—Figure 18-18b. An advanced brush seal technology was offered to the internal gland seal between the HP and IP steam path, but the customer gave preference for conventional labyrinth sealing technology because of its lower cost even despite the lower performance. With the existing outer casing being retained, it was found impossible to arrange a double-casing design of the cylinder's IP steam admission part, as commonly practiced in original steam turbines of ALSTOM—see Figure 3-14.

Pre-retrofit performance tests were carried out in 2002, immediately before the unit was shut down for the retrofit. The average internal efficiency values for the HP and IP steam path sections were found equal to 86.2% and 92.2%, respectively. The cylinder's post-upgrade internal

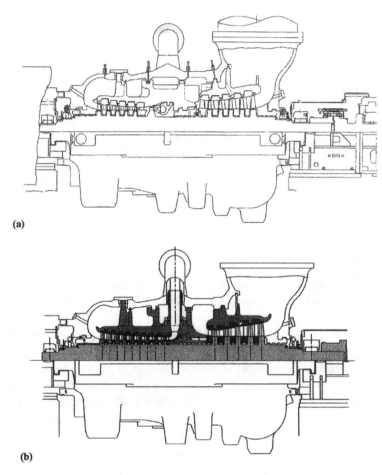

Figure 18-18. Original HP-IP cylinder of a 555-MW GE steam turbine of J.K. Spruce Unit 1 (a) and its refurbished version of ALSTOM (b). *Source: By courtesy of ALSTOM Power*

Steam Turbine Upgrade

efficiency was guaranteed at the levels of 93% for the HP steam path and 92.8% for the IP section. These data were completely confirmed by the acceptance guarantee tests carried out after recommissioning the turbine, with the IP section's efficiency equal 95.64%, much better than the guarantee. The derived turbine output was calculated from the test data to be 596 MW, which is 8 MW more than the guaranteed value of 588 MW. High levels of the HP and IP steam path efficiency was confirmed by subsequent performance tests after a year and then after 18 months of operation (Figure 18-19), with some degradation in efficiency caused by copper depositions on the blade surfaces. Analysis of samples scrapped out from the blade surfaces before the retrofit showed 39% copper content in composition of scrapings from the HP blades and 21% from the IP ones.

Being a successor of turbine design traditions from both GEC Alsthom on one hand and ABB on the other hand, ALSTOM Power seems to be the world's only major steam turbine producer that is able to offer both low-reaction (that is, impulse-type) and 50%-reaction technologies for newly designed and retrofitted turbines. What is more, they use reaction-type blading for retrofitting turbines initially designed with impulse-type steam paths, as done by ALSTOM Power divisions in Mannheim, Germany (former MAN Energie) and Elblag, Poland (former

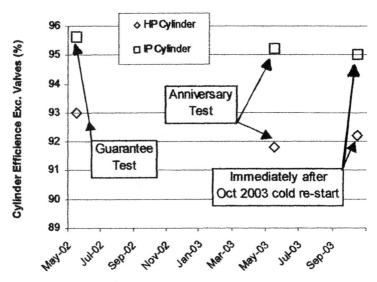

Figure 18-19. J.K. Spruce Unit 1 turbine's performance test data after its upgrading. *Source: By courtesy of ALSTOM Power*

Zamech Mechanical Works, later ABB Zamech Ltd.).

An example of an "ABB style" retrofit of the HP cylinder initially designed with an impulse-type steam path is shown in Figure 18-20 as applied to 500-MW subcritical-pressure steam turbines K-500-166-2 of LMZ.[49-51] By the time of retrofitting, the operating hours of these turbines ranged between 110,000 and 170,000. The original impulse-type steam path was replaced with up-to-date, three-dimensionally designed reaction-type blading, excepting the first impulse-type stage. Simultaneously, the original forged rotor was replaced with the welded one; the inner casing, casing rings, yokes of the guide vanes, and intermediate seal

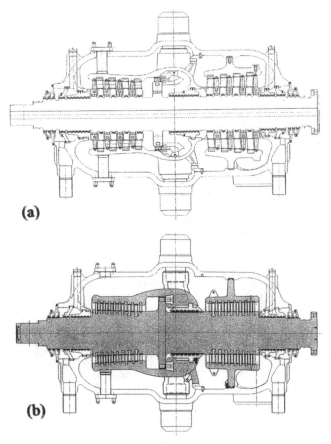

Figure 18-20. Original HP cylinder of a 500-MW steam turbine of LMZ (a) and its retrofitted version in an "ABB style" (b). *Source: H. Mandel et al.[51]*

Steam Turbine Upgrade 507

carrier were also replaced, with retaining the outer casing and end gland seals. In their original state, the HP cylinders of these turbines showed the internal efficiency varying between 76% and 80%. To compare, the HP cylinder's internal efficiency for the newly designed ABB turbines of an 800/900-MW-class at the Schwarze Pumpe, Boxberg, and Lippendorf power plants (see Figure 3-5) have lain at the level of over 90%. However with regard to the partial-arc steam admission and some other steam path peculiarities, the internal efficiency of the modified cylinder was assessed as equal to about 87.5% with the resulted increase of the power output about 10 MW and a reduction of the specific heat consumption of 140 kJ/kWh (133 Btu/kWh).

Turbines of this type have been operated at the power plants Kozhenice (Poland), Boxberg, and Jänschwalde (Germany), being installed in the late 1970s and early 1980s. Six turbines at the Jänschwalde power plant were retrofitted by the Mannheim division and the rest four turbines were retrofitted by the Elblang division. The LP cylinders were retrofitted with replacement of the original disc-type rotors with solid (in the first case) or welded (in the second case) rotors. The original 960-mm long LSBs of LMZ with the annular exit area of 7.48 m^2 per flow were replaced with shorter 900-mm LSBs with the finger-pin root and the exit area of 6.94 m^2 per flow. Nevertheless, the LP cylinder's internal efficiency was expected to be about 84.7%. Performance tests carried out at the Jänschwalde Unit E showed the internal efficiency values for the LP cylinders equal to 86.8%; whereas the internal efficiency of the IP cylinder after modernization increased by 2.2% up to 89.9%.[51]

Between 1982 and 1988, the Mechanical Works Zamech manufactured twelve 360-MW turbines under license of BBC for the Belchatow, the largest brown-coal-fired power plant in Poland and Europe, and in the 1990s another four turbines were installed at the power plant Opole. The turbine was designed in the 1960s for the steam conditions of 17.65 MPa, 535/535°C (2,560 psi, 995/995°F) in three cylinders: single-flow HP one and double-flow IP and LP ones. For the time being, retrofit of such a turbine has included upgrading only the LP cylinder by replacing its steam path with the use of blading of a new generation. Replacement of the welded rotor and stationary blade carriers occurs also necessary. The LSBs were made free-standing with hardening their leading edges for protection against WDE. The resulted increase in the unit efficiency made up over 3% and in the power output about 12 MW.

Besides the above mentioned ten 500-MW and sixteen 360-MW

508   *Steam Turbines for Modern Fossil-Fuel Power Plants*

turbines, by the early 2000s the list of steam turbines refurbished by the Elblang and Manheim divisions of ALSTOM Power had also included six 120-MW turbines manufactured by Zamech and 46 turbines with the single capacity of 200-215 MW designed and manufactured by LMZ or Zamech under license of LMZ, with a perspective to supplement this series with other turbines of these types.[49,50]

Diverse Versions for Upgrading Turbines of LMZ

Steam turbines with the single capacity ranged between 200 MW and 500 MW designed and manufactured by LMZ have been operated not only in Russia and other former Soviet republics, but also in many other countries. Some types of these turbines were also produced in other countries based on LMZ drawings. For a variety of reasons, these circumstances have resulted in diversity of proposals for upgrading these turbines, and there have appeared numerous publications devoted to possibilities of upgrading these turbines, which make it possible to compare different approaches practiced by diverse developers.

The most widespread type of steam turbines of LMZ has been its 200/210-MW turbine, initially designated as K-200-130. Later, there appeared its diverse modifications with the single capacity of up to 215 MW. By 2000, in Russia there were 76 turbines of this type in operation: 26 of them ran for less than 100,000 operating hours; for 21 the service time ranged between 100 and 170 thousand hours; for 13 it exceeded 170 thousand hours but did not reached 220 thousand hours, and 13 were operated beyond the threshold of the extended fleet lifetime of 220,000 hours.[53] Under license of LMZ, these turbines were also manufactured by Zamech in Poland, BHEL in India, and Shanghai Turbine Works in China. Originally designed for steam conditions of 12.8 MPa, 565/565°C (1,850 psi, 1,050/1,050°F), they have been mainly operated with the main and reheat steam temperatures of 540°C (1,004°F). The turbine is made in three cylinders, with single-flow HP and IP cylinders and double-flow LP one with the two-tier next-to-the-last stages (so called Baumann stages) and 765-mm (30-in) long LSBs, nozzle-group control and mechanical-hydraulic governing system. In its initial version, these turbines had single-casing HP and IP cylinders; some further modifications were designed with inner casings. The first turbine of this type was put into operation in 1958. The use of the Baumann stage in the LP steam path (Figure 18-21a) was stipulated by the fact that at that time LMZ had the LSBs not longer than 765 mm (30 in); years later, there appeared the LSBs

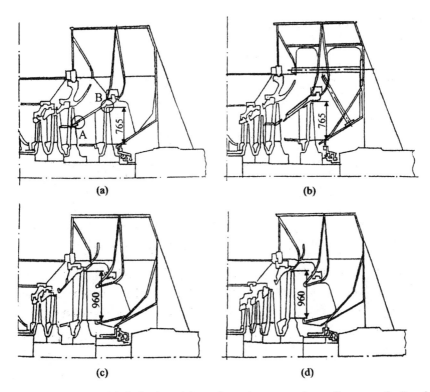

Figure 18-21. Initial design (a) and various versions (b, c, and d) of modernizing LP cylinders for 200-MW turbines of LMZ. *Source: V.G. Orlik et al.*[53]

of LMZ with the length of 960 mm and 1,200 mm (38 in and 47 in), but this happened later.

Along with a lower efficiency of the LP steam path, frequent breakages of the Baumann stage's rotating blades above the partition and increased erosion wear in the same zone, as well as at the last stage blades' tips (zones A and B in Figure 18-21a), have occurred the main disadvantages of the Baumann stage. Special investigations showed that the main origin of abundant coarse-grained moisture causing this erosion is an intense condensation on the surface of the partition conical visor of the Baumann stage's diaphragm under influence of intense heat removal by the colder exhaust steam of the upper tier. The leakage of supersaturated steam going through the lower tier under and along the partition and then running through the inter-tier seal into the main stream at the stage exit above the partition (zone A in Figure 18-21a and Figure 18-22a) significantly worsens the steam flow structure and intensifies WDE. Erosion of the trailing edge in the root section of the blade's upper portion forms stress concentration in this zone.[53]

LMZ in cooperation with Central Turbine and Boiler Institute of St. Petersburg and electric power utility of Tyumen' (Siberia) developed several versions of rather conservative modernization of the LP cylinder for 200-MW turbines in service at the Surgut power plant (12 x 200 MW)—see Figure 18-21b-d. The first version (Figure 18-21b) features a thin thermal sheet shielding the partition of the last stage's diaphragm; in addition, the diaphragm walls are provided with special grooves and catchers for water removal. In other respects, the steam path remains invariable. In the second version (Figure 18-21c), the last two stages (including the Baumann stage) are replaced with one stage with the rotating blade's length of 960 mm, the first two LP stages remaining the same. In the third version, additionally the first two stages are upgraded. And in the fourth version (Figure 18-21d), all the four stages are changed based on the last stage blade of 960 mm. In addition, the exhaust hood of the cylinder was initially designed with fins of rigidity. All the versions were considered with a possibility of replacing these fins with the rigidity bars

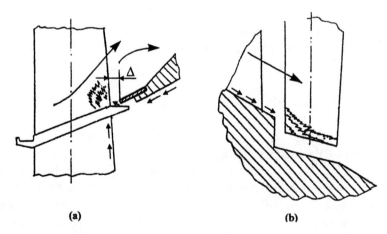

(a) (b)

Figure 18-22. Main places of erosion damages in the LP steam path of a 200-MW turbine of LMZ with the Baumann stage: for the next-to-the-last stage's rotating blades above the partition between the tiers (a) and at the tip of the last stage's rotating blades (b). *Source: V.G. Orlik et al.*[53]

Steam Turbine Upgrade

to decrease the energy losses in the exhaust hood. This replacement is especially effective when the turbine operates under partial loads.

The efficiency changes with relative volumetric steam flow amount were calculated for all the four above mentioned design versions. Results of these calculations are plotted in Figure 18-23. More complicated and laborious changes in the cylinder's design bring more effect, but it is understandable they are more expensive. According to estimations of the Surgut power plant, the cheapest first version pays itself for less than one year. Under conditions of limited financing, the second version (with replacement of only two last stages) seems most acceptable and for a short time period of 5-6 years brings maximum benefits. For long-term periods of seven years and more, the most expensive, fourth version seems the most profitable. According to estimations of LMZ, such a complete

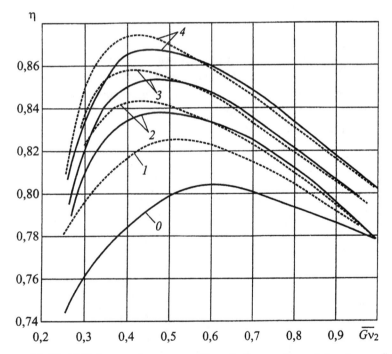

Figure 18-23. Efficiency changes with relative volumetric steam flow amount for the LP steam path of LMZ 200-MW turbines with the original design (0) and various modernization versions (1-4).
Solid lines refer to the exhaust hood with fins of rigidity and dotted lines refer to the case of replacing the ribs with rigidity bars. *Source: V.G. Orlik et al.*[53]

refurbishment of the LP cylinder, together with partial retrofit of the HP cylinder, allows increasing the turbine output to 225 MW with the same steam flow amount and decreasing the gross heat consumption for the turbine to 7,956 kJ/kWh (7,541 Btu/kWh) compared to 8,045 kJ/kWh (7,626 Btu/kWh) for the turbine version K-210-21.8-3.

It should be noted that all these modernization versions do not touch the LP rotor—it remains of the combined design with shrunk-on wheel discs.

Much more radical refurbishment of the LP cylinder was proposed and implemented by Skoda Energo at the Czech power plant Prunerzov II.[54] At this power plant, there are five steam turbines K-200-130-5 manufactured by Zamech. According to Skoda, all the main components of the cylinder are to be replaced: the steam path (with replacing four former stages with three new ones, including the last stage with the freestanding 840 mm long LSBs), rotor (with change to the forged design with a central bore), casing, and exhaust hood (with the use of rigidity bars instead of ribs). Practically, only the original foundation of the cylinder retains the same. As a result, the total turbine capacity under the nominal conditions rose by more than 5%. In the future, it is supposed to refurbish the HP cylinder of the turbine, too.

The most consistent and complete concept of refurbishing this tur-

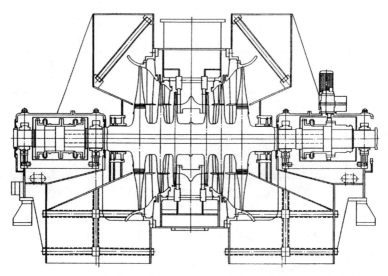

Figure 18-24. LP cylinder of 200-MW turbine of LMZ refurbished by Skoda Energo. *Source: L. Taich et al.*[54]

bine has been developed by ALSTOM.[49,50] In particular, it is completely materialized at Polish power plant Turow with ten 200-MW turbines and partially at some other power plants with these turbines of Poland, Estonia, Romania, and other countries.[49,50,55,56] According to ALSTOM, it is reasonable to conduct refurbishment of these turbines in three stages shown in Figure 18-25. In the first place, it is reasonable to modernize the LP cylinder. The second stage includes the retrofit of the HP cylinder, and the third stage refers to the IP cylinder. The second stage also embraces replacement of the mechanical-hydraulic governing system of the turbine with a contemporary electro-hydraulic control system, including replacement of the turbine valves, their steam-chests, and crossover pipes, too. This hierarchy of priorities looks quite reasonable, although more often than not it seems more advisable to combine the refurbishment of both the HP and IP cylinders. Accomplishment of the second and third stages of modernization greatly depends on economic circumstances and the degree of wear out of high-temperature components. For all three stages, the steam path is radically renewed with the use of advanced three-dimensionally designed blading. In particular, this refers to the next-to-last LP stage with the use of inclined and curved vanes, according to the concept of the advanced LP cylinder developed by ABB—see Figure 4-12. More often than not, the impulse-type blading is replaced with reaction-type one, even though in a few cases the impulse-type blading has been employed. All the rotors are replaced with welded ones. Both the high-temperature (HP and IP) cylinders are made double-casing, and the inner casing of the HP cylinder is made flangeless

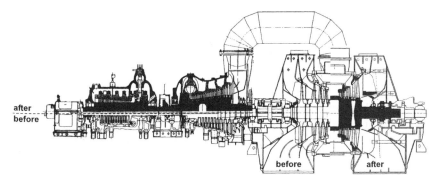

Figure 18-25. Concept of refurbishing 200-MW turbines of LMZ according to ALSTOM (as applied to the power plant Turow 10x200 MW). *Source: K. Linnebach.*[55]

514   *Steam Turbines for Modern Fossil-Fuel Power Plants*

with shrunk-on rings.

The complete refurbishment of the turbine, according to ALSTOM, brings the increase of the turbine output by about 35 MW and the decrease of the heat rate by approximately 7%. Refurbishment of the high-temperature components makes the turbine more flexible. With new rotors, casings, valve-steam chests, and steam-lines the turbine can be reliably operated for decades. Increased reliability of the turbine allows extending time intervals between the overhauls up to 10-12 years.

For the same turbines, design proposals developed by MEI on the basis of standard blading and design solutions of LMZ with complete upgrade of all the three cylinders promise the decrease of the gross specific heat consumption for the turbine to 7,737 kJ/kWh (7,334 Btu/kWh).[52] In this case, the turbine is equipped with LSBs of 960 mm with the total annular exhaust area equal to 14.96 m^2. Transition to the LSBs of 1,000 mm with the total annular exhaust area of 17.6 m^2 would allow decreasing the gross heat rate to 7,687 kJ/kWh (7,286 Btu/kWh). With raising the main and reheat steam temperatures to 575°C (1,067°F), this figure decreases to 7,519 kJ/kWh (7,128 Btu/kWh). According to this project, the turbine remains a three-cylinder machine, but the HP stop-valve steam-chests are relocated to the direct vicinity to the turbine to eliminate the HP crossover pipes; both the HP and IP cylinder are made double-casing, and all the turbine's bearing pedestals are fixed to the foundation frames, whereas the HP and IP cylinders slide along the supporting surfaces of the bearing chairs; the IP rotor in its front portion is provided with a forced cooling system delivering the cooling steam from the cold reheat steam-lines to the first stage's chamber. All these measures are intended to decrease the unsteady-state thermal stresses in the turbine's high-temperature critical elements, reduce the changes in the relative rotor expansions, and make the turbine more flexible for long-term operation in a semi-peak mood. All these changes can be carried out with retaining the existent turbine foundation.

Another widespread turbine of LMZ is K-300-240 with the output of 300 MW, steam conditions of 23.5 MPa, 540/540°C (3,414 psi, 1,004/1,004°F), and the rated heat rate of 7,710 kJ/kWh (7,308 Btu/kWh).[57] The turbine is made in three cylinders (see Figure 12-15); its steam path is single-flow in the HP and IP sections (the HP cylinder's steam path has a loop-flow configuration like those of the above mentioned turbines of LMZ K-800-240 and K-500-166—see Figs 2-4a and 18-20a) and triple-flow in the LP section, with settling one LP flow in the

Steam Turbine Upgrade 515

integrated LPIP cylinder. The first turbine of this type was manufactured in 1960. By 2000, there were 121 turbines of this type in operation at power plants of Russia and other countries.

Since the early 1990s, many of these turbines were modernized with refurbishment of the LP steam paths to make them more efficient. In doing so, the rotating blades of separate stages were replaced with integrally shrouded those of improved profiles, and fixed vanes of these stages were also refurbished. In the process of reconstruction, the meridional profile of the LP steam path, its root and peripheral outlines were improved, made smoother. In particular, such refurbishment was carried out in the mid-1990s at seven of eight 300-MW turbines installed at the Kostroma power plant.[58] In the same years, a significant number of steam turbines of this type in service at diverse power plants were furnished with systems for forced cooling of the IP rotors in their steam admission portion by steam from the cold reheat steam-lines; the system has been developed and implemented by TsKTI.[59] Such a measure lowers the highest metal temperature of the IP rotor, reducing in this way its creep rate, or makes it possible to raise the reheat steam temperature, heightening the unit's efficiency. This measure also reduces the maximum unsteady-state thermal stresses in the IP rotor, making possible to start up the turbine faster. In addition, lowering the temperature of steam going through the front end seal of the cylinder reduces the heat losses and prevents distortion of the seal carriers. (Similar systems for forced steam cooling of the rotors were also developed and employed for some other steam turbine types.)

Along with such works for partial refurbishment of these turbines, LMZ developed a project of their radical upgrade with replacing the entire steam paths of all the three cylinders, reconstruction of the HP steam admission organs and front end seal of the IP-LP cylinder, introducing the advanced moisture removal measures in the LP steam paths, and improving the exit diffusers of the LP exhaust hoods.[57] In particular, the refurbishment project also includes the use of hermetic (leakless) liquid-metal seals for stem of the HP control valves (see Figure 4-31). However, the most amazing feature of this project is transition in the HP cylinder from impulse-type blading, traditional for LMZ, to reaction-type one, which can be done with retaining the existent outer casing of the cylinder—like in the case shown in Figure 18-20. According to assessments carried out by LMZ, such a transition would increase the cylinder's internal efficiency by about 6-8%. An improvement of the existent im-

516   *Steam Turbines for Modern Fossil-Fuel Power Plants*

pulse-type IP steam path could increase its efficiency by approximately 1.8-2.2%, and improvement of the LP steam paths and exhaust hoods would raise their efficiency by 8-10%. As this takes place, the LP cylinder can be modernized either with retaining both the existent LSBs and LP rotor or with replacing the existent disc-type rotor for a solid one. The developed project can be carried into effect either for all the turbine cylinders at once or gradually, step by step, for individual cylinders.

It might be well to note that more detailed calculations executed by highly qualified specialists of MEI showed that the effect of transition to reaction-type blading in the turbine's HP cylinder would be somewhat less (not more that 3-4%, depending on the type of the used seals and heat recovery stage effects).[60,61] In addition, it is emphasized that the rate of efficiency degradation in the operation course caused by the wear of the labyrinth seals for reaction-type steam path would be greater than that for the impulse-type steam path, and restoring of seals for reaction-type steam path is more complex, laborious, and time consuming.

As an alternative, it is proposed to refurbish the existing 300-MW turbines of LMZ into three-cylinder 310-MW turbines with the integrated HP-IP cylinder and two double-flow LP cylinders, furnished with the same LSBs of the 960-mm length and total annular exhaust are of 29.92 m^2.[62,63] Such a turbine could be accommodated onto the existent turbine foundation. As compared to the guaranteed heat rate of 7,725 kJ/kWh (7,322 Btu/kWh) for the latest version of the existing 300-MW turbines of LMZ (K-300-240-3), such a turbine would provide the fuel saving of about 5%. With raising the main and reheat steam temperatures to 575°C (1,067°F), this gain would increase to 6.5%. For newly constructed power units of the 300-MW class, it is recommended to use a two-cylinder configuration with the integrated HP-IP cylinder and double-flow LP ones with the LSBs 1,200 mm (47 in) long, that is, with the annular exit area equal to 22.6 m^2.

References

1. "Estimating Steam Turbine Life Expectancy," *Turbomachinery International* 39, no. 5 (1998): 44.
2. MacDonald J. "Upgrading older power plants to improve their competitiveness," *Gas Turbine World*, March/April 2001: 32-34.
3. MacDonald J. "Upgrading steam turbine-generators, cost-effective solution for increasing plant efficiency," *Energy-Tech*, April 2006: 8-9.
4. Potter K. and D. Olear. "The Value of Steam Turbine Upgrades," *Power Engineering* 109, no. 11 (2005): 74-80.
5. Peltier R. "Steam turbine upgrading: Low-hanging fruit," *Power* 150, April 2006:

Steam Turbine Upgrade

32-36.

6. Schweimler J. "Experience with Long Reactor Runs without Main Revision, Taking Five Large Turbosets of a Lignite Power Plant as an Example," *VGB PowerTech* 79, no. 11 (1999): 33-40.

7. Krämer E., H. Huber, and B. Scarlin. "Low-pressure steam turbine retrofits," *ABB Review*, no. 5 (1996): 4-13.

8. Krämer E., N. Lannefors, and B. Scarlin. "Advanced Low-Pressure Steam Turbine Retrofits," *Advances in Steam Turbine Technology for the Power Generation Industry*, PWR-Vol. 26, New York: ASME International, 89-98.

9. Schimmoller B. "Steam Turbine Retrofits Upgrade Reliability and Performance," *Power Engineering* 102, no. 6 (1998): 31-34.

10. Longshore P. and M. Molitor. "Steam turbine control retrofit improves reliability," *Power Engineering* 99, no. 12 (1995): 41-43.

11. Thierfelder H.-G. "Advanced control technology for power plant refurbishments," *ABB Review*, no. 2 (1997): 15-24.

12. Suter F. and R. Sauter. "Upgrading the control and safety systems of reheat steam turbines," *ABB Review*, no. 2 (1999):

13. Weiss G.J. and A. Kopczynski. "Standard Packages for Steam Turbine Control Upgrades," presented at the 2001 International Joint Power Generation Conference, New Orleans, 2001.

14. Dünnwald J. and B. Walter. "Requirement for the Modernization of Governing and Protection Equipment at Steam Turbine Sets," *VGB PowerTech* 83, no. 1/2 (2003): 82-84.

15. Zwers J. "Uprating Controls Improves Unit Start-up," *Power Engineering* 108, no. 4 (2004): 46-48.

16. "As good as new," *Power* 148, March 2004: 10-12.

17. Varley J. "Modernizing Mehrum: 40 green MW from a coal plant," *Modern Power Systems* 24, no. 7 (2004): 11-13.

18. "More output, same input," *Power* 148, October 2004: 8-10.

19. "Enstedtværket competes after turbine upgrade," *Modern Power Systems* 18, no. 10 (1998): 29-33.

20. Hurd P., F. Truckenmueller, N. Thamm, et al. "Modern Reaction HP/IP Turbine Technology Advances & Experiences," presented at the 2005 International Joint Power Generation Conference, Chicago: 2005, PWR2005-50085.

21. Cheski J.R., R. Patel, K. Rockaway, et al. "A Large Steam Turbine Retrofit Design and Operation History," presented at the 2005 Power-Gen International Conference, Las Vegas: 2005.

22. Schimmoller B.K. "Turbine Touchups," *Power Engineering* 110, no. 1 (2006): 32-38.

23. "Bonanza unleashes 'stealth' capacity," *Power* 142, November/December 1998: 8.

24. Boss M.J., M. Gradoja, and D. Hofer. "Steam Turbine Technology Advancements for High Efficiency, High Reliability, and Low Cost of Electricity," presented at the 2005 Power-Gen International Conference, Las Vegas: 2005.

25. Kalyanaraman K. "Retrofitting the retro," *Turbomachinery International* 46, no. 3 (2005): 7-10.

26. Stephen D. "Optimized Plant Retrofits," presented at the 2001 International Joint Power Generation Conference, New Orleans: 2001, JPGC2001/PWR-19175.

27. Blachley S.R. and Foley M.E. "Testing for Turbine Degradation and Improving Performance with Seal Optimization," presented at the 2005 Power-Gen International Conference, Las Vegas: 2005.

28. Glynn B. "Undo the effects of aging in turbines," *Turbomachinery International* 47, no. 2 (2006): 34-35.

29. Brundle B. "World firsts for Yaomeng with vertical-tube low-mass-flow Benson unit," *Modern Power Systems* 22, no. 7 (2002): 24-28.
30. Spalding D. "Black, the new green?" *Modern Power Systems* 25, no. 9 (2005): 19-22.
31. Kawauchi A., K. Morikawa, and H. Arase. "Technologies for Repowering Existing Fossil Fuel Power Plants," *Hitachi Review* 44, no. 1 (1995): 45-50.
32. D'yakov A.F., A.F. Evdokimov, O.I. Demidov, et al. "Reconstruction of Old Thermal Power Stations Using a Combined-Cycle Technology," *Thermal Engineering* 44, no. 8 (1997): 663-669.
33. Zemtsov A.S. "Main Lines for Designing and Reconstructing Existing Power Stations Using Gas-Turbine and Combined-Cycle Technologies," *Thermal Engineering* 47, no. 10 (2000): 873-878.
34. Voronin V.P., A.A. Romanov, and A.S. Zemtsov. "Means for the Technical Upgrading of Electric Power Engineering," *Thermal Engineering* 50, no. 9 (2003): 701-705.
35. Smith D. "Great flexibility gained by Peterhead repowering," *Modern Power Systems* 18, no. 8 (1998): 59-63.
36. Hesketh J.A., H. Tritthart, and P. Aubry. "Modernisation of Steam Turbines for Improved Performances," *Proc. of the Symposium on Steam Turbines and Generators*, Monaco: GEC-Alsthom, 1994.
37. Smith D.J. "Retrofit Options Increase Steam Turbine Efficiency," *Power Engineering* 104, no 3 (2000): 34-36.
38. Hesketh A. and J. McCoach. "Fulfilling the Need for Turbine Retrofits Which Match Demand on 'Date and Duration' of Outage," *Proc. of the International Joint Power Generating Conference*, New York: ASME, 2002, 475-483.
39. Smith D.J. "Steam Turbine Upgrades Improve Reliability," *Power Engineering* 107, no. 6 (2003): 38-41.
40. Brummel H.-G. "Rehabilitation projects power Ukraine and India into the next millennium," *Modern Power Systems* 19, no. 10 (1999): 35-37.
41. Brummel H.-G. "The Ways and the Efficiency of Reconstruction of Thermal Power Stations," *Thermal Engineering* 46, no. 11 (1999):983-985.
42. Minami Y., N. Osaki, Y. Akaishi, and W. Newsom. "MHI Approach to Upgrading Old Steam Turbines," presented at the 2005 International Joint Power Generation Conference, Chicago: 2005.
43. Riolett G. "Outlook for the Large Steam Turbines of Tomorrow," *Modern Power Systems* 3, no. 1 (1983): 35-37
44. Brown R.D., F.Y. Simma, and R.J. Chetwynd. "Efficiency Improvement Features of Recent ABB-ALSTOM HP-LP Turbine Retrofit at Southern California Edison's San Onofre Nuclear Generating Station," *Proceedings of the International Joint Power Generating Conference*, New York: ASME, 2000, 85-93.
45. Leyzerovich A. *Wet-Steam Turbines for Nuclear Power Plants*. Tulsa (OK): PennWell Corporation, 2005.
46. McCoach J. "US utilities opt for steam turbine upgrades," *Modern Power Systems* 19, no. 4 (1999): 33-37.
47. Holms A. and A. Lord. "Turbine upgrading by design from survey data," presented at the 2001 International Joint Power Generation Conference, New Orleans: 2001, JPGC2001/PWR-19169.
48. Hogg. S. and D. Stephen. "ALSTOM-CPS San Antonio Retrofit of JK Spruce Unit 1 HP-IP Turbine—an Example of an Advanced Steam Turbine Upgrade for Improved Performance by a non-OEM Supplier," presented at the 2005 International Joint Power Generation Conference, Chicago: 2005, PWR2005-500227.

Steam Turbine Upgrade

49. Luniewicz B., R. Karpiuk, and B. Stasik. "10 Years Experience of Steam Turbine Modernization and Retrofit Activity at Elblag Works," presented at the 2001 International Joint Power Generation Conference, New Orleans, 2001, JPGC2001/PWR-19176.

50. Lunevich B.L., K.M. Ketlin'ski, J.A. Haskatt, and E.T. Kruger. "Experience of the Company Alstom Power in Modernizing Steam Turbines," *Thermal Engineering* 50, no. 6 (2003): 514-519.

51. Mandel H., G. Barth, W. Tanner, et al. "Retrofit Programme for the 500 MW Turbines Manufactured by LMZ," *VGB PowerTech* 83, no. 12 (2003): 75-79.

52. Trukhnii A.D., A.G. Kostyuk, and B.M. Troyanovskii. "Technical Proposals for Developing a Steam-Turbine Installation to Replace Obsolete 150-200 MW Power Units," *Thermal Engineering* 47, no. 2 (2000): 89-98.

53. Orlik V.G., M.V. Bakuradze, E.B. Dolgoplosk, et al. "Ways for Modernizing LP Cylinders of K-200-130 Steam Turbines" [in Russian], *Elektricheskie Stantsii*, no. 3 (2000): 29-34.

54. Taich L., Ya. Synach, L. Bednarzh, and A. Makarov. "The Retrofit of the LP Cylinder of a 210-MW Turbine," *Thermal Engineering* 48, no. 11 (2001): 959-964.

55. Linnebach K. "Technology Concepts for the EU Candidate Countries in Central and Eastern Europe," *VGB PowerTech* 83, no. 1/2 (2003): 45-50.

56. Uus M. "Clean power from Estonia's oil shale," *Modern Power Systems* 22, no. 2 (2002): 33-37.

57. Kondrat'ev V.N., A.S. Lisyanskii, Y.N. Nezhentsev, and V.D. Gaev. "A Project of Modernizing 300-MW Turbines of LMZ" [in Russian], *Elektricheskie Stantsii*, no. 7 (1999): 78-81.

58. Kuznetsov V.Y., O.E. Taran, A.P. Kurazhev, and D.L. Raginskii. "Experience of the Kostroma Power Station in Replacing the Low-Pressure Flow Paths of the Intermediate—and Low-Pressure Cylinders in the K-300-240-1 Turbines Manufactured by LMZ with Modernized Ones," *Thermal Engineering* 47, no. 11 (2000): 1023-1024.

59. Shargorodsky V.S., L.A. Khomenok, S.S. Rozenberg, and A.N. Kovalenko. "Raising the Technical Level of Steam Turbines with Implementing System of Forced Steam Cooling of the Rotors" [in Russian], *Elektricheskie Stantsii*, no. 1 (1999): 30-36.

60. Troyanovskii B.M. "Different versions of steam turbine flow paths" [in Russian], *Elektricheskie Stantsii*, no. 2 (2003): 18-22.

61. Kostyuk A.G. and A.D. Trukhnii. "A Comparison of the Impulse and Reaction HP Cylinders of Steam Turbines," *Thermal Engineering* 52, no. 6 (2005): 439-450.

62. Trukhnii A.D., A.A. Kalashnikov, A.G. Kostyuk, et al. "Turbine Installations for Technical Reequipment of Coal-Fired Power-Generating Units with K-300-23.5 Steam Turbines," *Thermal Engineering* 44, no. 7 (1997): 519-528.

63. Kostyuk A.G. "Certain Acute Problems of Designing and Modernizing Steam Turbines," *Thermal Engineering* 52, no. 4 (2005): 275-288.

Appendices

List of Abbreviations and Symbols

Abbreviated Titles of Institutions in the Power Industry

ABB	Asea Brown Boveri (Germany, Switzerland, and Sweden), merged with GEC Alsthom into ALSTOM Power
AEP	American Electric Power (the USA)
ANSI	American National Standard Institute (the USA)
ASME	American Society of Mechanical Engineers (the USA)
ASTD	American Society of Training and Development (the USA)
BBC	Brown Boveri Company (Germany, Switzerland), merged in ABB
BHEL	Bharat Heavy Electricals Ltd. (India)
CEM	Compagnie Electro-Mecanique (France), merged with Alsthom, presently in ALSTOM Power
DOE	Department of Energy (the USA)
EdF	Electricité de France
EPRI	Electric Power Research Institute (the USA)
GE	General Electric (the USA)
GEC	General Electric Company (the UK), merged with Alsthom into GEC Alsthom, presently in ALSTOM Power
KEMA	Commercial Enterprise for Technical Consultancy, Inspection, and Certification in Electric Power Industry (the Netherlands)
KhTGZ	Kharkov Turbine Works (the USSR, Ukraine), presently Turboatom
KWU	Kraftwerkunion (Germany), presently Siemens Power Generation
LMZ	Leningrad Metallic Works (the USSR, Russia), presently a branch of Silovye Mashiny (Power Machines)

MAN	MAN Energie, Maschinenfabrik Augsburg-Nürnberg (Germany), presently in ALSTOM Power
MEI	Moscow Power Engineering Institute –Technical University (the USSR, Russia)
MHI	Mitsubishi Heavy Industries (Japan)
NERC	North-American Electric Reliability Council (the USA)
NTPC	National Thermal Power Corporation (India)
ORGRES	Head Adjusting Institution for Power Plants (the USSR, Russia)
RPP NRW	"Reference power plant" for the North Rhine-Westphalia (Germany)
TMZ	Turbine-Engine Works (the USSR, Russia)
TsKTI	Central Turbine-Boiler Institute (the USSR, Russia)
TVA	Tennessee Valley Authority (the USA)
UDI	Utility Data Institute (the USA)
VGB	European Technical Association for Power and Heat Generation (initially Germany)
VTI	All-Russia (formerly All-Union) Thermal Engineering Research Institute (the USSR, Russia)
WEC	World Energy Council

Acronyms

ACC	advanced combined cycle
ACC	active control clearance (turbine seals)
AI	artificial intelligence
APCS	automated process control system
ASTD	automated system of technical diagnostics
BFP	boiler-feed pump
C&I	control and instrumentation
CAD	computer-aided design
CC	cross-compound (for double-shaft turbine)
CC	combined-cycle
CFD	computational fluid dynamics
CMS	control and monitoring system
CRH	cold reheat
CRT	cathode-ray tube, display
DACS	data acquisition and control system

DAS	data acquisition system
DCS	distributed control system
DCMS	distributed control and monitoring system
DDC	direct digital control
DSS	daily start/stop (for operation mode)
EFOR	equivalent forced-outage rate
ECW	erosion-corrosion wear
FATT	fracture appearance transition temperature
FEA	finite element analysis
FOR	forced-outage rate
FRF	fire-resistant fluid
FWH	feedwater heater
HHV	higher heat value (of fuel)
HP	high-pressure
HP-IP	high-pressure and intermediate-pressure (for integrated turbine cylinders)
HRH	hot reheat
HRSG	heat-recovery steam generator
HSR	high-speed reclosing
I&C	instrumentation and control
I/O	input/output
IP	intermediate-pressure
IP-LP	intermediate-pressure and low-pressure (for integrated turbine cylinders)
IT	information technology
LMS	liquid-metallic seal
LHV	lower heat value (of fuel)
LNG	liquefied natural gas
LP	low-pressure
LSB	last stage blade
LSDM-T	local subsystem of diagnostic monitoring for the turbine
M&D	monitoring and diagnostics
MCR	maximum continuous rate
MSV	main steam valve
NDE	non-destructive examination
NDT	non-destructive testing
O&M	operation and maintenance
ODA	octadecylamine ($C_{18}H_{37}NH_2$), amines-based surface-acting fluid

OEM	original equipment manufacturer
OMTI	fire-resistant oil of Thermal Engineering Institute (Russian abbreviation)
PC	personal computer
PCM	programmable control module
PLC	programmable logic controller
PU	processing unit
RH	reheater
RRE	relative rotor expansion
SCC	stress-corrosion cracking
SCS	supervisory and control system
SFC	static frequency converter
SH	superheater
SHP	superhigh-pressure
SPE	solid-particle erosion
SS	single-shaft (for CC units)
SSS	self-shifting and synchronizing (clutch)
TC	tandem-compound (for single-shaft turbines)
TSC	turbine stress controller
TSE	turbine stress evaluator
TTD	terminal temperature difference
USC	ultra-supercritical (for steam pressure)
USI	ultrasound inspection
WBT	web-based training
WDE	water drop erosion
3D	three-dimensional

Symbols

Symbol	Units	Definition
a	m/s	acoustic velocity
a	m^2/s	thermal conduction
a	-	factor of influence
c	m/s	steam velocity
c	kJ/(kg×°C)	specific heat capacity
d, D	m	diameter
E	m^4/(s×kg)	water-drop erosion criterion
F	m^2	annular exhaust area

f	m²	cross-section area
f	Hz	grid frequency
g	m/s²	free fall acceleration
G	kg/s, t/h	steam mass flow amount
H	kJ/kg	enthalpy
H	m	wall thickness, characteristic size
k	-	characteristic coefficient
l	m	blade length
L	m	distance, characteristic size
L	-	Laplace operator
M	kg	mass of metal
n	1/s, rpm	rotation speed
N	MW	electric output, load
p	MPa	steam pressure
r, R	m	radius
s	-	Laplace transfer parameter
S	-	corrosion-erosion rate
t	°C	temperature
T	s	time constant
u	m/s	circumferential rotation speed
V	kPa	vacuum (in condenser)
v	m³/kg	specific volume
v	mm/s	rate of crack growth
w	m/s	relative steam velocity
W	-	transfer function
W_t	°C/min	rate of ramp temperature change
W_N	MW/min	rate of ramp load change
x, y, z	m	spatial coordinates
X	-	steam dryness
$y = 1\text{-}X$	-	steam wetness
α	W/(m²×°C)	convection heat transfer coefficient
β	°C⁻¹	coefficient of thermal expansion
δ	m	wall thickness
Δt	°C	temperature difference
$\overline{\Delta t}$	°C	effective temperature difference ($\overline{\Delta t} = t_s - \overline{t}$)
ζ	-	energy loss factor
η	-	efficiency
θ	-	dimensionless temperature
λ	-	relative steam velocity ($\lambda = c/a$)

λ	W/(m×°C)	thermal conduction
μ	kg/(m×s)	dynamic viscosity
ν	-	Poisson's ratio
ν	m^2/s	kinematic viscosity
ξ	-	dimensionless length, radii ratio
ρ	kg/m^3	specific density
ρ	-	reaction degree
ρ	-	relative radius
σ	MPa	stress
$\sigma_{0.2}$	MPa	yield strength (yield limit)
τ	s, min, h	time
ψ	-	separation efficiency factor
ω	1/s	angular rotation speed ($\omega = 2\pi n$)

Subscripts and Superscripts

0	steam conditions at the stage (section, turbine) inlet, main steam
1	steam conditions between the nozzle and blade rows
2	steam conditions at the stage (section, turbine) outlet
a	axial, ambient
c	(in) the condenser
el	electric
ex	exit
ext	external (surface, efficiency)
f	force
fl	across the flange width (temperature difference)
fl-b	flange-bolt (temperature difference)
in, 0	initial
ins	insulated (surface)
m	mass
m	medium
m	metal
max	maximum
min	minimum
opt	optimum
r	radial (constituent), root
rh	reheat (steam)

s	surface (heated)
sat	saturation
set	set, given
sh	superheated (main steam)
st	stationary
st	steam
t	tip
t	temperature (stress)
t-b	top-bottom
th	thermal (efficiency)
u	tangential (constituent)
w	across the wall thickness
z	axial (direction, constituent)
θ	tangential, circumferential (constituent)

Criteria of Similarity

$$Bi = \frac{\alpha h}{\lambda} \left(\text{or } \frac{\alpha R}{\lambda}\right) \qquad \text{- Biot number}$$

$$Fo = \frac{at}{h^2} \left(\text{or } \frac{at}{R^2}\right) \qquad \text{- Fourier number}$$

$$Gr = \frac{g \times \beta \times \Delta\theta}{\nu^2} \times R^3 \qquad \text{- Grashoff number}$$

$$M = \frac{c}{a} \left(\text{or } \frac{w}{a}\right) \qquad \text{- Mach number}$$

$$Nu = \frac{\alpha d}{\lambda_{st}} \left(\text{or } \frac{\alpha x}{\lambda_{st}}\right) \qquad \text{- Nusselt number}$$

$$Pr = \frac{c_p \mu}{\lambda} \qquad \text{- Prandtl number}$$

$$Ra = Gr \times Pr \qquad \text{- Raleigh number}$$

$$Re = \frac{pwd}{\mu} \left(\text{or } \frac{pwx}{\mu}\right) \qquad \text{- Reynolds number}$$

Conversion Table for Main Units Used

Physical qualities	SI to US Customary System	US Customary System to SI
Length	1 m = 3.28 ft = 39.36 in	1 ft = 0.305 m; 1 in = 0.0254 m
Area	1 m^2 = 10,76 ft^2	1 ft^2 = 0.093 m^2
Volume	1 m^3 = 35.3 ft^3	1 ft^3 = 0.0283 m^3
Velocity, speed	1 m/s = 3.28 ft/s	1 ft/s = 0.305 m/s
Density	1 kg/m^3 = 0.0624 lb$_m$/ft^3	1 lb$_m$/ft^3= 16.02 kg/m^3
Force	1 N = 0.225 lb$_f$	1 lb$_f$ = 4.45 N
Mass	1 kg = 2.205 lb$_m$	1 lb$_m$ = 0.454 kg
Pressure, stress	1 MPa = 10^6 N/m^2 = 10 bar = 145.0 psi (lb$_f$/in^2) 1 kPa = 0.145 psi	1 psi = 6.9 kPa = 0.0069 MPa 1 in. Hg = 3.39 kPa
Energy, work, heat	1 kJ = 0.948 Btu	1 Btu = 1.055 kJ
Power, heat flow	1 W = 1 J/s = 3.41 Btu/h	1 Btu/h = 0.293 W
Heat flux per unit area	1 W/m^2 = 0.317 Btu/(h×ft^2)	1 Btu/(h×ft^2) = 3.154 W/m^2
Energy per unit mass, enthalpy	1 kJ/kg = 0.43 Btu/ lb$_m$	1 Btu/ lb$_m$ = 2.33 kJ/kg
Temperature	t$^\circ$C = (t$^\circ$F − 32)/1.8 T$^\circ$K = (t$^\circ$F−32)/1.8 + 273.16	t$^\circ$F = 1.8× (T$^\circ$K-273.16) + 32 t$^\circ$F = 1.8×t$^\circ$C + 32
Temperature difference	1°C = 1°K = 0.556 $^\circ$F	1°F = 1.8°C
Specific heat capacity, entropy	1 kJ/(kg ×$^\circ$C) = 0.239 Btu/ (lb$_m$×$^\circ$F)	1 Btu/ (lb$_m$×$^\circ$F) = 4.187 kJ/(kg×$^\circ$C)
Thermal conduction	1 W/(m×$^\circ$C) = 0.578 Btu/(h×ft×$^\circ$F)	1 Btu/(h×ft×$^\circ$F) = 1.73 (W/m×$^\circ$C)
Heat transfer coefficient	1 W/(m^2×$^\circ$C) = 0.176 Btu/(h×ft^2×$^\circ$F)	1 Btu/(h×ft^2×$^\circ$F) = 5.68 W/(m^2×$^\circ$C)
Dynamic viscosity	1 kg/(m×s) = 0.672 lb$_m$/(ft×s)	1 lb$_m$/(ft×s) = 1.49 kg/(m×s)
Temperature conduction, cinematic viscosity	1 m^2/s = 10.76 ft^2/s	1 ft^2/s = 0.0929 m^2/s

Index

Symbols

3D aerodynamic calculations 148

A

abbreviations 521

abradable coating 126

abradable seals 127

absolute thermal expansion 170

acronyms 522

active clearance control (ACC) 119

  seals 89

advanced main steam

  temperatures 20

advanced steam conditions 23, 26

advanced steam temperatures 38

aging steam turbine in service 5,

  458

altering the thermal conditions of

  the power unit 492

annual number of shut-downs 320

annual number of start-ups 305,

  310

annular exhaust area 49

arrangement of thermal expansion

  169

assessment 403

ASTDs 367, 395

attachment bases 141

automated systems of technical

  diagnostics 363, 430

axial and lateral side exhaust ports

  150

axial exhaust 73, 89

  and axially arranged condensers

  79, 85

  ports 152

B

backpressure 73, 252

barrel-type outer casing 52, 85

bar charts 373

Baumann stages 508

bearings 171

  fatigue cracking and plastic

  distortion 318

  mutual for the adjacent rotors 56

  pedestals 169, 174

between overhauls 489

blades 144

  attachment 336

blading 89

blocks 55

boiler-turbine control mode 301

brittle fractures 461

  and burst rupture 326

  of rotors 326

brushing and rubbing 334

brush bristle(s) 122

brush seals 394, 503

bypasses 202

C

calculated optimization of start-up

  diagrams 286

CC 15

  steam turbines 137

  turbines 45

  units 74

CFD 127

chief manufacturers of large power

  steam turbines 6

cockpit-type control room 356

cold, warm, and hot start-ups 194,

256
cold start-ups 250, 272, 323, 325, 326
combined LP rotors 57
combined rotors (welded or solid) 495, 500
combined stop-and-control valves 130
common inner casing 62
compact simulators 441, 445
compact two-cylinder turbine 73
competitions of the operational personnel 443
complex refurbishment of steam-turbine power plants 494
computational fluid dynamics (CFD) 101
concept of ordinary 68
condensers, consequentive connection of 28
condensers connected consecutively 391
conservative steam conditions 37
conservative steam parameters 69
control and monitoring systems 351
control valves 128
conversion into CC turbines 494
cooling-down constant 238
    for major turbine components 241
cooling down characteristics 238
cooling of rotors 515
cooling system 514
copper depositions on blading surfaces 335
covering 295
    power consumption variations 312

the variable part of power consumption graphs 14
cracks 463
    in steam-lines 461
criteria of similarity 527
criterion 341
critical 264
    crack size 459
    design elements 216
    elements 258, 322, 405
    turbine elements 195, 283, 287
cross-compound (CC) 43
    configuration, steam turbines 11
crossover pipes 46, 55, 219, 226, 247, 332
CRT-based arrangement of operational control and monitoring 353
CRT-based data presentation 443
curved-entry fir-tree roots 141
cycling 21, 198, 496
    operation 305, 314, 318, 330
cylinder number 43

D
DACSs 362, 370, 484
daily start/stop, stop/start 77, 198
data acquisition and control systems 315, 351
DCS 357, 361, 489
deaeratorless scheme 69
definition of start-up types 197
dense pack 491
    steam turbine design 118
design materials 35, 66
deviation 383
    of the operating conditions from the rated 455
diagnosibility 366

Index

Diagnostic Expert System for Turbomachinery 418
diagnostic functions 365
diagnostic monitoring of power equipment 362
DIGEST 367, 382
distributed control systems 352, 484
double-flow 51, 58
double-reheat 20, 23, 49
  scheme 21
double-shaft 43
downtime corrosion 345
DSS operation 312

E

effective temperature differences 283, 286, 297, 322, 408
efficiency 25, 27, 28, 32, 34, 37, 49, 77, 82, 333, 405, 456, 487, 491, 494, 504
efficiency gains 477, 482, 486
elevated main steam temperatures 19
embrittlement phenomena 460
emergency lube reservoirs 182
end energy losses 116
energy drops 127
energy loss 151
  with steam leakages 119
equivalent forced-outage rate 318
estimating thermal stresses 235
evaluation of the control quality at the transients 433
exhaust hoods 500, 510
exhaust port efficiency 151
extending time intervals between the overhauls 514
extensions to the overhaul

inspection intervals 483
extension of overhaul interval periods 502

F

fatigue 430
  cracks 332
field of velocities 149
field tests 274
fire-resistant fluids as lubricant 178
fix-point 170
fixed and rotating blades 105
flange joints 60
fleet resource 470
flexibility characteristics 201, 312
floor-mounted 88
  steam turbines 85
FOR 333
forced 191
  cooling of the rotor surfaces with steam 57
  outages (and rates) 16, 316, 320, 456
forged 56
  bored rotors 57
fossil-fired steam-turbine power greatest single capacity for 15
fracture appearance transition temperature (FATT) 197, 250, 326, 461
freedom of thermal expansion 171
free standing 113, 135, 486
  LSBs 144
fuel and energy expenditures for start-ups 315

G

gain in efficiency 491
gland seals 112, 318, 334

532   *Steam Turbines for Modern Fossil-Fuel Power Plants*

governed load change range 305
governed range 200
gross efficiency 9, 11, 21, 32

H
heat-rate 14
   monitoring 426
   performance field tests 333
   tests 380, 394, 490
heat-resistant high-chromium-
   percentage steels 23, 67
heating steam temperature 216, 217
heat consumption 16, 479
heat rate 9, 379
   under partial loads 315
heat transfer conditions 216
hetero-material rotor 89
hetero-material welded 67
higher heat value 11
horseshoe-shaped vortices 111
hot start-ups 198, 275, 306, 312, 330
human-machine interface 369, 440,
   484
human interface 355
hybrid, or combined, program 301

I
identifying the operating
   conditions 424
impulse-type 505
   blading 46
   stages 113
   turbines 118
individual resource 470
influence of a blading type on the
   cylinder efficiency 116
influence of main steam
   parameters 11
informative support 353, 369, 392,

404, 422
for the operational personnel
   320
for the turbine operators at
   transient 418
of the operator 365
initial matching of the steam and
   metal temperatures 202
instruction diagram 272, 310
   warm start-up 274
integrally formed tie-bosses 144
integrally shrouded 113, 144
integrated 85, 515
   HP-IP cylinder 45, 58, 61, 68, 73,
   82, 123, 127, 488, 498, 501, 502
      largest 46
   IP-LP cylinders 82, 88
intelligent tutoring systems 440
interlocked 145
internal efficiency 28, 97, 117, 333,
   379, 394, 478, 481, 491, 507
intrachannel water separation 162
IP-LP 85

J
journal bearings 175

L
largest TC steam turbines 45
last stage blades 23, 43, 135
   protectionagainst water drop
   erosion 155
lateral exhaust ports 152
leading indices 286, 322, 431
leaf 89
   seal 124
length-to-mean-diameter ratio 137
lifetime duration 455
lifetime expenditures 418, 431

Index

lifetime extension strategy 470
lifetime limitations 459
life expenditure 403
liquid-metallic seal 130
loading duration 289
loading time 285
loading up 194
load change amp rate 310
load changes 305
   within and beyond the
   governed range 295
load variations within the
   governed range and beyond 469
losses in control valves 128
losses with leakages of steam 97
low-flow operating conditions 146,
   158
lower heat value 11
LP exhaust ports 150
LSBs 49, 86, 113, 141, 148, 155, 481,
   488, 489
   annular exit area 135
   length 136

M

maintenance diagnostics 364, 429
main and reheat steam-lines 68
main operating parameters and
   indices of the equipment state
   373
mathematical modeling 411, 413
   of the heating up process for the
   rotors 283
meridional profiling of the nozzle
   channel 107
meridional profiling of the steam
   path 108
metal temperature fields forturbine
   design elements 234

mimic diagrams 373, 424, 426
   and trend graphs 448
minimum continuous load 305
minimum load 312
minimum stable load 201
moderate 68
Mollier diagram 9, 10, 218
monitoring of heat-rate
   performances 379
monitoring of turbine temperature
   403
monitoring temperature
   differences 322
monitoring thermal-stress states
   403

N

nomograms 278
non-bladed steam path areas 127
nozzle-box configuration 502
nozzle-group control 331
nozzle-group steam admission
   control 46, 219, 302, 388, 407
nozzle-group steam control 52
number cylinders 70
number of cylinders 49, 68, 72
number of LP cylinders 135

O

octadecylamine 344
oil fires 178
   in turbine halls 178
once-through 268
   boilers 13, 77, 203, 209, 329
   HRSG 77
   subcritical-pressure boilers 493
one-bypass 69
   start-up systems 203, 205, 247
operating conditions 189, 375

operating instruction 372
operation quality 375, 433, 459
operative diagnostics 364
optiflow configuration 60, 500
optimal intensity of cycling 318
optimization of the backpressure 388
overboring 466
overhaul intervals 458, 491
overheating of blading 340

P
parallel condensing systems 391
parasitic steam flows within the stage 112
PC-based local subsystem of diagnostic monitoring 362, 411, 430
PC-based LSDM-T 415, 422
performance test code for steam turbines 379
periods between overhauls 458
peripheral water separation 160
physical modeling 416
   of the rotor temperature state 412
planning the fuel consumption and power generation 392
plastic distortion of casings and other stator elements 325
post-operating analysis 375
   of operating conditions 429
power capacity and generation 3
power generating capacities 74
pre-start warming up 194, 247, 326
predictive, or condition-oriented, maintenance 380
profile 116
   energy losses 97

prong-and-finger roots 141
protection 344

R
raising the main and reheat steam temperatures 30, 198, 202
ramp rate 298
Rankine cycle 13
reaction 505
reaction-type 116, 515
   blading 52, 501, 506
   machines 46
   turbines 125
reaction degree 116
refurbishment by non-OEMs 495
reheater's steam bypass 208
relative rotor expansions (RRE) 61, 170, 251
remote monitoring-and-diagnostic center 368
renewal of the turbine's control and protection systems 483
replacement 148
replacing LSBs 491, 492
replacing original LSBs 486
residual lifetime 403
   assessments 459, 465
resource counters 403
restoring the cracked turbine casings or valve steam-chests 467
revelation of turbine steam path damages 394
reverse flows and vortices 344
reverse motion and vortices 146, 158
rolling up 194, 277
   for units with two-bypass start-up systems 255

Index

the turbine 247
rotors 56, 67, 326, 461
RREs 174, 197, 286, 334, 431, 433

S

saber-like 104, 105
 vanes 110, 147
SCC 470, 500
scheduled 191
scroll 56
scroll-type steam inlets 89
seals, abradable, brush, and
 retractable 490
seals with retractable packings 119
secondary losses 97
self-shifting and synchronizing
 (SSS) clutch 82, 265
sensitized seals 122
separation efficiency 162
shrinking rings 52
side condensers 86, 88
simulators 439
single-cylinder turbine 70
single-flow 51
 cylinders 58
 solution 58
single-seat angle valves 46, 55
single-shaft (SS) CC 265
single capacity of turbines 11
situation diagnostics 365
skin peeling 325, 466
sliding 388
 pressure 301
 steam pressure 37, 300, 315
snubbers 144
softening 460
solid 326
 (or monoblock) forged rotors 57
 particle erosion 202, 318, 327

sources of heat-rate performance
 deterioration 394
SPE 502
spray attemperators 208, 331, 386,
 493
stage, impulse-type 116
start-ups from the hot stand-by
 state 195
start-up diagrams 277
start-up duration 261, 282
start-up field tests 215
start-up for CC units 261
start-up fuel expenditures 283
start-up instructions 271, 282
start-up loading and raising the
 steam conditions 258
start-up number 311
start-up operations 190
start-up systems 69, 201
start-up technology 85
stationary 189
statistical processing of the
 operation 430
steam-bypass systems of
 combined-cycle (CC) units 209
steam-metal mismatch 281
steam-metal temperature
 mismatch 272
steam cooling 407
steam heating systems 61
steam leakage 117
 along the valve stems 130
 losses within the stage 113
steam pressure 388
steam temperature difference
 lengthwise of the steam transfer
 path 223
steam temperature drop
 lengthwise of the hot reheat 258

steam temperature excursions 208, 324, 330, 469

steam turbines 15, 43, 66, 68

steam velocity fields 100

stellite laminas 156

storing 373

stress-corrosion cracking 13, 57, 336, 464, 492

stress concentrators 324, 405, 462

subcritical-pressure 13, 14, 77, 335

subcritical main steam pressure 9, 15

subcritical steam pressure 77

subscripts and superscripts 526

supercritical-pressure 19, 28, 37, 335, 493

  steam turbines 130

supercritical- and USC-pressuresteam turbines of the 1960s-80s 14

surface-acting fluids (surfactants) 344

surface-type water trap 162

surface steam cooling 407

symbols 524

T

tandem-compound (TC), or single-shaft, turbines 11

TC 68

  (single-shaft) high-speed steam turbine (largest) 15

  turbine configuration 44

teaching and training the operational personnel 320, 374, 429, 436

technical diagnostics of power equipment 362

temperature differences 60, 325, 405

  top-bottom 239, 334, 410

temperature fields 222

temperature measurements 332, 404, 411, 434, 483

thermal 430

  expansion 169

thermal fatigue 420, 430, 461, 466

  of thick-wall turbine design elements 321

thermal stress analysis of rotors 435

thermodynamic diagnostics 379

three-dimensional design 89

  CFD design of the turbine steam path 107

  computational models 102

  of the steam path 104

throttle control 502

throttle steam admission control 52, 331, 405

throttle steam control 219

tilted stationary blading 53

tip circumferential speed 137, 138, 155, 159

titanium 49, 86

  -alloy blades 139

  LSBs 44, 73, 110, 140, 149, 155, 159, 162

top-bottom temperature differences 433

torsional oscillations 418, 461

  of the turbine-generator shaft 326

total number of start-ups 306

transient 189, 191, 404

  operating conditions 374, 421

trend graphs 373, 424

turbines 202

bypasses 258
casings 461
efficiency 495
elements 264
gland seals 119
largest 46
stress controller 417, 420, 469
stress evaluator 264, 403, 408, 417
transients, mathematical modeling and calculated optimization of 215
with subcritical main steam pressure 9
two-bypass 203
start-up systems 200, 340
two-shift daily start/stop (DSS) operation 306
typical patterns of WDE 156
typology of operating conditions 189

U

ultra-supercritical (USC) pressure 19
units with one-bypass systems 256
units with two-bypass start-up systems 254
upper and lower admissible values of the turbine's load 420
USC-pressure 37

V

vacuum 256, 272
valve steam-chests 461
variable reaction blading 117
variable reaction degree 116, 489
variations of the relative rotor expansion 58
varying the cooling water flow 390
velocity field 150

W

warming the hollow nozzle vanes 164
warm start-ups 198, 281, 305
water-drop erosion 340, 474
water induction 55, 196, 205, 208
into the turbine 318, 331
WDE 140, 486, 510
wear 318
wear out of gland seals 333
web-based training 442
wedge-shaped edges 144
welded 56
welded hetero-material 86
welded rotor 67, 71, 501
what-if scenarios 392
Wilson (phase-transition) region 57
Wilson zone 336
windage 205, 340, 418